The Urban Book Series

Series Advisory Editors

Fatemeh Farnaz Arefian, University College London, London, UK
Michael Batty, University College London, London, UK
Simin Davoudi, Newcastle University, Newcastle, UK
Geoffrey DeVerteuil, Cardiff University, Cardiff, UK
Karl Kropf, Oxford Brookes University, Oxford, UK
Karen Lucas, University of Leeds, Leeds, UK
Marco Maretto, University of Parma, Parma, Italy
Fabian Neuhaus, University of Calgary, Calgary, Canada
Vítor Oliveira, Porto University, Porto, Portugal
Christopher Silver, University of Florida, Gainesville, USA
Giuseppe Strappa, Sapienza University of Rome, Rome, Italy
Igor Vojnovic, Michigan State University, East Lansing, USA
Jeremy Whitehand, University of Birmingham, Birmingham, UK

Aims and Scope

The Urban Book Series is a resource for urban studies and geography research worldwide. It provides a unique and innovative resource for the latest developments in the field, nurturing a comprehensive and encompassing publication venue for urban studies, urban geography, planning and regional development.

The series publishes peer-reviewed volumes related to urbanization, sustainability, urban environments, sustainable urbanism, governance, globalization, urban and sustainable development, spatial and area studies, urban management, urban infrastructure, urban dynamics, green cities and urban landscapes. It also invites research which documents urbanization processes and urban dynamics on a national, regional and local level, welcoming case studies, as well as comparative and applied research.

The series will appeal to urbanists, geographers, planners, engineers, architects, policy makers, and to all of those interested in a wide-ranging overview of contemporary urban studies and innovations in the field. It accepts monographs, edited volumes and textbooks.

More information about this series at http://www.springer.com/series/14773

Ransford A. Acheampong

Spatial Planning in Ghana

Origins, Contemporary Reforms
and Practices, and New Perspectives

Springer

Ransford A. Acheampong
Department of Planning and Environmental
  Management, School of Environment,
  Education and Development
The University of Manchester
Manchester, UK

ISSN 2365-757X   ISSN 2365-7588 (electronic)
The Urban Book Series
ISBN 978-3-030-02010-1   ISBN 978-3-030-02011-8 (eBook)
https://doi.org/10.1007/978-3-030-02011-8

Library of Congress Control Number: 2018957058

© Springer Nature Switzerland AG 2019
This work is subject to copyright. All rights are reserved by the Publisher, whether the whole or part of the material is concerned, specifically the rights of translation, reprinting, reuse of illustrations, recitation, broadcasting, reproduction on microfilms or in any other physical way, and transmission or information storage and retrieval, electronic adaptation, computer software, or by similar or dissimilar methodology now known or hereafter developed.
The use of general descriptive names, registered names, trademarks, service marks, etc. in this publication does not imply, even in the absence of a specific statement, that such names are exempt from the relevant protective laws and regulations and therefore free for general use.
The publisher, the authors and the editors are safe to assume that the advice and information in this book are believed to be true and accurate at the date of publication. Neither the publisher nor the authors or the editors give a warranty, express or implied, with respect to the material contained herein or for any errors or omissions that may have been made. The publisher remains neutral with regard to jurisdictional claims in published maps and institutional affiliations.

This Springer imprint is published by the registered company Springer Nature Switzerland AG
The registered company address is: Gewerbestrasse 11, 6330 Cham, Switzerland

My mother, Florence Adwoa Sekyiwaa Hackman

Preface

In 2012, I had the unique opportunity to be involved in a new wave of institutional and legal reforms that would reintroduce and establish a new spatial planning system in Ghana. These reforms had actually begun in 2007, while I was still an undergraduate student of planning, under the Land Use Planning and Management Project (LUPMP). LUPMP, one of the sub-component initiatives under the umbrella of the Land Administration Project (LAP), was an ambitious attempt to resurrect a hitherto weak and under-resourced apparatus for land use planning and management in Ghana. Having freshly graduated from a master's degree programme in 2012, I had the privilege of working in one of the teams of consultants that had been tasked, to among other things, provide technical assistance in the preparation of the very first set of spatial development plans. Our task was quite significant in that it would help experiment with the various spatial planning models initiated by the reforms and pave the way to institutionalize the new spatial planning system.

For myself and for spatial planning—a discipline and profession that I love and have been passionate about—this was a historic moment. For the first time, since the first national physical development plan for Ghana was formulated around 1963, spatial planning was receiving the much needed attention in both the public policy and academic discourse. This was certainly a great impetus for the professional practice of spatial planning and the roles it could play in addressing some of the key developmental challenges of the country. Students of spatial planning, including myself, beamed with hope and optimism not only about the career opportunities a well-resourced spatial planning system could provide, but also about the opportunity this would afford us to put our skills to use to address some of the challenges facing our towns and cities.

I really enjoyed the experience of working with my team. I also received practical knowledge and invaluable insights from working closely and interacting with practitioners in the public sector (i.e. Town Planning Officers) who had spent decades doing what they possibly could to salvage a profession that was essentially at the brink of demise. That said, on the scholarly side of things, I found a major gap. Historically, spatial planning as a discipline has attracted limited scholarship in

Ghana. This may partly be as a result of the historically weak land use planning system and the limited attention that the profession has received in public policy discourse in the past. Yet another reason may be the shift in the focus and content of planning education in the early 1990s, which emphasized the social and economic dimensions of planning and development to the neglect of their physical expression and distributional consequences within and between geographical entities (i.e. neighbourhoods, towns, cities and regions).

The motivation to write this book, the first of its kind on the subject in Ghana, therefore, stems partly from my passion for spatial planning as a profession and the practical knowledge and first-hand experience I obtained through my involvement in the delivery of some of the initiatives under LUPMP, which would contribute to the establishment of a new spatial planning system in Ghana. I have always believed that one of the most effective ways by which the future cadre of planners would learn from these significant developments that have shaped spatial planning would be if we can document the experiences and lessons while reflecting on ways of improving the system and the associated practices to deliver its objectives. Towards this end, I conceived the idea to write this book.

Furthermore, this book is the result of my desire as a researcher to contribute to the scholarship needed to expand the frontiers of knowledge in the discipline while searching for ways of making spatial planning useful in addressing the country's developmental challenges. Indeed, traditionally, books on spatial planning have been written from Global North perspectives. In my student days, both in Ghana and abroad, the key textbooks I read on the topic were mostly written from either European or North American experiences. These sources have offered the opportunity to engage with the experiences and practices from their respective contexts. However, we know that the experience in one context may not necessarily be relevant in other contexts. Similarly, what has been proven to work in one context might not necessarily work in another. This is particularly true for spatial planning scholarship and practice because planning education and planning systems tend to be deeply rooted in the sociocultural, legal, political and economic situations of the countries of origin. I see this book on spatial planning in Ghana as being a part of an ongoing effort to advance scholarship and knowledge on the subject from contexts that have received limited attention in the past.

By focusing on Ghana, the book aims to provide an in-depth and comprehensive discussion of the spatial planning system and associated practices from this context. It hopes to take its audience on a journey to discover and understand spatial planning from a unique social, cultural, legal and political-economy context. Readers will engage with the wide-ranging issues covering the systems, processes, procedures, mechanisms and practices that characterize the spatial planning system and shape outcomes in this context. The book has been written with undergraduate and graduate students, academic researchers and practitioners and policy-makers in the multidisciplinary field of spatial planning in mind. In particular, this book aims at readers who seek an international perspective in spatial planning systems and practices.

Preface

Many individuals have helped me in cultivating the knowledge, skills and the motivation that would enable me to write this book. I am particularly thankful to my teachers for the inspiring lectures I have received on the subject. They passed on useful knowledge and nurtured in me the passion that would ultimately culminate in writing this book. I must thank Prince Aboagye Anokye, my long-time mentor from my undergraduate days at the Kwame Nkrumah University of Science and Technology. On his recommendation, I got the opportunity to get closer to the action and to experience first-hand, some of the major initiatives that would help establish the new spatial planning system in Ghana—an experience that would also be indispensable in writing this book. I also want to thank Priscilla Ankomah-Hackman for helping me in the digital production of most of the maps in this book.

I am grateful for the support I have received from the editorial team at Springer, especially from Juliana Pintanguy, Selvaraj Divya, Sanjievkumar Mathiyazhagan and Komala Jaishankar towards the publication of this book.

Finally, I could not have completed this book without the support of my wife, Nana Adwoa Nyarko, who kept the distractions to the barest minimum and also proofread most of the initial draft chapters of this book. What will I do without your unconditional support and encouragement, Nana?

Dublin, Ireland Ransford A. Acheampong, Ph.D.
July 2018

Acknowledgements

I would like to thank Stephen Asabere, Priscilla Ankomah-Hackman and Mohammed Abdul-Fatawu for providing, free of charge, the built-up land-cover datasets used for the analyses of urbanization trends in Ghana. Stephen provided the datasets on Accra; Priscilla provided the datasets on the Sekondi-Takoradi Metropolis, and Mohammed provided the datasets covering the Greater Kumasi sub-region.

Some of the chapters in this book build on the evidence from initial works I have done and/or co-published with other colleagues. Notable among these works are: (i) 'Housing for the urban poor: towards alternative financing strategies for low-income housing development in Ghana' (with P. A. Anokye) published in *International Development Planning Review* (37(4), 445–465, 2015); (ii) 'One nation, two planning systems? Spatial planning and multi-level policy integration in Ghana: Mechanisms, challenges and the way forward' (with Ibrahim Alhassan) published in *Urban Forum* (27 (1) 1–18, 2016); (iii) 'Quantifying the spatio-temporal patterns of settlement growth in a metropolitan region of Ghana' (with Felix S. K. Agyemang and Mohammed Abdul-Fatawu) published in *GeoJournal* (82(4), 823–840, 2017); and (iv) 'Examining the determinants of utility bicycling using a socio-ecological framework: An exploratory study from the Tamale Metropolis in Northern Ghana' (with Siiba Alhassan) published in the *Journal of Transport Geography* (article in press, 2018). I would like to thank my co-authors for allowing me to build on some of the ideas and information initially presented in these works.

I would also like to thank Mr Kofi Amedzro Kekeli (TCPD Head Office, Accra) for helping me to obtain the necessary permission from the TCPD to use some of the diagrams and illustrations in this book. Finally, I want to thank the two anonymous individuals who reviewed and provided feedback on the book proposal. I am particularly grateful for their very useful comments and suggestions and, above all, for their encouragement which further motivated me to write this book.

Contents

Part I Setting the Scene

1 Introduction 3
   References 10

2 The Concept of Spatial Planning and the Planning System 11
   2.1 Introduction 11
   2.2 What Is Spatial Planning? 12
   2.3 Spatial Planning and Political-Economy Contexts 15
        2.3.1 State-Market Interactions: Market Failure,
               the Welfare State and Spatial Planning 15
   2.4 Normative Theories of Spatial Planning: An Overview 19
   2.5 The Spatial Planning System: Key Features and Policy
        Instruments 21
        2.5.1 Governance Structure and Spatial Planning 22
        2.5.2 Policy Instruments Deployed to Achieve Spatial
               Planning Objectives 22
   2.6 Conclusions 25
   References 26

**3 Historical Origins and Evolution of Spatial Planning
   and the Planning System in Ghana** 29
   3.1 Introduction 29
   3.2 Origins of Spatial Planning in Ghana 30
        3.2.1 Pre-1945 Colonial Planning: Essentially Spatial Plans
               in Character 31
        3.2.2 The 1945 Town and Country Planning
               Act (CAP 84) 32

xiii

	3.3	Landmark (Spatial) Plan-Making Projects in the Era of Independence....................................	34
		3.3.1 National Perspectives—Five-Year Plan (1951–1956) and Seven-Year Development Plan (1963/64–1969/70)............................	35
		3.3.2 The National Physical Development Plan (1963–1970).................................	36
		3.3.3 The Master Plan for Accra.....................	37
		3.3.4 The Master Plan for Tema.....................	41
		3.3.5 Planning Scheme for Kumasi, 1963................	43
	3.4	Decentralization and Contemporary Planning in Ghana.......	45
	3.5	The Age of Reform: The Land Use Planning and Management Project and the New Spatial Planning System..............	48
		3.5.1 The Land Use and Spatial Planning Act (Act 925), 2016......................................	49
		3.5.2 Institutional Framework for Spatial Planning Under the Land Use and Spatial Planning Act (Act 925)....	50
		3.5.3 The Three-Tier, Hierarchical System of Spatial Planning Instruments..........................	51
		3.5.4 The Land Use and Spatial Planning Development Fund......................................	52
	3.6	Conclusions..	53
	References...		56
Part II	**Modern-Day Planning Systems and Scales of Spatial Planning**		
4	**Contemporary Traditions of Planning and Spatial Planning at the National Level**.....................................		59
	4.1	Introduction.......................................	60
	4.2	Three Decades of National Medium-Term Development Planning..	60
	4.3	The Systems Act and the New Land Use and Spatial Planning Act: A Recipe for Institutional Conflicts and Duplications?....	65
	4.4	National-Level Spatial Planning Under the New Spatial Planning System....................................	68
		4.4.1 The Ghana National Spatial Development Framework (2015–2035)................................	69
		4.4.2 The National Urban Policy Framework.............	76
	4.5	Realizing Long-Term Spatial Development Imperatives Within a Culture of Medium-Term Development Planning....	77
	4.6	Conclusions..	79
	References...		81

5	**The Inception of Regional Spatial Planning**	83
	5.1 Introduction	83
	5.2 The Concept of Regions and Rationale for Regional Spatial Planning	85
	5.3 Spatial Planning for Regions in Ghana: A Previously Uncharted Territory	88
	5.4 Contemporary Planning Systems and Authority and Competences for Regional Spatial Planning	90
	5.5 Regional Spatial Planning in Practice Under the New Spatial Planning System	93
	5.5.1 Western Region Spatial Development Framework (2012–2022)	93
	5.5.2 Greater Kumasi Sub-Regional Spatial Development Framework (2013–2033)	95
	5.6 The Emerging Paradigm of Special Development Zones and Development Authorities	99
	5.7 Towards Effective Regional Spatial Planning	100
	5.8 Conclusions	105
	References	106
6	**Local-Level Spatial Planning and Development Management**	107
	6.1 Introduction	108
	6.2 Local-Level Spatial Planning Instruments	109
	6.2.1 District Spatial Development Framework	109
	6.2.2 Structure Plan	109
	6.2.3 Local Plan	110
	6.3 The Context of Local Spatial Planning in Ghana	111
	6.3.1 Four Decades of Severed Connection Between National Policy Frameworks and Local Land Use Plans	111
	6.3.2 A Complex Land Tenure System and the Separation Between Ownership Rights and Determination of Use of Land	113
	6.3.3 Legal and Institutional Arrangements for Contemporary Local Spatial Planning and Development Management	114
	6.4 Local Plans and Development Management in Practice	118
	6.4.1 Plan-Led Development versus Non-Plan Development-First and the Permitting Process	119
	6.4.2 Rezoning and Change of Use	126

	6.5	Non-Compliance with Local Plans and Development Management Regulations.............................	129
	6.6	Towards Effective Local Spatial Planning and Development Management......................................	132
	References		134

Part III Issues in Spatial Planning

7 Policy Integration in Spatial Planning: Mechanisms, Practices and Challenges............................... 139
  7.1 Introduction 139
  7.2 The Concept of Integration in Spatial Planning............. 140
    7.2.1 Types and Dimensions of Policy Integration 141
  7.3 Integration in Ghana's Spatial Planning System: Mechanisms, Practices and Challenges 142
  7.4 Towards Effective Policy Integration in Spatial Planning...... 147
  7.5 Conclusions 148
  References 148

8 Public Engagement in Spatial Planning: Statutory Requirements, Practices and Challenges............................... 151
  8.1 Introduction 151
  8.2 Public Participation in Theory 152
  8.3 Spatial Planning and Public Participation in Ghana: Practices and Challenges 156
    8.3.1 Public Participation Requirements and Practices Under the 1945 Town and Country Planning Ordinance (CAP 84)......................... 156
    8.3.2 Democratization, Decentralization and Public Participation in Spatial Planning: Just More of the Same? 157
    8.3.3 Public Participation Under the New Spatial Planning System: Practices and Emerging Issues 159
  8.4 Towards Effective Stakeholder Engagement: The Case for Technology-Mediated Participation Practices 166
  8.5 Conclusions 168
  References 168

9 Urbanization and Settlement Growth Management............. 171
  9.1 Introduction 171
  9.2 Global Urbanization Trends and the Processes of Urban Change... 173
  9.3 Urbanization Trends in Ghana 176
    9.3.1 Urban Population Growth...................... 177
    9.3.2 Spatio-temporal Settlement Expansion Trends 179

	9.4	Urbanization and Urban Growth Management	192
		9.4.1 Regulatory Instruments	193
		9.4.2 Fiscal Instruments—Taxes, Exactions and Fees	197
		9.4.3 Incentive-based Instruments	198
	9.5	Conclusions	199
	References		201

10 The Spatial Planning System and Housing Development: Prospects and Models ... 205
- 10.1 Introduction ... 205
- 10.2 The Housing Situation in Ghana ... 207
  - 10.2.1 Channels of Housing Supply and Supply-Side Challenges ... 208
  - 10.2.2 Demand-Side Challenges ... 211
- 10.3 Policy Responses in the Past ... 215
- 10.4 The Planning System and Housing Delivery: Models and Approaches ... 217
  - 10.4.1 Model 1: Supplying Land for New Housing Development ... 218
  - 10.4.2 Model 2: Development Exactions and Affordable Housing Quotas ... 222
- 10.5 Conclusion ... 226
- References ... 227

11 Integrated Spatial Development and Transportation Planning ... 231
- 11.1 Introduction ... 231
- 11.2 Conceptualizing the Spatial Development and Transportation Nexus ... 233
  - 11.2.1 Spatial Development and Transportation at the Strategic Level ... 235
  - 11.2.2 Spatial Development and Transportation at the City and Neighbourhood Levels ... 238
- 11.3 Urban Spatial Structure and Mobility Patterns in Ghana ... 241
  - 11.3.1 A Conceptual Model of the Structure of Metropolitan Regions ... 242
  - 11.3.2 Exemplifying the Generalized Model of Urban Spatial Structure the Case of the Greater Kumasi Metropolitan Region ... 245
  - 11.3.3 The Influence of Prevailing Spatial Structures on Commuting Patterns and Travel Behaviours ... 249

		11.4	Towards Integrated Spatial Development and Transport Policy and Planning	251
			11.4.1 Land Use and Transport Integration at Policy and Institutional Levels..........................	252
			11.4.2 Urban Planning and Development to Promote Sustainable, Non-Motorized Mobility	258
			11.4.3 A Portfolio of Complementary Policy Strategies: Public Transit, Shared-Mobility and Intermodal Integration	261
			11.4.4 The Role of Integrated Land Use and Transport Decision-Support Systems.....................	263
		11.5	Conclusion	264
		References		265
12	**Spatial Planning and the Urban Informal Economy**			269
		12.1	Introduction	269
		12.2	Informal Urbanism and the Urban Informal Economy: An Overview of Perspectives...........................	271
		12.3	Historical Origins and Contemporary Manifestations of the Informal Economy	273
		12.4	Location Decisions and Challenges of Urban Informal Economy Businesses and Workers......................	275
			12.4.1 Home-Based Informal Economic Activities	276
			12.4.2 Itinerant Informal Economy Workers..............	277
			12.4.3 Sedentary Sales/Commercial and Artisanal Activities Located Outside the Home	278
		12.5	Planning to Accommodate the Informal Economy...........	280
			12.5.1 Shifting Attitudes and Contemporary Policy Responses	281
			12.5.2 Translating National Policies into Local Strategies and Actions	283
		12.6	Conclusion	286
		References		287
13	**Epilogue: Perspectives on Pathways Towards a Responsive Spatial Planning System**			289
		13.1	A Synopsis of the Story so Far	289
		13.2	Strengthening the Spatial Planning System	291
		13.3	Responding to Developmental Challenges Through Spatial Planning...	294
			13.3.1 Local-Level Spatial Planning and Development Control	294
			13.3.2 Urbanization Impact Monitoring and Settlement Growth Management	296

		13.3.3	Housing Supply—The Role of Spatial Planning	297
		13.3.4	Integrated Spatial Development and Transport Planning for Sustainable Mobility	298
		13.3.5	Supporting Livelihoods and Local Economic Development—Spatial Planning and the Urban Informal Economy	299
		13.3.6	Public Engagement for the Co-production of Urban Transformations	300
	13.4	Conclusion		301
References				302

About the Author

Ransford A. Acheampong *Ph.D., M.Phil. Cambridge, UK*; *B.Sc. KNUST, Ghana* is a Presidential Academic Fellow at the University of Manchester, UK. Before joining the University of Manchester, he was a Postdoctoral Research Fellow at Trinity College Dublin, The University of Dublin, Ireland. His research focuses on the nexus between spatial development and transportation. Themes covered in his research include comparative study of spatial planning systems; land use and transport interaction modelling; socio-spatial impacts of emerging mobility paradigms (e.g. shared-mobility and autonomous transport); and travel behaviours, focusing on the socio-ecological determinants of sustainable mobility options such as bicycling and walking in cities. He has contributed to planning-related projects for national and international organizations, including as a spatial planning consultant for the Town and Country Planning Department (TCPD) in Ghana and a research consultant for the Organization for Economic Co-operation and Development (OECD).

Abbreviations

AMA	Accra Metropolitan Area/Assembly
AUN	Aflao Urban Network
BAU	Business Advisory Unit
BRT	Bus Rapid Transit
CBD	Central Business District
CCUN	Cape Coast Urban Network
CEDECOM	Central Region Development Commission
CSGs	Civil society groups
DACF	District Assemblies Common Fund
DEPP	Development Policy and Planning
DPA	District Planning Authorities
DPU	Development Planning Unit
DVLA	Driver and Vehicle Licencing Authority
ESPON	European Spatial Planning Observation Network
GDP	Gross Domestic Product
GKSR	Greater Kumasi Sub-Region
GKSR-SDF	Greater Kumasi Sub-Regional Spatial Development Framework
GLSS	Ghana Living Standards Survey
GPRS	Growth and Poverty Reduction Strategy
GREDA	Ghana Real Estate Developers Association
GSGDA	Ghana Shared Growth and Development Agenda
HHUN	Ho-Hohoe Urban Network
HIPC	Heavily Indebted Poor Countries
ICT	Information and Communication Technology
IDA	International Development Association
ILO	International Labour Organization
IMF	International Monetary Fund
IPEP	Infrastructure for Poverty Eradication Project
JICA	Japan International Cooperation Agency

KMA	Kumasi Metropolitan Area/Assembly
KNUST	Kwame Nkrumah University of Science and Technology
LAP	Land Administration Project
LP	Local Plan
LUPMP	Land Use Planning and Management Project
LUSPA	Land Use and Spatial Planning Authority
LUTI	Land Use Transport Interaction
MDGs	Millennium Development Goals
MESTI	Ministry of Environment, Science, Technology and Innovation
METLOMP-SIM	Metropolitan Location and Mobility Patterns Simulator
MLGRD	Ministry of Local Government and Rural Development
MMDA	Municipal, Metropolitan and District Assembly
MTDP	Medium-Term Development Plan
NCA	The National Communications Authority
NDC	National Democratic Congress
NDPC	National Development Planning Commission
NEUN	North-East Urban Network
NGOs	Non-Governmental Organizations
NHP	National Housing Policy
NIP	National Infrastructure Plan
NPF	National Policy Framework
NPP	New Patriotic Party
NRSC	National Road Safety Commission
NSDF	National Spatial Development Framework
NSPS	National Spatial Planning System
NTP	National Telecommunications Policy
NTP	National Transport Policy
NUP	National Urban Policy
OECD	Organization for Economic Co-operation and Development
OfD	Oil for Development
PIP	Public Investment Programmes
RCC	Regional Co-ordinating Council
RPCU	Regional Planning Co-ordinating Unit
RSD	Regional Spatial Development Framework
RSPC	Regional Spatial Planning Committee
SADA	Savannah Accelerated Development Authority
SAP	Structural Adjustment Programme
SDF	Spatial Development Framework
SDGs	Sustainable Development Goals
SP	Structure Plan
SPC	Statutory Planning Committee
SPRING	Spatial Planning for Regions in Growing Economies
STMA	Sekondi-Takoradi Metropolitan Area/Assembly
STUN	Sekondi-Takoradi Urban Network
SUN	Sunyani Urban Network

TAZs	Traffic Analysis Zones
TCDP	Town and Country Planning Department
TDR	Transfer of Development Rights
TOD	Transit-oriented development
TUN	Tamale Urban Network
UBG	Urban growth boundary
USB	Urban service boundary
WRSDF	Western Regional Spatial Development Framework
WUN	Wa Urban Network

List of Figures

Fig. 2.1	Constituents of a typical spatial planning system............	21
Fig. 3.1	Proposed residential densities in the 1958 Master Plan for Accra. *Source* Reproduced digitally from analogue maps obtained from the TCPD, Accra.........................	38
Fig. 3.2	Distribution of major open spaces in the 1958 Master Plan for Accra. *Source* Reproduced digitally from analogue maps obtained from the TCPD, Accra.........................	39
Fig. 3.3	Designated as slums and areas proposed as suitable for lowering of construction standards in the 1958 Master Plan for Accra. *Source* Reproduced digitally from analogue maps obtained from the TCPD, Accra.........................	40
Fig. 3.4	A typical neighbourhood plan accompanying the 1958 Master Plan for Accra for private development at Kaneshie, Accra. *Source* 1958 Master Plan for Accra, TCPD, Accra..........	40
Fig. 3.5	Underlying principles of spatial organization in the Tema Master Plans. Adapted from Doxiadis Associates (1962)......	41
Fig. 3.6	Proposed road system in the Tema Master Plan. *Source* Reproduced digitally from analogue map obtained from the TCPD, Accra.......................................	42
Fig. 3.7	Distribution of communities based on income groups in the Tema Master Plan. *Source* Reproduced digitally from analogue map obtained from TCPD, Accra.......................	43
Fig. 3.8	Planning Scheme for Kumasi, 1963. *Source* TCPD, Kumasi...	44
Fig. 3.9	Governance structure and institutional competences under the National Development Planning (System) Act (Act 480, 1994) and the Land Use and Spatial Planning Act (Act, 925, 2016). *Source* Acheampong and Alhassan (2016, p. 7)............	46
Fig. 3.10	Governance structure and institutional competences for spatial planning under the 2016 Land Use and Spatial Planning Act (Act 925)..	51

Fig. 3.11	Three-tier model of spatial development frameworks and derivative plans established by the new decentralized spatial planning system	52
Fig. 4.1	Institutional arrangements for planning under two main planning laws in Ghana	65
Fig. 4.2	Composition of the governing body of the Land Use and Spatial Planning Authority (LUSPA)	67
Fig. 4.3	Urbanization trends in Ghana. *Source* Based on Population and Housing Census data, Ghana Statistical Services (For more information see: http://www.statsghana.gov.gh/statistics.html)	71
Fig. 4.4	Regional population growth rate (1960–2010). *Source* Based on Population and Housing Census data, Ghana Statistical Services	72
Fig. 4.5	'Integrated spatial development concept' proposed by the NSDF. *Source* Ghana National Spatial Development Framework	75
Fig. 5.1	Composition of the Regional Spatial Planning Committee established by the Land Use and Spatial Planning Act (Act 925)	91
Fig. 5.2	Diagram of the Western Region spatial development framework. *Source* Town and Country Planning Department, Accra Head Office	96
Fig. 5.3	Diagram of the Greater Kumasi Sub-Regional Spatial Development Framework. *Source* Town and Country Planning Department, Kumasi Ghana	98
Fig. 6.1	Physical development and development control under normative plan-led versus non-plan scenarios	121
Fig. 8.1	Arnstein's (1969) ladder of participation	153
Fig. 8.2	Pretty's typology of participation. Adapted from Pretty (1995)	154
Fig. 9.1	Percentage of the population living in urban areas (1950–2050). *Source* World urbanization prospects: based on world urbanization prospects data, United Nations (2014)	173
Fig. 9.2	Built-up area expansion for a globally representative sample of 200 cities (1990–2014). *Source* Based on data obtained from Atlas of urban expansion, 2016 edition (Atlas report can be downloaded at: http://www.lincolninst.edu/sites/default/files/pubfiles/atlas-of-urban-expansion-2016-volume-1-full.pdf. Raw data files were downloaded at website: http://www.atlasofurbanexpansion.org/data)	175
Fig. 9.3	National population growth trend (1950–2050). *Source* Based on world urbanization prospects data, United Nations (2014)	177

List of Figures

Fig. 9.4	Percentage of urban population in Ghana, West Africa and the African Continent (1950–2050). *Source* Based on world urbanization prospects data, United Nations (2014)	178
Fig. 9.5	Urban population growth rate in Ghana, West Africa and the African Continent (1950–2050). *Source* Based on world urbanization prospects data, United Nations (2014)	179
Fig. 9.6	Spatio-temporal built-up land-cover changes in the Accra metropolis and surrounding areas	182
Fig. 9.7	Spatio-temporal built-up land-cover changes in the Greater Kumasi sub-region	187
Fig. 9.8	Spatio-temporal built-up land-cover changes in the Sekondi-Takoradi Metropolis	190
Fig. 11.1	A three-level conceptual illustration of the links between spatial development and transportation	234
Fig. 11.2	Road and railway transport infrastructure networks in Ghana	237
Fig. 11.3	Fibre-optic networks for ICT in Ghana	239
Fig. 11.4	A conceptual illustration of the spatial structure of metropolitan regions	242
Fig. 11.5	Map of the GKSR showing local government administrative areas	246
Fig. 11.6	Three broad urban zones in the KMA	247
Fig. 11.7	Distribution of broad land use activity functions in the KMA	248
Fig. 11.8	A conceptual illustration major land use activity functions in the KMA and road distances between them	249
Fig. 11.9	Institutional setting for integration between spatial development and transportation policy and planning	255

List of Tables

Table 3.1	Periodizing significant developments in the history of spatial planning in Ghana.	54
Table 4.1	Spatial planning-related issues captured in national policy frameworks under the tradition of development planning	63
Table 4.2	Historical population size and growth rate.	70
Table 4.3	Historical population distribution in the ten administrative regions of Ghana.	71
Table 4.4	Proposed hierarchy of settlement in the NSDF	73
Table 4.5	City-regions and urban networks identified and delineated in the NSDF	74
Table 5.1	Proposed hierarchy of settlement in the WRSDF.	95
Table 5.2	Spatial Development Zones.	101
Table 9.1	Historical population growth in major towns and cities in Ghana.	179
Table 9.2	Land-cover changes in Ghana between 1975 and 2013.	181
Table 9.3	Built-up land changes in Accra and surround districts between 1984 and 2015.	183
Table 9.4	Aggregated built-up land changes in Accra and surrounded districts between 1984 and 2015	184
Table 9.5	Built-up land changes in Greater Kumasi sub-region between 1986 and 2014	186
Table 9.6	Population distribution in the Greater Kumasi sub-region	186
Table 9.7	Built-up land changes in Sekondi-Takoradi between 1990 and 2016.	190
Table 9.8	Types of growth management instruments.	193
Table 10.1	Annual household incomes housing expenditure ratio	212

List of Boxes

Box 6.1	Functions of the District Spatial Planning Committee and Technical Subcommittee under the Land Use and Spatial Planning Act (Act 925, 2016)	116
Box 6.2	Composition of District Spatial Planning Committee and Technical Subcommittee under the Land Use and Spatial Planning Act (Act 925, 2016)	116
Box 6.3	Development Zones in the Zoning Guidelines and Planning Standards	122
Box 6.4	Conditions for Change of Use	127
Box 6.5	Procedure for Processing Rezoning and Change of Use Applications	128
Box 8.1	Stages and Methods of Consultation in SDF and Local Plan Preparation	160
Box 9.1	The Greater Kumasi Urban Growth Boundary	195
Box 10.1	Examples of Properties Supplied in the Private Real-Estate Market in Accra, Ghana	214
Box 10.2	Housing Policy Statements in Ghana	216
Box 11.1	Ministries and Agencies in the Transportation Sector	252
Box 11.2	The National Transport Policy: Sector Goals and Strategies for Land Use and Transport Integration	254
Box 11.3	National Transport Policy: Policy Statement and Strategies for Non-motorized Transport	260
Box 12.1	Informal Business Zone (BL)	282

Part I
Setting the Scene

Chapter 1
Introduction

Abstract This chapter provides a general introduction for the book. It sets out the aim, scope and structure of this book. It provides a brief account of the story of spatial planning in Ghana, highlighting landmark events in spatial planning from the mid-1940s when the planning system was first established to the contemporary times. Following from this, it introduces the three main parts of the book and provides an overview of the themes in spatial planning that are subsequently explored in the remaining twelve chapters of the book.

Keywords Spatial planning · Institutions · Legal systems · Socio-economic Development · Global south · Ghana

Planning systems are deeply rooted in the sociocultural, legal, political and economic situations of the countries that use them to achieve specific development goals. In view of this, scholarship on the subject tends to take a country-specific focus and/or a comparative approach by addressing spatial planning issues and experiences from multiple countries. Historically, scholarship on spatial planning has based their narratives on perspectives and experiences from the Global North. The Global North perspectives and experiences have therefore not only made significant contribution to our understanding of the evolving meaning, scope, purpose and practices of spatial planning, but also shaped spatial planning theory development and the models that are currently applied globally. In recent years, however, there have been calls to explore, analyse and document knowledge and experiences from different contexts with respect to spatial planning. This, in turn, has generated a huge interest in and demand for authoritative perspectives on contemporary spatial planning from other contexts, particularly in countries in the Global South.

Efforts are currently underway to respond to the gaps in the existing academic literature by documenting spatial planning practices and experiences from previously unexplored contexts. While some of these initial attempts have focused exclusively on spatial planning as it relates to urban issues in multiple countries, others written as compendiums, due to their relatively wider scope, have understandably been unable to offer detailed discussions on the wider context of national governance structures and political-economy ethos that shape the spatial planning systems, practices and outcomes within the individual countries they cover.[1] Thus, in addition to the aforementioned endeavours, it is vital that we continue to have scholarly engagements with spatial planning from unique sociocultural, legal and political-economy contexts on wide-ranging issues, including the systems, processes, procedures, mechanisms and practices that characterize the spatial planning system and shape outcomes in particular national contexts. This book is intended to be one of the responses to this imperative.

This book is about spatial planning in Ghana. Although the main focus is on Ghana, it situates the discussion within the wider contemporary discourse on the evolving meaning, scope and purpose of spatial planning as both an established academic discipline and a public-sector activity for addressing a plethora of social, economic and environmental challenges. The mandate and expected outcomes of spatial planning may be similar across different contexts. Yet, the actual scope of issues embraced by the activity could be wide-ranging, depending on the prevailing sociopolitical, economic, legal and cultural systems within which the activity takes place and from which it derives its mandate and relevance.

The aim of this book is not to address every possible issue that may be delineated and addressed through spatial planning. Instead, the core of this book engages with some of the dominant themes in spatial planning globally, with a particular focus on detailed analyses of the experiences and practices from the unique context of Ghana. It provides a comprehensive and critical discussion on the changing socio-economic conditions, institutional settings, legal arrangements and practices which have shaped and defined Ghana's spatial planning system for more than seven decades. The book traces the history of spatial planning in Ghana to identify the landmark initiatives that have been responsible for introducing, reforming and institutionalizing spatial planning as we know it today. One of the major themes addressed in this book is the geographical scale of spatial planning, where we examine the role of the spatial planning system in formulating programmes, policies and plans at the national, regional and local levels of governance and decision-making. The issue of policy integration across spatial scales and between policy domains in spatial planning is also addressed in this book. Adopting a problem-based approach and drawing on primary and secondary data sources, the book also engages with important themes in the field, including: public participation in spatial planning; urbanization and sustainable growth management; spatial planning and housing development; spatial planning and the transition towards sustainable urban transport; and spatial planning and the informal economy.

Spatial planning in Ghana has had a chequered history. In the first decade following independence from British rule, physical planning, as it was then called, became one of the most important machineries for articulating and realizing the ambitious programme of socio-economic transformation envisioned by the socialist government of Dr. Kwame Nkrumah. Over this period, landmark spatial plan-making projects including the formulation of the first National Physical Development Plan

---

[1] See, e.g., Silva (2015) and Bhan et al. (2018).

(1963–1970), and derivative Master Plans for major cities including Accra, Tema and Kumasi would be completed. The National Physical Development Plan would provide a blueprint for the development of strategic infrastructure to drive the country's ambitious industrialization programme. The accompanying Master Plans would also initiate programmes and strategies to transform major towns and cities in the country.

In the history of a new Ghana, these developments were symbolic, but as would become of the vision, hope and aspiration that fuelled the struggle for independence and the significant developments that followed it, the glory days of planning in all its forms were short-lived. Being a public-sector-driven activity, spatial planning thrives where there is political stability. It was therefore not a coincidence that the national planning system ceased to exist between the mid-1960s and the late-1980s when successive military interventions became pervasive. The ensuing chaos essentially dismantled the governance apparatus needed to sustain all forms of public-sector activities, including spatial planning.

At the beginning of the twentieth century, the reversion to constitutional rule and the ensuing decentralization reforms offered a unique opportunity to entrench spatial planning at all levels of governance. However, the reforms would rather install a new ethos of planning that would become known as the 'development planning system', which was supported by the newly established system of local governments, a new legislation—the 1994 National Development Planning (System) Act (Act 480) and a reinvigorated National Development Planning Commission. As we will later discover in this book, the System Act of 1994 envisaged and entrenched a new planning tradition that was inherently aspatial from the national level to the local government level. Also, the focus and content of planning education would for many years be shaped by the prevailing ethos of development planning.

It is not entirely clear how a country, which begun its transformational journey after independence with a planning tradition, which explicitly recognized the need to anchor socio-economic development visions and goals spatially for optimum distributional outcomes, would later entrench a new tradition that essentially neglected this imperative. What we now know, however, is that as a result of these reforms, the legislations, institutions, financial resources, personnel and logistics needed to effectively carry out the mandate of spatial planning were missed on the nation's radar. The spatial planning system was weakened as a result: for nearly five decades after the first landmark spatial planning projects initiated in the early 1960s were curtailed, spatial planning hardly featured in the national development agenda.

With the new tradition of development planning neglecting the doctrines, ideals, practices, legal and institutional arrangements needed to support a well-functioning spatial planning system, it became almost impossible to discern the existence and impact of the activity of spatial planning at all levels. Where it existed, its functions were limited to the preparation of local sub-division plans and development control by under-resourced district TCPDs, which had no connections legally and/or administratively to the local governments they were mandated by 1945 Town and Country Planning Ordinance (CAP 84)—an obsolete piece of legislation—to work for.

It would take another reform under the Land Use Planning and Management Project (LUPMP)—a three-year initiative implemented between 2007 and 2010—to begin to reintroduce spatial planning into mainstream national policy-making. While the reforms under LUPMP are commendable, it has opened up a new Pandora's box of technical and strategic challenges with respect to the legal and institutional arrangements entrenched by the tradition of development planning on the one hand and those established by the new spatial planning system on the other hand. The emergent conflicts between the two planning systems existing side by side, which has been referred to elsewhere as a situation of *'one nation, two planning systems'*, are a major theme in this book.

It is within the context of the brief history on the evolving traditions of planning and the associated practices outlined above that this book has been written. It is worth mentioning that the terminology of *spatial planning* in this book is intentionally chosen to refer to a public-sector-driven activity that explicitly addresses the geographical and distributional implications of socio-economic development imperatives. Adopting this view, spatial planning, as used in this book, embraces an integrated and coherent approach to formulating and implementing programmes, policies, plans and projects at multiple spatial scales by multiple actors. The multi-scale view also implies that spatial planning is not only a local-level activity as traditionally conceived and practised in the past. Instead, it refers to various interventions that affect the distribution of population, economic activities and infrastructure that may be articulated in national, regional, sub-regional and local planning instruments. Inherent in this meaning of spatial planning is a rejection of a dichotomy between what may be referred to as 'development planning' and 'spatial planning' as it currently exists in Ghana. As we will later discover in this book, such a distinction does not only lack compelling intellectual justification, but also results in needless compartmentalization and duplication of functions and resources in practice.

Aside engaging with the governance structures, legal frameworks, policy instruments and practices that define planning in Ghana, this book has been written with a genuine conviction that spatial planning constitutes one of the pathways towards sustainable socio-economic transformation. It offers a platform to bring together expertise from different academic and policy domains to address the growing social, environmental and economic challenges facing the country. In strategic terms, spatial planning and policy emphasizes the distributional implications of development and thus, is an effective vehicle to address the development imbalances that exist between and within regions. At the settlement level, we see the problems of haphazard, uncoordinated physical development and the associated problems of perennial flooding, lack of basic amenities and services in residential neighbourhoods, slum formation, congestion and environmental degradation. All these problems are place-based, which require conscious, spatially focused interventions to address them. Devising innovative solutions to these problems and implementing them to improve the health of the population is one of the major priorities of spatial planning.

The scope of issues addressed in this book therefore reflects the need to understand and engage with the evolving institutional, legal and political-economy settings within which spatial planning is embedded. Following the underlying conviction that spatial planning is indispensable in the country's transformational agenda now and in the future, this book is written with an intent to engage with the major developmental problems we face today and the opportunities that spatial planning presents in addressing them. These two broad themes of the book have been addressed in 12 chapters, organized into three parts. Below, an overview of each of the chapters is presented.

Part I of the book comprises three chapters. Following the introduction chapter (i.e. Chap. 1) which gives an overview of the book, we set the foundation for the subsequent chapters of the book in Chap. 2. This chapter explores the meaning of spatial planning, identifying the nature, scope and purpose of this activity. It situates spatial planning within the wider context of the debate around the relative importance of free markets and the welfare state to provide justifications for the activity as a form of public-sector intervention. Normative planning theories which provide meta-narratives about what spatial planning ought to be and underpin planning practice are also outlined and briefly discussed. The key features of a typical spatial planning system are identified, outlining the institutional and legal milieu, and the various policy and development management instruments embodied in the planning system to: deliver land use planning and development control objectives, achieve resource allocation and (red) distributive outcomes, promote economic growth and pursue environmental protection imperatives.

The spatial planning system and the accompanying practices in Ghana has evolved since it was formally introduced in 1945 under the Town and Country Planning Act (CAP 84). Chapter 3 of the book traces the historical origins of spatial planning from the colonial era to modern-day Ghana. It reviews landmark spatial planning projects implemented in the early years after independence to identify the scope and purpose of the activity then. The inception of decentralization at the beginning of the twentieth century entrenched a new tradition of 'development planning' in Ghana: the impacts of this shift in the ethos of planning on spatial planning are examined in this chapter. Finally, the new spatial planning system instituted in the twenty-first century, following the renewed interest in spatial planning is presented, discussing the accompanying legal, institutional and planning instruments. This chapter also engages critically with the complexities in the planning arena which have resulted from the existence of two separate planning systems—the development planning system and the spatial planning system.

In Part II, we focus on the modern-day planning systems in Ghana and the scales of spatial planning. Chapter 4 begins the discussion on the tiers of planning. This chapter focuses on spatial planning at the national level. As has been highlighted previously, national-level planning in Ghana today is realized under two legal instruments—the 1994 National Development Planning (System) Act (Act 480) and the 2016 Land Use and Spatial Planning Act, (Act 925). These instruments have established two seemingly different planning systems, the 'Development Planning System' and the 'Spatial Planning System' all of which draw on the governance structure established by the decentralization law (i.e. 1993 Local Government Act (Act 462)). This chapter examines the focus and scope of three decades of national-level planning under the System Act (Act 480). With the promulgation of the new spatial planning law and the establishment of a new spatial planning system, which operates alongside the development planning system, the potential sources of institutional conflicts and duplications at the national level will be examined. The newly formulated National Spatial Development Framework (NSDF) will be presented as evidence of the renewed emphasis on spatial planning in mainstream strategic planning at the national level. This chapter also engages with the challenges associated with realizing long-term national spatial development imperatives within a culture of medium-term national development planning, which has historically been dictated by the tenure of governments stipulated in the national constitution.

Spatial planning at the level of regions is necessary for realizing national development goals at sub-national levels. However, what exactly constitutes a region is contested, implying that there could be different types of regions. Chapter 5 of this book focuses on spatial planning at the scale of regions. We first define what a region is, identify the different types of regions that could be delineated and outline the rationale for planning at this scale. Narrowing down to the Ghana context, the chapter outlines the story of regional planning in Ghana and examines critically the effect on contemporary regional spatial planning of what has previously been described as a situation of 'one nation, two planning systems' in Ghana. Ongoing efforts at institutionalizing regional spatial planning will be discussed, presenting examples of spatial plans formulated for different types of regions in the country. The chapter also reflects on the current structures for regional spatial planning as well as the experiences gained so far, and suggests ways in which regional spatial planning could become effective in the future.

In Chap. 6, the attention will shift from regional spatial planning to spatial planning at the local level. In Ghana, local governments are by law, authorities of development for their respective metropolitan, municipal and district assemblies, and as such are mandated to perform spatial planning functions. This chapter examines local-level spatial planning in Ghana. It outlines the various instruments intended under the new spatial planning system to translate national and regional development policies to district-wide frameworks and subsequently to cities, towns and rural areas within the districts. The contextual issues influencing planning and development management at the local level are discussed. The main issues discussed are: the historical disconnect between national-level planning and local land use planning; the complex indigenous landownership systems that spatial planning must grapple with; and the institutional and legal structures established formally to mediate the separation between landownership rights on the one hand, and the determination of land use and development control on the other hand. The gap between the established development management systems in normative terms as opposed to its functioning in practice is also examined. Based on the challenges of local spatial planning and development management identified, ways in which planning at this level could be made effective are recommended.

In Part III, we address a number of important themes in spatial planning. We begin the discussion of issues in planning in Chap. 7 by focusing on policy integration in spatial planning. Effective integration across policy domains and between spatial scales is indispensable in dealing with the inherently complex process of policy formulation and implementation at all levels. The purpose of this chapter therefore is to examine the instruments and mechanisms of vertical and horizontal integration with respect to spatial planning in Ghana. We will explain the concept of policy integration. Following from this, we will outline and discuss the various mechanisms and instruments adopted to achieve coherence between various policies in spatial planning in Ghana, identifying the gap(s) between integration as embedded in the design of the planning system and integration in action. Based on these, the challenges of policy integration are identified. Possible ways in which the system could be improved to deliver its objectives in an effective and efficient manner are put forward.

Spatial planning is about people and as such the success of spatial planning, to a larger extent, depends on how best it deals with the varying interests, views, preferences and aspirations of individuals, groups, communities and businesses. Chapter 8 is dedicated to the subject of public engagement in spatial planning in Ghana. It discusses public engagement and its value from the perspective of the normative theories of planning. Public engagement in practice is also examined, highlighting the statutory procedures and methods adopted, the shortcomings and the theory–practice gap in stakeholder participation in spatial planning in Ghana. On the basis of the challenges, ways in which public engagement could be improved in the spatial planning process are discussed.

The rapid pace of urbanization and its link with emerging challenges of climate change, environmental degradation and resource depletion poses serious challenges for urban growth management and sustainable development. Chapter 9 of this book focuses on the interface between urbanization and spatial planning. We will begin with a discussion of the urbanization—sustainable development nexus globally. We will then examine historical population growth patterns and discover spatio-temporal urban expansion trends in major metropolitan areas in Ghana based on satellite data. Moreover, we will outline the various tools and strategies deployed through spatial planning to achieve growth management objectives. The specific growth management tools and strategies that are currently being used in Ghana will be identified. Strategies towards sustainable growth management are also discussed in this chapter.

Chapter 10 addresses the nexus between housing supply and spatial planning. In this chapter, the housing situation in Ghana is examined, highlighting the demand and supply-side challenges of the housing market. Various policy responses in the past and the resulting challenges are discussed. We will also discover in this chapter the various approaches that can be advanced through the spatial planning system to contribute to addressing the country's housing crises. To this end, two models of housing supply, which are designed to be deployed through the spatial planning system, will be put forward in this chapter.

In Chap. 11, we will focus on one of the central themes in spatial planning, which is the nexus between spatial development and transportation. This chapter draws on the established evidence of the two-way dynamic relationship that exists between spatial development, observed as the distribution of land use activities on the one hand, and transportation, taken as the system of infrastructure and travel modes that facilitates travel/movement necessitated by the distance separation between activity locations on the other hand. The chapter explores the links between spatial development and transportation at the strategic level (i.e. national and regional scales) and the city and neighbourhood scales. A generalized model illustrating the spatial structure of metropolitan regions in Ghana will be advanced and applied to understand the structural forces that underline patterns of spatial interaction in metropolitan areas. Various ways in which spatial development policies and transportation policies could be integrated to bring about sustainable urban development outcomes will be identified and discussed in this chapter.

The informal economy is an enduring phenomenon in town and cities in Ghana. In Chap. 12, we will focus on the urban informal economy in relation to spatial planning. We will develop a typology of urban informal economy activities that is based on the spatial dimensions of activities in the informal economy. We will then examine the location preferences and strategies of the various activity types identified and highlight the key challenges informal economy workers and enterprises face as they attempt to utilize the urban space for their livelihood activities. We will discuss the extent to which contemporary urban policy has sought to address support the informal economy and explore pathways towards accommodating the informal economy in spatial planning and the urban development process.

In the final chapter, we will synthesize the key issues presented in this book and highlight the perspectives on pathways towards making spatial planning responsive to the evolving challenges of towns, cities and regions. In doing so, we will bring together all the perspectives and insights which are discussed in relation to the spatial planning system in Parts I and II as well as those that are advanced under each of the single thematic issues presented in Part III of the book.

References

Bhan G, Smita S, Watson V (eds) (2018) The routledge companion to planning in the global South. Routledge, New York, p 424

Land use and spatial planning Act (2016) Act 925 The nine hundred and twenty fifth act of the parliament of the Republic of Ghana

Local Government Act (1933) Act 462 The four hundred and sixty two act of the parliament of the Republic of Ghana

National Development Planning (System) Act (1994) Act 480 The four hundred and eightieth Act of the parliament of the Republic of Ghana. http://urbanlex.unhabitat.org/sites/default/files/urbanlex//gh_national_development_planning_act_1994.pdf. Accessed on 24 Jul 2018

Silva CN (ed) (2015) Urban planning in Sub-Saharan Africa: colonial and post-colonial planning cultures. Routledge, New York, p 306

Town and Country Planning Division (1963); National Physical Development Plan 1963–1970. Prepared by Town and Country Planning Division Ministry of Lands

Chapter 2
The Concept of Spatial Planning and the Planning System

Abstract This chapter explores the evolving meaning, scope and purpose of spatial planning. It situates spatial planning within the wider context of the debate around the relative importance of free markets and the welfare state in the generation of growth allocation of the gains of development and protection of public health to provide a justification for the activity. An overview of normative planning theories, which provide meta-narratives about what spatial planning ought to be and underpin practice, is also provided in this chapter. Conceptualizing spatial planning as a form of the multi-level governance activity, this chapter identifies the key features of a typical spatial planning system, examines the national governance structures that shape the institutional and legal arrangements for spatial planning and identifies the various policy instruments deployed by spatial planning to deliver its objectives.

Keywords Spatial planning · Planning theory · Governance structure
Political-economy · Welfare state · Policy instruments

2.1 Introduction

It is estimated that over 7 billion human beings currently live on planet Earth, and by the first half of the twenty-first century, global population is expected to reach over 9.5 billion inhabitants (United Nations 2014). Over this period, many places in the Global South are expected to have a significant proportion of their populations living in urban areas. On the African continent, for example, it is estimated that about 55% of the inhabitants will be living in cities by 2050 (United Nations 2014). Within these societies, there is a growing need to ensure equitable distribution of opportunities, such as employment, housing, transportation and social services, and to promote social justice and inclusivity. In addition, systems are needed to formulate and implement innovative strategies to deal with the almost overwhelming challenges of environmental sustainability induced by historical socio-economic development trajectories. Indeed, these goals underpin socio-economic transformation. But without the accompanying system of institutions, legislative frameworks and various policy instruments deliberately created to translate them into action, such goals, however ambitious and well-intentioned they may be, would merely constitute wishful articulations.

This book is about one of the well-established systems deployed by many modern societies to achieve specific socio-economic development goals. It is about *spatial planning*. Spatial planning anchors national visions, goals, programmes, policies and plans to human settlements of varying sizes at different spatial scales. In most countries, the goals of co-ordinating the physical manifestations of sectoral policies and ensuring equity in the distribution of the outcomes of development between places while resolving the essential tensions between socio-economic development and environmental protection imperatives are articulated and pursued through spatial planning. As an activity, spatial planning has a profound impact on the internal layout and functional organization of land uses and their regulation at the level of towns and cities and ultimately shapes the emergent distribution of population and economic activities observed in human settlements.

The purpose of this chapter is fourfold. Firstly, we introduce the concept of spatial planning and explore the nature, scope and purpose of this activity. Secondly, we will operate from the premise that public policy decisions and outcomes are the product of complex interactions between market forces and a strong welfare state and that spatial planning exists as a form of state intervention in free markets. In view of this, we will situate spatial planning within the wider context of political-economy and draw on the principles of welfare economics to further justify the need for this activity. Within the overarching framework of political-economy perspectives that justify spatial planning, we will also briefly outline the normative theories that set out what spatial planning as an activity ought to be and underpin its practice. In addition, we will draw on the forgoing discussions to conceptualize the key features of a typical spatial planning system, identifying the institutional structures, legal frameworks and the different policy instruments that constitute a spatial system. In the end, we will bring together all the concepts, theories and principles to highlight some of the practical contributions of spatial planning to national development by highlighting the links among land use planning and development, business promotion, economic growth, resource redistribution and social justice and environmental sustainability.

2.2 What Is Spatial Planning?

The use of the term *spatial planning* to describe the activities, processes, practices and the accompanying legal and institutional milieu described in this book is quite recent. Since its inception, the activity of planning has been known and continues to be known by other related terminologies including: '*land use planning*', '*physical planning*', '*urban planning*', '*town and country planning*', '*regional planning*' and even just '*planning*'.

2.2 What Is Spatial Planning?

As we will show shortly, just as societies in general continue to evolve and the prevailing socio-environmental and economic challenges of the day continue to assume different degrees of complexity, so has planning, which seeks to confront these complex challenges evolved in terms of the underlying theories and the nature and scope of the activities associated with it in practice. In general, the nature and scope of planning within any given society, indicative of the terminology used, reflect the priorities of that society, which in turn determine the focus and core functions embraced by the activity. In some instances, the terminologies used also indicate the spatial scale (i.e. whether town, city, regional or national level) at which the activity of planning is undertaken. With this recognition at the background, we will in the sections that follow, review the meanings of some of these terminologies, learning in the process why we have come to use the term spatial planning.

Going back to the earliest civilizations such as Egypt, Benin, Rome and Greece, the need to organize the layout and the built form of human settlements has featured prominently in the history of human existence. From the original ideals of designing and building impressive towns and cities, arose the activity of planning as we know it today. Consequently, the meanings implied by terminologies such as '*land use planning*', '*town planning*' and '*physical planning*' are deeply rooted in the desire to control not only the location of human activities such as housing, education, industry, retail and agriculture to ensure harmony among them, but also to shape the intensity, form, amount, spacing and the interlinkages between various land use activities. From this fundamentally architectural and design origins, emerged principles and standards, including efficiency, aesthetics, economy, harmony and health that underlined the design and development of towns and cities (see, e.g. Chapin 1965; Keeble 1969; Cullingworth 1972). One of the most succinct descriptions of the nature and scope of planning in its traditional sense is perhaps demonstrated in the definition of town planning by Keeble (1969). Keeble defined town planning as:

> the art and science of ordering the use of land and siting of buildings and communication routes so as to secure maximum practicable degree of economy, convenience and beauty.

The term '*urban planning*', while embodying the original design and development control focus of planning outlined above, also reflects the geographical scope and spatial scale of the activity of planning. As Couch (2016) delineates:

> the main focus of concern is the urban area, ranging from large agglomerations to free-standing cities and smaller towns, and the different dimensions of urban planning from the city-region, through the city scale down to the planning of districts, neighbourhoods, individual sites and buildings.

From the mid-1960s, the concept of '*regional planning*' gained prominence in both the theory and the practice of planning, beginning in Europe and North America and spreading its doctrines to other parts of the world. This type of planning, which is still undertaken in many countries, today aims to transcend the limitations imposed by boundaries administered by local authorities by embracing planning issues for *regions* as spatial units larger than a single town, city or metropolitan area, but smaller than the state. A regional approach is largely justified by the need to achieve functional integration between places as the basis to improving linkages, to promote balanced development between places and to co-ordinate the actions of different administrative bodies, for example, in the planning and development of major infrastructure projects (see, e.g. Murphy 1984; Lichfield 1964; Friedmann and Weaver 1979). It is uncommon these days to find the term *'urban and regional planning'* being combined, the rationale of which reflect a type of planning that embraces the fundamental idea of thinking and acting within and across spatial scales. We will revisit the topic of regional (spatial) planning later in Chap. 4.

From the original meanings implied by the terminologies discussed above have emerged the concept and doctrines of *spatial planning*. Spatial planning as a concept has its origins in Europe, where the term has been coined and used in the generic sense to refer to the established machinery used by governments to influence the distribution of activities in space and to manage spatial development. Healey (1997) defines spatial planning as: 'a set of governance practices for developing and implementing strategies, plans, policies and projects, and for regulating the location, timing and form of development'.

From the above definition, we can begin to unpack the nature, scope and purpose of spatial planning. Firstly, spatial planning, while embracing the traditional functions of design, regulation and development control, attempts to engage with the wider sociopolitical processes and practices affecting the development of villages, towns and cities. Moreover, proponents of the doctrines of spatial planning present it as new approach to planning which is meant to engage with the distributional implications of allocating land for various activities and how these might be delivered with other public policy outcomes (Morphet 2010). In essence, it embraces social, economic and environmental issues across multiple spatial scales (i.e. national, regional and local levels) and between different sectors of public policy. Spatial planning, therefore, implies not only an evolving ethos of planning but also a shift in the culture of planning practice: it is comprehensive in nature and emphasizes the need for policy integration, co-ordination and collaboration among multiple actors (Healey 2004; Morphet 2010; Albrechts 2004). Also, spatial planning embraces not only a regulatory posture but a promotional one that supports businesses and developers to deliver economic development as well as the livelihoods of individuals and households by removing unnecessary regulations, prohibitions and costs.

Some scholars have questioned the claim that spatial planning constitutes a new ethos of and approach to planning that is any different from scope of issues embraced and addressed by terminologies preceding it—that is, land use planning, town planning or urban and regional planning. The central argument of critics is that planning has always been about proposing preferred forms of development, controlling development and engaging with different actors affected by planning decisions (see, e.g. Couch 2016). Others, while pointing out a lack of common understanding of what spatial planning means in practice, do also acknowledge that the ambiguous nature and malleability of the term give it a neutral and unifying appeal, allowing for universal acceptance (Allmendinger and Haughton 2009).

In summary, we have learnt from the forgoing discussion of the following: from an initial emphasis on land use designation, design, regulation and development control, the activity of planning has evolved to embrace more responsibilities in accordance with the prevailing challenges and priorities of modern societies. Spatial planning is about visioning and how they are expressed in spatial terms. It is about place-making and so embraces the economic, social and environmental dimensions of development, while acknowledging and engaging with the politics needed bring about sustainable transformations. Spatial planning is supposed to serve both regulatory and promotional purposes to bring about desired development outcomes. Spatial planning is also meant to be proactive, promotional and interventional in it not only being able to anticipate socio-economic development trends but also formulating and implementing policies to attract, promote and bring about desired changes. Finally, it emphasizes the need to think and act within and across spaces, thereby ensuring effective integration of policies between spatial scales and across different institutions in the public policy arena.

2.3 Spatial Planning and Political-Economy Contexts

So far, we have focused on defining what spatial planning is and to understand the nature, scope and purpose of the activity of spatial planning. Among other things, we have identified that spatial planning constitutes one of the universally established systems used by governments to influence the distribution of activities and their implications for the social, economic and environmental dimensions of development. Spatial planning has become an integral part of the way societies are organized and how resources are allocated, thereby having far-reaching distributional consequences. Thus, although spatial planning is largely a public-sector activity, the outcomes of any planning activity affect individuals, households and private business. With such far-reaching consequences on individuals, groups and communities, it is important beyond knowing what spatial planning is, to also understand why it is necessary. Thus, this section addresses the following questions: what is the justification for spatial planning? How does the need for spatial planning fit within the broader ideas and philosophies that shape social, economic and political systems in modern societies? And what theories and principles underpin contemporary spatial planning thinking and practice?

2.3.1 *State-Market Interactions: Market Failure, the Welfare State and Spatial Planning*

The fundamental principles of state-market interaction underpin the functioning of many modern, democratic societies including Ghana—the country we focus on in this book. Although the state-market nexus is widely recognized, different societies have different philosophies with respect to how much of decisions should be left to the free market and how much should be directed by state interventions. It is possible in one context to find a strong welfare state and to find in another where free market reign supreme. The evolving balance between these forces and how they are operationalized in practice within any given context is what we call political-economy.

In Ghana, for example, the contemporary political-economy is that of a mixed economy where the free market operates alongside public-sector interventions. In other words, the private and public sectors play important roles to bring about socio-economic development. Indeed, this conception is supported by the prevailing ideologies of the two major political parties—the National Democratic Congress (NDC) who present themselves as '*social democrats*' and the New Patriotic Party who by virtue of their emphasis on development through the private sector have been branded as '*property-owning democrats*'. Thus, across this divide, we can expect at least in theory, different emphasis on the role of the state as against the role of private markets. In principle, socialists tend to favour a strong welfare state whereas governments with a leaning towards the capitalist mode of production tend to favour private enterprise and the individual liberties associated with it and, thus, focus on programmes and policies that support the principles of free markets. It is within this contemporary thinking of state-market interaction that we situate the discussion of the relevance of spatial planning.

Several years ago, socialism and *laissez-faire* economics stood as opposing ideologies. While the former advocated for dominant state involvement, the latter advocated no role for government at all (Stiglitz and Rosengard 2015). The state-market interaction framework, in contemporary times, is grounded in the notion that markets, under certain conditions such as perfect competition, generate outcomes that are efficient or socially optimal and that under such conditions free markets should dictate production and distribution. This philosophy, deeply rooted in the 'invisible hand' aphorism of Adams Smith, has been given a variety of labels such as '*laissez-faire economics*', '*neoclassical economics*' and '*neoliberalism*'. Within this framework of thinking, the state should only put in measures that will aid markets rather than hinder markets. Unnecessary interference in the market is therefore considered a nuisance with potential distortionary and inefficient consequences.

In the real world, however, markets do not operate under perfect conditions, which imply that they do not always yield efficient or socially optimal outcomes. Markets fail. Market failure manifests, for example, in the inequitable distribution of wealth, leading to a wide gap between a few who have a lot of resources/wealth and many others who have limited means to sustain their lives. In urban land and property markets, for example, the lack of decent and affordable housing; homelessness and the existence of slums are all evidence of market failure. These problems show that private markets are largely supplying housing for the affluent in society to the neglect of the shelter needs of the poor and marginalized in society. Market failures of this kind provide the rationale for government intervention: one of the tools at the disposal of the welfare state to correct such distortions in private markets is spatial planning.

Markets fail for several reasons, and a detailed discussion of these reasons is beyond the scope of this book. Instead, we will briefly discuss the sources and manifestations of market failure that justify spatial planning as a form of government intervention. In relation to spatial planning, we discuss two main sources of market failure which justifies its relevance. These are externalities and public goods. The meaning of these concepts and the underlying principles from which we can make a case for the activity of spatial planning at all levels are discussed as follows.

Firstly, although societies grant different levels of individual liberties, the actions and inactions of individuals utilising those freedoms do not only affect themselves but affect others as well. The external effects of our actions and inactions, whether positive or negative, are known as *externalities*. Positive externalities benefit others and are called '*economies of scale*', while negative externalities impose cost on others and are called '*external diseconomies*'. Externalities provide some economic justification for the way in which planning operates (Evans 2008). For example, by having the freedom to congregate in bigger settlements, we enjoy several benefits, including being able to: meet and interact with others, match businesses with labour and vice versa, attain population thresholds required to support various businesses offering various services, and provide amenities for residents (Glaeser 2010; Duranton and Puga 2004).

Notwithstanding the benefits associated with many people concentrating in cities, there are many negative externalities too. The freedom to own and use cars in cities, for example, can have negative impact on the health of others and the natural environment through pollution. Firms, in their quest to produce for profit, may end up polluting the urban environment resulting in public health consequences. Another example of negative external effects of individuals' actions is what some have termed as '*abattoir effect*'—imagine spending a lot of money to either buy or build a house and waking up to find an abattoir sited right next to your property. Another way to look at this is having a landfill or waste disposal site sited right next to your property. Given that land and property values are so strongly linked to the actions of owners of adjoining and nearby properties, land markets, if left unregulated, would exhibit serious problems of market failure (Cheshire and Vermeulen 2009), hence the need for planning. This justification for spatial planning is well articulated by MacLoughlin, one of the early thinkers in the field as follows:

> planning seeks to regulate or control the activity of individuals and groups in such a way as to minimize the bad effects which may arise, and to promote better performance of the physical environment in accordance with a set of broad aims and more specific objectives in the plan. (MacLoughlin 1969, p. 59)

Another way of looking at externalities is the effect of slums on the people who live in them and those in the wider city. Slums in themselves are manifestations of market failure, but given that environmental sanitation in these areas is often poor, they could as well become epicentres of communicable diseases such as cholera, which often affects slum dwellers and non-slum dwellers. Thus, spatial planning, through its regulatory and development control tools such as zoning and development and building permitting, endeavours to secure the interest of the public by, for example, prohibiting land use incompatibilities to eliminate the 'abattoir effect' of others' actions. Area-based interventions such as slum-upgrading programmes could also be implemented to improve sanitation conditions, thereby protecting public health.

Yet another instance where markets fail to deliver efficient outcomes, for which reason government intervention, and spatial planning for that matter is needed, is the supply of *public goods*. Public goods have two unique characteristics: firstly, they are non-excludable, meaning that individuals cannot be excluded from using them. It is generally difficult or impossible to exclude individuals from the enjoyment of a pure public good. Secondly, they are non-rivalrous, meaning that one individual's use of it does not limit its availability to others, and in most instances, it costs nothing for an additional individual to enjoy the benefits of such goods. The unique characteristics of public goods (i.e. non-excludability and non-rivalry) imply that private markets will either not supply them at all or when it does, only supply a limited amount. It is also difficult to know what the optimal quantity to provide is for public goods, since markets do not always provide the necessary signals of need or demand (Cheshire and Vermeulen 2009).

For the reasons outlined above, it becomes necessary that the government steps to provide public goods. A lot of urban amenities exhibit one or both characteristics of public goods. The provision and preservation of open spaces, the supply of parks and recreational amenities, provision of street lights, conservation of architecturally significant areas, and the demarcation and protection of environmentally sensitive areas are some examples of public goods delivered through spatial planning (Cheshire and Vermeulen 2009). Another closely related type of goods is *merit goods*. These are goods that the government compels or encourages individuals to consume because they are considered beneficial for society's overall well-being (Stiglitz and Rosengard 2015; Lipsey and Chrystal 1995). The provision of decent and affordable housing and upgrading of slums and/or relocation of slum dwellers from hazard-prone locations to safe locations, for example, may be justified as forms of government intervention to ensure people have access to decent shelter that improves their health and well-being. In many countries across Europe, spatial planning constitutes one of the key vehicles for delivering affordable housing for the poor and vulnerable in society, who without such support would be unable to secure decent housing. We address the nexus between spatial planning and housing development later in Chap. 10 of this book.

In summary, by drawing on the contemporary thinking of the roles of the welfare state and private markets in addressing various problems in society, we have provided justifications for spatial planning as a form of government intervention. We have demonstrated that although free markets are necessary in the way societies are organized, markets do actually fail. Two conditions under which markets fail, namely the presence of externalities and the supply of public and merit goods, underpin the need for spatial planning. Through land use allocation decisions, regulations and development control, spatial planning brings together complementary uses while eliminating different sources and forms of negative externalities. Similarly, public goods, such as open spaces, parks and conservation, and merit goods, such as affordable housing for which private markets have little or no incentive to supply, are delivered through spatial planning for the welfare of society.

2.4 Normative Theories of Spatial Planning: An Overview

In the forgoing sections, we have defined spatial planning, identifying its purpose, scope and goals. The relevance of spatial planning has also been justified by drawing on the principles of state-market interactions and the role of the welfare state. In addition to these, it is important that we identify and understand the various philosophies, ideas and thinking that set out what spatial planning is and how it is to be carried out in practice. Thus, we now turn our attention to some of the normative theories underpinning spatial planning.

Planning theory is a whole field of knowledge on which many scholars have written extensively (see, e.g. Allmendinger 2017; Faludi 2013; Healey 2003; Taylor 1999; Fainstein and Campbell 1996). The theories of planning have co-evolved with planning practice and the prevailing realities in the societies from which these ideas and philosophies were initially espoused. It is therefore not feasible, in the context of this chapter or even this book to cover the entire terrain of planning theory. In view of this, the goals in this section are not to offer a detailed discussion of planning theory, but instead to provide an overview of the territory, highlighting some of the key principles and thinking that continue to shape planning scholarship and practice to date. In the paragraphs that follow, we provide an overview of: systems and rational comprehensive, pragmatism, advocacy, and communicative planning and collaborative planning theories.

One of the enduring traditions of how spatial planning should be approached is the *systems and rational comprehensive theories.* This school of thought, which gained prominence in the 1960s, combines ideas from systems theory and rational decision-making theory. The former, propagated by planning theorists such as Chadwick (1971) and Mcloughlin (1969), is grounded in systems thinking in the biological sciences. A systems approach views cities and regions as complex entities with several interconnected parts. The systems view may be applied to the physical city where we treat the land uses and the infrastructures supporting them as sub-systems, which together, constitute the city as a complex system. This view could also extend beyond the physical city to encompass the people and the more intangible aspects of socio-cultural and political systems that shape the existence of cities. Almost always, the systems perspectives encompass both the physical and tangible and the aspatial and intangible forces that make up the city. Thus, the systems approach to spatial planning emphasizes the need to discern and address these complex and often dynamic interrelationships in order to address any given problem of interest.

The rational comprehensive approach, which draws partly on the systems view, also argues that planning should be concerned with scientific and objective methods that can be applied to all aspects of planning practice (Allmendinger 2009). In so doing, proponents argue that planners should separate 'means' from 'ends', meaning that planning should be separated from political processes and approached purely from a technical and rational perspective. Consequently, planning practices underpinned by systems and rational comprehensive theories focus on evaluating comprehensively, all possible courses of action and their consequences, and selecting the best possible course of action. The planner in this setting is a technician who relies on cognitive skills, tools, models and administrative expertise in carrying out the activity of planning. Contemporary planning ideals and practices including environmental impact analysis, transport demand modelling and traffic impact analysis, urban growth modelling and simulations all have roots in the systems and rationale comprehensive theories of planning.

The traditional conception of planning as a procedural, technically rational and politically neutral activity, concerned with unitary public interest, has been criticized for being undemocratic, failing to engage with the wider political processes involved in decision-making (Campbell and Marshal 2002; Healey 1997). Consequently, theories that recognize the highly political nature of spatial planning, its (re)distributive consequences and the need therefore for democratic principles to underpin spatial planning practice have emerged. One of these theories is *pragmatism*. As implied in the dictionary meaning of the word, pragmatic planning theory advocates a highly practical and incremental approach to the activity of spatial planning. This approach to planning prioritizes intellectually practical methods of getting things done by going for what works best in a given situation over philosophising and theorising (Allmendinger 2009). Leading exponents of pragmatism including Hoch (1984, 2002) argue that unnecessary bureaucratic command disguised as scientific and objective knowledge should give way to planning based on socially shared and democratic means. Within this framework, rather than reproducing and perpetuating inequalities, spatial planning should pursue social justice by using communication and negotiation as means of bringing together different interests (Campbell and Marshal 2002).

Yet another politically sensitive approach to spatial planning that takes a pluralistic and inclusive view of the activity is *advocacy* planning, formulated in the 1960s by Paul Davidoff. Planning practice underpinned by advocacy theory seeks to engage with and represent the interests of various groups in society. The planner is seen as an advocate who embraces social justice. Moreover, under advocacy planning, the notion of value neutrality and instrumental rationality is replaced with the need for planners to be aware of their own biases and to be transparent about the values that have informed their decisions. Indeed, in a bid to demystify planning as a technical activity, Davidoff goes as far as suggesting that multiple plans could be put forward to reflect the interests of different groups and using dialogue to resolve conflicting plans.

In recent years, the role of discourse as a means of determining the relative outcomes of planning for those with differential access to power (Morphet 2010) has been explored, giving birth to new approaches to planning discussed under slightly different labels including *transactive planning* (Friedmann 1973), *communicative planning* (Forester 1988; Healey 1997; McGuirk 2001) and *collaborative planning*. The central tenet of these discourse-based approaches to spatial planning essentially is an activity concerned with inclusiveness and dialogue among stakeholders, based on a deliberate attempt to pool together resources to solve problems. These theories advocate for place-making practices and strategies that are deeply rooted in the ideals

2.4 Normative Theories of Spatial Planning: An Overview

of co-production: promoting equal and reciprocal relationships among stakeholders to identify and evaluate and solve problems affecting communities.

As the brief overview of some of the theories of planning has shown, these theories help us to identify the normative principles that ought to underpin planning practice, the roles of and power relations among those involved in the planning process (e.g. planners and communities) and the expected and/or actual outcomes of the planning. Most importantly, the theories also expose the gap between what might be considered ideal in terms of the processes, practices and outcomes of spatial planning on the one hand and what is being practiced and may be achievable in reality on the other hand. As such, as the discussion has shown, these theories have evolved to point practitioners and scholars to how the status quo should and could be changed in order to improve the distributive outcomes of spatial planning among different members of society.

2.5 The Spatial Planning System: Key Features and Policy Instruments

The evolving institutional and legal frameworks and various policy instruments used to articulate visions, policies and strategies that affect the distribution of population and land use activities together constitute a typical spatial planning system. As discussed previously and further illustrated in Fig. 2.1, the relative importance attached to free markets and the state, at any given point in time, demarcates the role and scope of spatial planning and legitimizes the activity as a form of intervention by the welfare state to bring about desired socio-economic transformations. Moreover, normative planning theories provide the meta-narratives that set out what spatial planning should be and underpin how planning is carried out in practice.

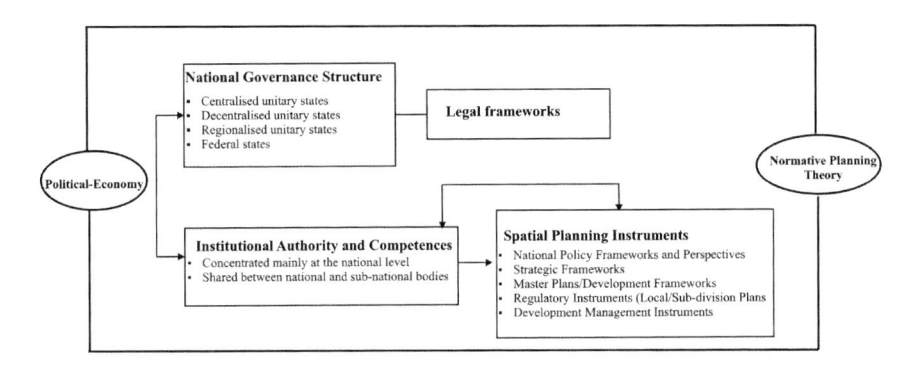

Fig. 2.1 Constituents of a typical spatial planning system

2.5.1 Governance Structure and Spatial Planning

Since spatial planning is largely a public-sector activity, the spatial planning system in any country reflects the formal national governance structure in that country. Broadly, the formal governance structure of countries within which their planning systems are situated falls within one of four main typologies, namely the centralized unitary state, decentralized unitary state, regionalized unitary state and federal state (see Silva and Acheampong 2015). Whereas decision-making powers are concentrated mainly in the central government and may seldom be delegated to subnational institutions in centralized unitary states, in decentralized unitary states, local authorities wield substantial powers delegated from the central government through a deliberate programme of devolution and decentralization. In countries operating the federal system of government, partially self-governing states under a central federal government have a high degree of constitutionally entrenched autonomy.

From the prevailing formal national governance/administrative structure derives the institutions with authority and competences in spatial planning. The institutional arrangements therefore reflect the tiers of government, so that a country running a three-tier decentralized unitary system, for example, for its spatial planning system, would have specific institutions mandated to carry out planning at the national, regional and local levels of administration, respectively. Moreover, depending on the system of governance, the authorities and competences for spatial planning are either concentrated mainly in central government departments, agencies and/or shared between similar departments and agencies at the national and sub-national levels.

All institutions at various levels of public administration derive their spatial planning powers and mandate from legislative instruments in the form of acts, ordinances and decrees. Legislative instruments promulgated specifically for the activity of planning spell out the procedures for spatial planning, define the role of stakeholders involved and give legal backing to different policies and plans that result from the activity. The spatial planning system also sources powers from other sector-specific legal instruments bordering on issues such as decentralization and devolution, land administration and property rights, building and housing construction codes, and environmental protection and biodiversity conservation.

2.5.2 Policy Instruments Deployed to Achieve Spatial Planning Objectives

Within the planning system, the various institutions and the legal frameworks defining their mandate realize the goals and objectives of the activity using various policy instruments. These policy instruments are formulated and deployed at different spatial scales and across different sector agencies to achieve specific development goals. In broad terms, there are two main types of instruments deployed to accomplish the

2.5 The Spatial Planning System ...

goals of spatial planning. The first category of instruments is those deployed in the form of *policy frameworks* and *plans*, while the second category is those deployed as *development management* instruments. As we will show shortly, these categories of instruments are mutually linked in that while policy frameworks and plans articulate spatial planning visions, goals, objectives and strategies, development management instruments function to translate the content of policies and plans into their spatial manifestations by shaping the type, timing, location and intensity of development.

Under the first category of spatial planning instruments, a four-tier system of policy frameworks and plans, namely (i) national policy and perspectives; (ii) strategic or regional plans; (iii) master plans and frameworks and (iv) local/subdivision plans, are formulated and implemented in many countries (see Commission for European Communities 1997; Silva and Acheampong 2015). We briefly describe these instruments as follows:

National policy and perspectives are used to articulate governments' vision at the national level. They may be issued as general guidelines, setting out central government priorities in different areas such as economic growth, infrastructure development and environmental conservation with or without specific spatial references. In democracies, national policy frameworks and perspectives are either short- to medium-term frameworks formulated to overlap with the tenure of governments or may be long-term plans that go beyond the tenure of any one government. As such, their content may or may not be legally binding.

National policy frameworks and perspectives often rely on lower-level instruments to translate and anchor strategic spatial visions and goals to sub-national spatial units. In countries where regional planning exists as an activity immediately below national-level planning, *strategic or regional plans* are formulated to further translate national visions into regional development priorities. Geographically, their coverage is a well-defined region or sub-region, which transcends the boundaries of single local administrative authorities. They combine social and economic policy statements with the allocation of specific quantities of growth and development for the various spatial units within their coverage. Strategic or regional plans may be indicative or legally binding.

Master plans and frameworks follow from regional plans and are therefore concerned with specific locations. They may cover single local administrative units and are used to set out the criteria for the regulation of land uses within the areas they cover. They specify broad land use zoning policies, determining where various land uses such as residential, recreational, educational, manufacturing, parks and nature reserves would be permitted and areas where such activities are prohibited. Master plans and frameworks tend to be legally binding and provide the basis for land use regulation and development control.

At the local level, the spatial planning system deploys *local land use plans or subdivision plans*. These instruments tend to cover specific locations demarcated in Master Plans as local planning areas. They may cover areas ranging from one site, a neighbourhood or several neighbourhoods. These instruments are in most cases detailed subdivision schemes showing individual parcels of land designated for specific uses (e.g. residential, commercial and recreational) and streets connecting them. They specify the codes and standards to which proposed developments should conform (e.g. building heights, dwelling types). They are therefore legally binding and provide the basis to regulate development through predetermined development and building permitting procedures which grant permits for uses that conform to the plan and rejects planning applications that they prohibit.

As the discussion in the previous paragraphs has shown, the hierarchy of planning instruments deployed by the spatial planning system articulates national visions and priorities, offers the blueprints of what the desired distributions of activities would be and provides the standards and codes to regulate development at the local level. These sets of instruments in themselves are not sufficient to bring about desired developments. Consequently, as we mentioned previously, development management instruments are deployed through the spatial planning system in most countries to contribute to the realization of strategic goals expressed in various policy frameworks and plans.

Development management instruments could be deployed to achieve specific regulatory objectives. In this case, *containment instruments* and policies such as greenbelts, development moratoria, rate-of-growth controls, urban growth boundaries and urban service boundaries are deployed to affect the timing and extent of development by directing activities to areas where development is intended to occur. *Fiscal instruments* such as taxes, exactions and fees that draw their legitimacy from the imperatives of capturing the positive externalities accruing from public investments or mitigating the negative externalities resulting from the development process are instituted as part of the spatial planning system to raise revenues through the development process (Evans-Cowley 2006).

Moreover, incentive-based development management instruments are delivered through the spatial planning system to stimulate markets and to encourage and attract more desirable activities to locations of strategic interest (Tiesdell and Allmendinger 2005). Such instruments include Brownfield Redevelopment Incentives provided to incentivize developers to offset the costs and challenges involved in redevelopment of brownfield sites while averting unsustainable urban expansion (McCarthy 2002); Historic Rehabilitation Tax Credits aimed at providing incentives for the public to preserve and rehabilitate historic places and cultural heritage (McCleary 2005) and Transfer of Development Rights a market-based incentive programme intended to reduce or eliminate development potential in places that should be preserved by increasing development potential in places where growth is wanted (Pruetz and Standridge 2008).

Summarizing, the spatial planning system, comprising institutions and legal frameworks, and policy instruments in the form of hierarchical policy frameworks and plans as well as development management instruments, is used by governments to articulate their visions and priorities and bring about sustainable development outcomes. Its contribution to socio-economic development can therefore not be overemphasized. The regulatory function of spatial planning eliminates negative externalities, provides the basis for defining and enforcing property right and protects private investments by providing certainty for business (Tiesdell and Allmendinger 2005). It supports economic growth by making developable land available and co-ordinating the provision of infrastructure (Couch 2016). Environmental protection objectives are also realized through the spatial planning system. In many countries, spatial plans are used to create the preconditions for harmonizing socio-economic development goals with environmental protection imperatives (Silva and Acheampong 2015). Plans not only designate special areas for conservation, but also contribute to their continued existence through the development control process and enforcing practices such as Strategic Environmental Assessment (SEA) and Environmental Impact Assessment (EIA). It provides the platform for stakeholders to participate in the democratic processes affecting the distribution of resources, ultimately having distributional consequences individuals, households, groups and communities. The spatial planning system provides the vehicle to provide essential public and merit goods and to protect public health. Moreover, it contributes to revenue mobilization to finance additional development and provide a wide range of services through fiscal instruments such as property taxes, impact fees and exactions. Last but not least, incentive-based instruments are deployed through the spatial planning system to stimulate markets and to promote and attract desirable activities such as historical preservation in human settlements.

2.6 Conclusions

In this chapter, we have shown that the nature, scope and purpose of the activity called spatial planning have evolved from an initial emphasis on land use designation, design and development control to embrace the economic, social, political and environmental dimensions of development. It is about place-making at different spatial scales and thus requires planners to think and act across multiple scales and between different sectors of public policy. Spatial planning is intended to influence the distribution of activities and population and to manage spatial development towards sustainable development outcomes.

Within the broader context of political-economy, the role and scope of spatial planning are determined and legitimized by the relative importance attached to free markets and the state at any given moment in time. While free markets are essential in the development process, we have demonstrated that markets do fail and that market failure provides some justification for spatial planning. Spatial planning regulates and seeks to eliminate negative externalities endemic in human settlements. It is a vital vehicle for governments to provide essential public and merit goods for the well-being of society.

Our discussion of normative planning theories also demonstrated a shift in the traditional conception of planning as technical, apolitical and value-free activity to renewed emphasis on imperatives of planning practice underpinned by theories that recognize the highly political nature of the activity and thus promote the democratic principles of dialogue, citizen participation and social justice. The purpose of spatial planning, as reflected in the various concepts, theories and principles, is accomplished through an evolving system of institutional legal frameworks and policy instruments in the form of hierarchical policy frameworks and plans as well as development management instruments which together comprises the spatial planning system.

The themes covered in this chapter, therefore, set the scene for the remaining chapters of this book. With the background knowledge provided by this introductory chapter, we will proceed in Chap. 2 to begin to examine spatial planning in Ghana, discovering the historical origins of the activity and examining the key features of the contemporary spatial planning system and practices in the country.

References

Albrechts L (2004) Strategic (spatial) planning reexamined. Environ Plan B: Plan Des 31(5): 743–758
Allmendinger P (2009) Critical reflections on spatial planning. Environ Plan A 41(11): 2544–2549
Allmendinger P (2017) Planning theory. Macmillan International Higher Education
Campbell H, Marshall R (2002) Utilitarianism's bad breath? A re-evaluation of the public interest justification for planning. Plan Theory 1(2): 163–187
Chadwick G (1971) A systems view of planning: towards a theory of the urban and regional planing process. Pergamon Press, New York
Chapin FS (1965) Urban land use planning. University of Illinois Press, Urbana, IL
Cheshire P, Vermeulen W (2009) 6 Land markets and their regulation: the economic impacts of planning. Int Handb Urban Policy: Issues Dev World 2: 120
Commission of the European Communities (CEC) (1997) The EU compendium of spatial planning systems and policies. Regional Development Studies, 28 (Luxembourg: CEC)
Couch C (2016) Urban planning: an introduction. Macmillan International Higher Education
Cullingworth JB (1972) Town and country planning in Britian. Allen and Unwin, London
Duranton G, Puga D (2004) Micro-foundations of urban agglomeration economies. In Handbook of regional and urban economics, vol 4, Elsevier, pp 2063–2117
Evans AW (2008) Economics and land use planning. John Wiley & Sons
Evans-Cowley J (2006) Development exactions: process and planning issues. Lincoln Institute of Land Policy, Cambridge
Fainstein SS, Campbell S (eds) (1996) Readings in urban theory. Blackwell, Oxford, pp 216–245
Faludi A (2013) A reader in planning theory, vol 5. Elsevier, Amsterdam
Friedmann J (1973) Repacking America. A theory of transactive. Doubleday, New York
Friedmann J, Weaver C (1979) Territory and function: the evolution of regional planning. University of California Press
Forester J (1988) Planning in the face of power. Univ of California Press
Glaeser EL (2010) Introduction to "Agglomeration Economics". In Agglomeration economics. University of Chicago Press, pp 1–14
Healey P (1997) The revival of strategic spatial planning in Europe. In Making strategic spatial plans. UCL Press, London, pp 3–19
Healey P (2003) Collaborative planning in perspective. Planning Theory 2(2): 101–123
Healey P (2004) The treatment of space and place in the new strategic spatial planning in Europe. Int J Urban Reg Res 28: 45–67
Hoch C (1984) Doing good and being right the pragmatic connection in planning theory. J Am Plan Assoc 50(3): 335–345
Hoch CJ (2002) Evaluating plans pragmatically. Plan Theory 1(1): 53–75
Keeble LB (1969) Principles and practice of town and country planning. Estates Gazette

References

Lichfield N (1964) Cost-benefit analysis in plan evaluation. Town Plan Rev 35(2): 159

Lipsey RG, Chrystal A (1995) Economics. Oxford University Press, Oxford

McCarthy L (2002) The brownfield dual land-use policy challenge: reducing barriers to private redevelopment while connecting reuse to broader community goals. Land Use Policy 19(4): 287–296

McCleary RL (2005) Financial incentives for historic preservation: an international view. Thesis University of Pennsylvania

McGuirk PM (2001) Situating communicative planning theory: context, power, and knowledge. Environ Plan A 33(2): 195–217

McLoughlin JB (1969) Urban & regional planning: a systems approach. Faber and Faber

Morphet J (2010) Effective practice in spatial planning. Routledge

Murphy PA (1984) Regional planning: purpose, scope and approach in New South Wales. Australian Planner 22(4): 16–19

Pruetz R, Standridge N (2008) What makes transfer of development rights work?: Success factors from research and practice. J Am Plan Assoc 75(1): 78–87

Silva EA, Acheampong RA (2015) Developing an inventory and typology of land-use planning systems and policy instruments in OECD countries. OECD environment working papers, No. 94

Stiglitz JE, Rosengard JK (2015) Economics of the public sector: fourth international student edition. WW Norton & Company

Tiesdell S, Allmendinger P (2005) Planning tools and markets: towards an extended conceptualisation. Planning, public policy & property markets, pp 56–76

Taylor N (1999) Anglo-American town planning theory since 1945: three significant developments but no paradigm shifts. Planning Perspectives 14(4): 327–345

United Nations (2014) World urbanization prospects, the 2011 revision. Population Division, Department of Economic and Social Affairs, United Nations Secretariat

Chapter 3
Historical Origins and Evolution of Spatial Planning and the Planning System in Ghana

Abstract The spatial planning system and the accompanying practices in Ghana have evolved since the inception of the activity in 1945 through the Town and Country Planning Act (CAP 84). This chapter traces the historical origins of spatial planning from the colonial era to modern-day Ghana. It reviews landmark spatial planning projects implemented in the early years after the declaration of independence to identify the scope and purpose of the activity then. The return to democratic governance at the beginning of the twentieth century, and the decentralization programmes that followed entrenched a new system of planning in Ghana, called the development planning system. The ethos and scope of planning under the development planning system as well as its impacts on spatial planning are examined. Reforms in recent years that are meant to institutionalize a new spatial planning system and strengthen the activity at all levels are also discussed. The notion of '*a one country two planning systems*' will be introduced in this chapter to set the stage for further analyses in the subsequent chapters of the implications of having two competing planning systems in Ghana (i.e. the development planning system and the new spatial planning system).

Keywords Physical planning · Comprehensive plan · Master plan · Democracy Decentralization · Development planning · Planning systems
Ghana Spatial planning

3.1 Introduction

In this chapter, we begin the discussion of the spatial planning system and the activity of spatial planning in Ghana. As we discovered in the previous chapter, the activity called spatial planning in this book has been known and continues to be known by other related terminologies. In Ghana, the use of the term *spatial planning* is recent. Since its inception, the activity of planning has been known by three main terminologies; '*physical planning*', '*land use planning*' and '*town and country planning*'. Indeed, the public-sector department that was established in 1945 and mandated to perform spatial planning functions is still known today as the Town and Country Planning Department (TCPD).

In the last decade or so, spatial planning has formally entered the academic and policy lexicon, being used more frequently alongside its predecessor terminologies by government institutions, scholars and practitioners. Indeed, as we will see shortly, the new legislative framework accompanying reforms of the activity use the compound terminology '*land use and spatial planning*', ostensibly to reflect both the traditional regulatory function and the emerging strategic role that spatial planning is expected to play. Throughout this book, we use the term spatial planning to connote both the old and the new understanding of what the nature, scope and purpose of this activity encompasses.

The spatial planning system and the activity of spatial planning has evolved since it was first introduced in Ghana in the mid-1940s when Ghana was still under British colonial rule. We will therefore begin this chapter by tracing the historical origins of spatial planning in Ghana. Landmark spatial planning projects, especially those implemented in the early years after the declaration of independence in 1957, will be reviewed, identifying the scope and goals of spatial planning then. The inception of decentralization in the late 1980s ushered in a new national planning system which institutionalized a new planning tradition that was not explicitly spatial in terms of scope, focus and outcomes. This new tradition of planning and the associated practices would become known as the development planning system. We will examine the nature of this ethos of planning and the impact it has had on contemporary spatial planning in Ghana. Emerging reforms in the early part of the twenty-first century (i.e. 2007–2010), following a renewed interest in spatial planning has entrenched a new spatial planning system. We will examine the features of the new planning system, focusing on the legal and institutional context within which it is meant to operate as well as the set of policy instruments it would use to deliver its objectives.

3.2 Origins of Spatial Planning in Ghana

The story of spatial planning in Ghana can be divided into four broad timelines of landmark events. The first is the period before 1945 when the then Gold Coast was still under British rule. The second major event worth noting in the history of spatial planning is the promulgation of the Town and Country Planning Ordinance (CAP 84) in 1945. Later on, from 1957 when Ghana became an independent country, this ordinance would provide the legal basis for a series of plan-making projects at the national level and for the country's major towns and cities. The period between 1966 and the late 1980s marked the dark days in the history of independent Ghana as successive military interventions saw the overthrow of governments and a dismantling of the state apparatus, including any form of sustained long-term strategic planning at the national and sub-national levels. After close to four decades of political and economic instability, the reversion to constitutional rule in 1992 and the decentralization reforms that gained impetus immediately after would bring into existence a new tradition of planning. As we will later discover, this new tradition of planning almost eliminated spatial planning from policy discourse and rendered the institutions with competences in the field weak to carry out their mandate effectively. The last momentous period in the history of spatial planning is the reforms that were implemented between 2007 and 2010 under the Land Use Planning and Management Project (LUPMP). These reforms would once again bring spatial planning into mainstream policy and academic discourse and initiate a number of projects to experiment with various spatial planning models, introduce a new legislation and entrench a new system of hierarchical spatial planning at the national, regional and local levels of political administration. In the sections that follow, we will examine the nature, scope and purpose of spatial planning during these significant periods in history.

3.2.1 Pre-1945 Colonial Planning: Essentially Spatial Plans in Character

Modernist planning systems in Africa including Ghana are deeply rooted in their European doctrines of planning which were introduced during the days of colonialism. The modern state of Ghana, then Gold Coast, was shackled under British imperialism for a period of 113 years from 1844 to 1957 when the struggle for self-rule and liberty culminated in political independence. While the overarching motive of British occupation of the Gold Coast was that of natural resource exploitation to propel the industrial revolution underway back home (Nkrumah 1973), a handful of significant events happened in the latter part of the twentieth century that in many respects, signified the earliest forms of spatial planning in Ghana.

The ten-year Development Plan of Gordon Guggisberg (9 October 1919–24 April 1927) and the four-year Development Plan of Alan Burns (29th June 1942–2nd August 1947), both agents of British exploitation of the Gold Coast, sought to develop infrastructure and social services. These plans targeted improvements in transportation infrastructure, development of hydroelectric projects, delivery of public services such as water, sanitation, schools and hospitals and towns improvement in the form of housing development schemes. They defined and mapped specific geographic areas where these major physical developments would be implemented to achieve specific strategic objectives.

Except for the social services such as schools and hospitals that benefited the masses, a significant share of the infrastructure proposals implemented under these plans was built in the resource-rich southern parts of the country. Notable among them are the Takoradi Harbour, the existing railway lines in the southern part of Ghana, the development and improvement of some 5340 km of roads and the opening of diamond mines in the south-eastern and south-western parts of Ghana. Clearly, the geographical distribution of these projects shows that they were implemented to further strengthen the infrastructure base needed to facilitate resource exploitation in the Gold Coast. Notwithstanding, the physical infrastructure development focus of these plans and their clear expression in spatial terms meant that they were essentially spatial plans. These could therefore be duly credited as the earliest examples of spatial planning projects in Ghana.

3.2.2 The 1945 Town and Country Planning Act (CAP 84)

Spatial planning as we know it today was formally introduced in Ghana through the 1945 Town and Country Planning Act (CAP 84). By 1945, World War II had ended, ushering in a new dawn of a post-war reconstruction programme in Britain, which then still occupied the Gold Coast. New planning legislations were introduced between 1945 and 1952 in Britain to guide the post-war reconstruction programme (Greed and Johnson 2014). Thus, the planning system that existed in Britain then was exported to Ghana, one of its colonies in West Africa. The 21 April 1945 Act established the TCPD in Ghana and laid the foundations for the modern planning system. The Act which was the first of its kind to define the scope, purpose and legal basis for spatial planning set the ambitious objective to:

> …[provide] for the orderly and progressive development of land, towns and other areas, to preserve and improve their amenities and for related matters (Town and Country Planning Act 1945 (CAP 84)).

In addition, the Act of 1945 defined development in relation to land to include:

> …a building or re-building operations and the use of the land or a building on it for a purpose which is different from the purpose for which the land or building was last being used (Town and Country Planning Act 1945 (CAP 84)).

The regulatory and development control functions of spatial planning and the role it would play in the agenda of modernization and social transformation were clearly expressed in its intent to bring about orderly and progressive development of human settlements in the country. In practice, however, the outcomes of planning were far from what appeared to be the selfless progressive agenda proclaimed by the 1945 Act in Ghana and the other colonies in Africa where similar legislations were introduced. Instead, the planning ordinance would become instrumental in facilitating, strengthening and perpetuated the exploitative and spatially biased development agenda initiated by the pre-1945 colonial development planning system.

While the planning system in its home origin was deployed to rebuild the British economy and societies in general crippled by the war (Greed and Johnson 2014), the type of planning instituted in the colonies, including the Gold Coast, in terms of outcomes did not measure up by any imaginable standards to the welfare objectives that were aggressively being pursued back home in Britain. Instead, spatial planning was limited by scope, covering few urban centres. It largely parochial in its interests, objectives and outcomes as it was directed towards securing the well-being of the representatives of the colonial representatives and the few numbers of local elites. It was essentially British officials and their families first; all other benefits to the masses were merely trickle-down effects. In its worse expression, segregationist objectives,

3.2 Origins of Spatial Planning in Ghana

as was seen in its worst form in South Africa and other colonies, were pursued and realized, using planning and zoning policies (Mabogunje 1990). As noted by Mabogunje concerning the scope, intent and outcomes of planning in the colonies:

> In its colonial origin, therefore, it was improvement in community health and general sanitary conditions for the white population who were the resident agents of colonial capitalism that provided the rationale for undertaking limited planning of urban areas... All over Africa, therefore, colonial urban planning had its origin not so much in the state effort at resolving class conflicts or struggles but in its direct commitment to ensure adequate health protection for its agents and those of international capitalism. (Mabogunje 1990, pg. 137–139)

In addition to establishing the TCPD, the 1945 Act instituted the Town and Country Planning Board, whose powers were vested in the TCPD to perform the spatial planning functions stipulated in the Act. Subsequent amendments—The Town and Country Planning Act of 1958 (Act 30) and Town and Country Planning (Amendment) Act of 1960 (Act 33)—abolished the Town and Country Planning Board. In its place, planning powers were vested with the Minister responsible for Town and Country Planning. The TCPD operated purely as a Civil Service Department with the head office in Accra and branch offices at the administrative regions and the districts. The cadre of professional planners were trained in Britain and carried out the task of planning under British Town Planning Advisors. This institutional setup whereby powers were vested with the Minister for planning prevailed until 1993 when administrative reforms set in motion a decentralization programme.

The approach and processes conceived and entrenched by the 1945 planning ordinance, by which places in the country would qualify for a plan, were deeply woven into the bureaucratic structures of the day. The 1945 Act and the subsequent amendments (i.e. Act 30 of 1958 and Act 33 of 1960) instituted a procedure referred in the Acts as '*Designation of Planning Areas*'. Under this procedure, before a planning scheme was prepared for an area, the Minister responsible for town planning would consult with the relevant local governing authority. If after the consultation, the Minister was of the opinion that a scheme should be prepared for an area, the Minister may, by executive instrument declare that area as planning area. In other words, the decision for extend the benefits of land use planning to any village, community, neighbourhood, town or city was reached through the discretionary powers granted the Minister. The executive instrument would require gazetting and remain in force for a period of three years after which it ceased to have effect if a scheme was not prepared and approved for the designated area. It was also a legal requirement that a copy of the instrument be posted at places within the planning area to elicit concerns and views from the relevant stakeholders.

Moreover, on the publication of the planning instrument, a Planning Committee was appointed by the Minister for the area concerned. The Planning Committee, as the Acts stipulated, would then be required to provide the particulars and information that the Minister may require with regard to the present and future planning needs of and the probable direction and nature of the development of its area. Moreover, the instruments declaring the planning area, once published, prohibited any form of development including construction, demolition, alteration, extension and repair or renewal of a building until a final scheme was approved for the area.

From the forgoing, the key features of the pre-independence planning system formally instituted in 1945 could be summarized as follows:

i. The planning system was highly centralized. This is evidenced by the fact that decisions regarding the preparation of a plan for an area under the jurisdiction of local government authorities were taken at the highest level of decision-making by a Minister responsible for town planning. Despite being a highly centralized activity, spatial planning in this era had an overall positive impact in terms of bringing about strategic infrastructure development, distributing national resources across regions and shaping the growth and development of major towns and cities (see e.g Fuseini and Kemp 2015; Adarkwa 2012);
ii. Planning was carried out through a piecemeal approach where the practice of declaring planning areas meant that not all areas qualified for and benefited from planning interventions. Instead, decisions as to whether a plan would be prepared for an area or not depended on the discretionary powers of the Minister. This also implied that planning could easily be directed towards achieving strategic objectives of the colonial government;
iii. While the approved plan provided the legal basis for development control, a moratorium instrument was imposed to halt development over the period of plan preparation;
iv. Stakeholder participation was largely by information. The legal requirement to post instruments declaring an area a *planning area* supports this assertion. This also meant that only the few educated members of the population then were the primary targets and could probably participate in the process.

From the aforementioned features of the planning system, it becomes apparent the implications of the design and functioning of the planning system on the outcomes of spatial planning and the distributional consequences of the benefits of planning over different groups in society. It is also worth noting that the planning system introduced then has remained entrenched and continues to shape planning practice even today. Later in Chap. 5, these legislative instruments and the planning systems they instituted would be revisited, assessing the impacts that they have had on development control processes and spatial development outcomes today.

3.3 Landmark (Spatial) Plan-Making Projects in the Era of Independence

At the peak of the struggle for independence in the 1950s, which would eventually bring an end to over a century of British oppression in the Gold Coast on 6 March 1957, the Dr. Kwame Nkrumah government initiated an ambitious programme of socio-economic transformation in the new Ghana. Nkrumah's vision firmly grounded in the philosophy and ideals of socialism sought to build the productive structures needed for economic growth and to develop infrastructure and social services to improve the well-being of the population. To translate this grand vision into action, a series of planning-making projects at the national and city level, both socio-economic and spatial in scope, were formulated. A brief review of these initial plans is presented in the sections that follow.

3.3.1 National Perspectives—Five-Year Plan (1951–1956) and Seven-Year Development Plan (1963/64–1969/70)

In 1951, the first National Perspective, the Five-Year Development Plan (1951–1956) was formulated. Before the five-year development, The Plan for the Economic and Social Development of the Gold Coast, a ten-year plan to further cement British rule of the Gold Coast from 1 April 1950 had been formulated by the colonial administration. Nkrumah, then Leader of Government Business, on the basis of the prevailing socio-economic conditions, found it expedient to begin the transformation agenda that he had envisage with midterm national perspective. This resulted in the government collapsing the ten-year plan into a five-year National Perspective. By independence in 1957, which would see Nkrumah become the first president of Ghana, the landmark achievements from the first National Plan included the completion of the Volta Valley Scheme; completion of the Aluminium Smelter Plant; the establishment of the Industrial Development Corporation to oversee the formation of industrial estates and provide funding for industries; and the establishment of the Tema Development Corporation (TDC) to provide housing for the newly created Tema industrial city.

In October 1961, the National Planning Commission was set up to formulate the second National Perspective/Plan, the Seven-Year Development Plan for National Reconstruction and Development (1963/64–1969/70). The seven-year plan put forward a strong case to replace the hitherto narrower conception of planning with a tradition of national-level comprehensive planning in Ghana. This would become the medium through which the government would articulate its socialist policies and help build the public and cooperate sectors in the productive economy that it had envisaged. The need for a comprehensive approach, despite the prevailing institutional challenges was well articulated in the introduction of the plan document:

> In spite of the small size [of the National Planning Commission] and relative inexperience of the apparatus of planning in Ghana, it was early on decided to attempt a comprehensive Plan along the lines set out in the in this document rather than to restrict our planning to the drawing up of a public investment programme as has been done hitherto in this and other developing countries. (The Seven-year Development Plan 1963/64–1969/70, p. vii)

The resulting plan set out an ambitious modernization agenda with the aim of creating a society:

> ..in which the individual Ghanaian will be able to enjoy a modern standard of living in his home supplemented by an advanced level of public services outside. (The Seven-year Development Plan 1963/64–1969/70, p. 1)

3.3.2 The National Physical Development Plan (1963–1970)

The Seven-Year Development Plan for National Reconstruction and Development (1963/64–1969/70) was not only unique for its comprehensiveness, but also for being the first National Perspective to have been accompanied by a truly spatial plan—the National Physical Development Plan (1963–1970). The plan, prepared by the Town and Country Planning Division of the then Ministry of Land, emphasized the need for an integrated planning, which would take into consideration the economic, social political and physical aspects of development in a single coherent framework. Considered the first of its kind in Africa, the physical development plan was deployed as the first of a series of short-term physical plans to guide the realization of the long-term goals of National Perspective/Plan.

Moreover, the recognition that development and development planning are multi-dimensional and complex was evident in the formulation of the first physical plan. It recognized that for maximum welfare benefits, the comprehensive planning tradition instituted in the 1960s needed to move a step further to consider the timing, numbers and types of productive activities and social facilities to be provided, and to identify areas where these facilities would be located.

Consequently, the aims of the physical plan were to:

i. Guide the annual phasing and budgeting of the seven-year investment programme;
ii. Solve the locational problems in connection with all investment project from 1963 to 1970;
iii. Guide investment assumption for the preparation of the perspective plan;
iv. Guide the elaboration of regional development plans;
v. Guide the elaboration of local development plans for cities and urban settlements; and
vi. Guide the establishment of a practical system of survey and analysis.

The making of the plan followed the technical process beginning with data collection and analysis to understand the problems for which interventions were needed, with proposals and strategies for implementation coming out at the end of the process. As demonstrated in the analyses of the background socio-economic challenges that had necessitated its formulation, the physical plan recognized that to achieve the goals of the National Perspective would require profound changes in the demographic and employment structure of the country, which in turn, would affect growth and possibly decline of various settlements. In other words, the plan understood well that real impacts of the national vision would be felt at the level of villages, towns and cities, hence the need to address the micro-, spatial distribution implication of the national vision. The overall spatial strategy of the physical development plan, comprising the patterns of individual settlements, their organization and functions within the national space economy envisaged by the plan, was to provide answers to two basic problems:

i. The determination of the number of people, employment structure, the level of social facilities and the general living conditions in particular localities and regions; and
ii. The determination of the need as well as the possibilities and economic potentialities of particular localities and regions, which in the long run should either conform the feasibility of the development programme or lead to its revision.

In corollary of the aforementioned strategic imperatives, about eight thematic areas, covering topographic and natural elements, population growth and distribution, agriculture, industry, infrastructure, urban settlement pattern, service centres and social services, were identified in the plan. Under each of these thematic areas, background situational analyses fed into future forecast and projections of needs and demand, which were, in turn, translated spatially by identifying and mapping the locations of all proposals. A comprehensive schedule detailing the construction period, estimated costs and projected employments to be generated by individual projects accompanied the physical plan. Given its national coverage, estimates of the land requirements for the proposals were, however, left to be detailed out in the local plans that were to be developed from the national physical plan.

3.3.3 The Master Plan for Accra

By the late 1950s, national- and city-level spatial planning had been established as one of the major vehicles of the government to deliver its wider socio-economic transformation goals. Following the promulgation of the Town and Country Planning Act 1945 (CAP 84) and the introduction of comprehensive planning at the national level, a series of lower-tier instruments were prepared. These instruments called Master Plans were to translate the visions of the National Perspectives and the National Physical Development Plan into action and provide the blueprint for the future development of major towns and cities in Ghana.

The Master Plan for Accra officially captioned '*Accra: A Plan for the Town*', which was formulated in 1958, constituted one of the first landmark spatial planning projects at the city level. Prior to the 1958 Master Plan, an overall plan for Accra had been prepared in 1944, although this initial plan never became a statutory one. Attempts were, however, made to fit overall pattern of development of Accra to the 1944 Plan. Thus, to a larger extent, the 1944 plan had to a larger extent dictated desired patterns of physical development in the capital city despite never becoming a statutory plan. Recognizing the impacts that the 1944 plan had had in shaping development of the capital city, the then Minister of Housing in the introductory section acknowledged:

> …all development in the town, since [1944] has been legally subject to planning control…The result of this form of control of the innumerable buildings which have been erected over the years, each one insignificant in itself but gradually forming a whole town, can now beginning to be appreciated. Accra already has the definite making of a fine city". (Minister of Housing, Accra: A Plan for the Town—1958)

The 1958 Master Plan, therefore, was to build on the gains of the 1944 plan. The Master Plan in its conception was to reflect the ambitious modernization and transformation agenda initiated by the National Perspectives. Indeed, Dr Kwame Nkrumah's foreword accompanying the Master Plan reiterated the national vision that

> it is fitting that we should improve our main towns alongside our rural and industrial development and that our capital city should offer improved amenities and standards of living. (Prime Minister Kwame Nkrumah, Foreword Accra: A Plan for the Town—1958)

The Master Plan itself was modest in its objectives and proposals, with the aim to 'improve present conditions and to give future development to be carried out in a manner fitting to the dignity of a capital city and which is also practically and economically feasible'. Some of the key features and proposals in the plan are summarized below:

- A series of traffic flow studies formed the basis for road improvements and development of new roads to serve the new development areas. The existing and proposed road system therefore defined the city structure and delimited the extent of planned expansion.
  A residential zoning plan specified the new areas planned for future expansion of the capital city. A density zoning policy specified three main residential development zones, namely high-density, medium-density and low-density housing

Fig. 3.1 Proposed residential densities in the 1958 Master Plan for Accra. *Source* Reproduced digitally from analogue maps obtained from the TCPD, Accra

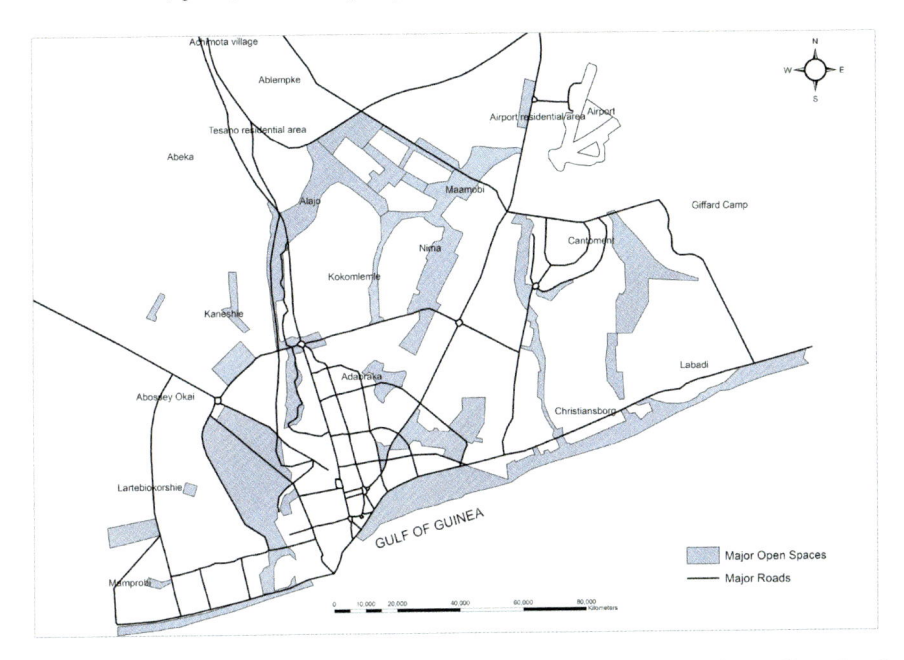

3.3 Landmark (Spatial) Plan-Making Projects in the Era of Independence

Fig. 3.2 Distribution of major open spaces in the 1958 Master Plan for Accra. *Source* Reproduced digitally from analogue maps obtained from the TCPD, Accra

development zones. High-density development was planned for the central areas of Accra, except for the areas designated as earthquake-prone zone (see Fig. 3.1)
- A network of parks, gardens and recreational centres were also demarcated in the Master Plan (see Fig. 3.2).
- Recognizing existing areas that exhibited conditions of slum, the plan contained proposals to address the emerging slum formation. Three main areas reflecting proposed interventions were designated. These were 'slums for remedial treatment'; 'slums for demolition'; and 'areas proposed for reduced structural standards' (see Fig. 3.3). Within the central area alone, about five slum settlements were proposed for demolition.
- Detailed neighbourhood layouts for private development specified the types of houses to be built, their arrangement in space as well as the locations of ancillary amenities (see Fig. 3.4)

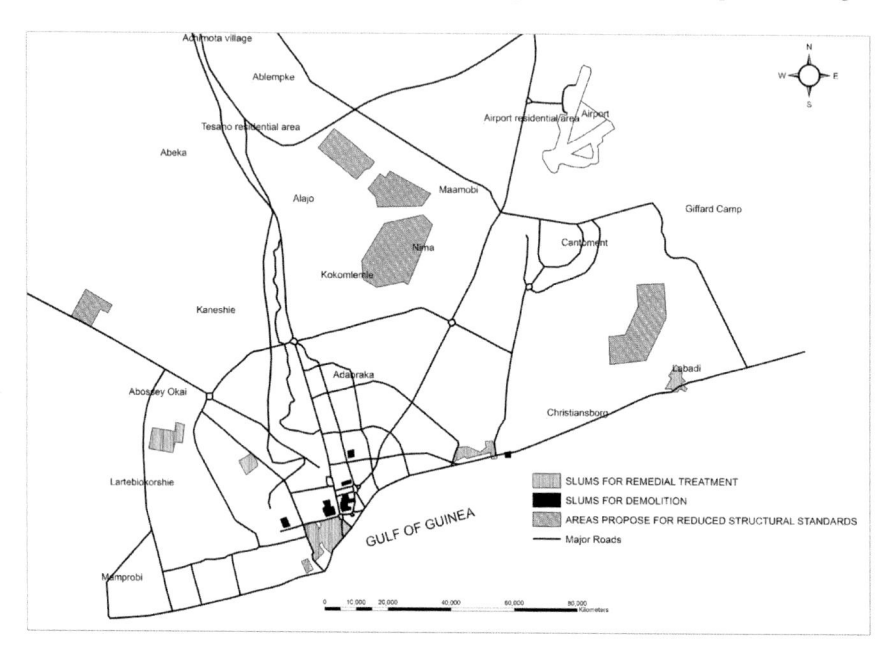

Fig. 3.3 Designated as slums and areas proposed as suitable for lowering of construction standards in the 1958 Master Plan for Accra. *Source* Reproduced digitally from analogue maps obtained from the TCPD, Accra

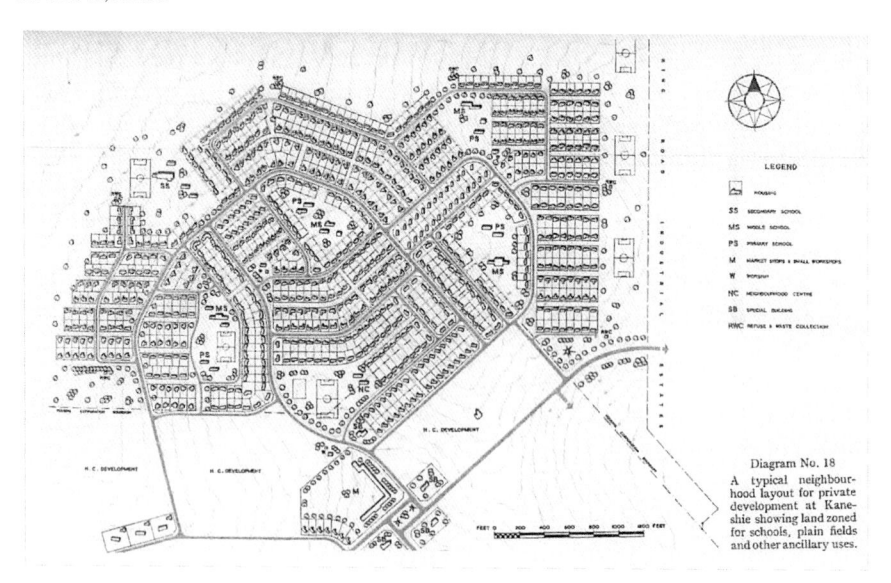

Fig. 3.4 A typical neighbourhood plan accompanying the 1958 Master Plan for Accra for private development at Kaneshie, Accra. *Source* 1958 Master Plan for Accra, TCPD, Accra

3.3.4 The Master Plan for Tema

The national industrialization and infrastructure development policies articulated by the post-independence comprehensive planning projects necessitated the preparation of the Master Plan for Tema. In the broader scheme of things, Tema would become a major industrial hub, accommodate a major port surrounded by thriving communities of some 80,000 people on a 5000-acre land based on an initial plan prepared by the Tema Development Corporation.

Construction of the Town of Tema started in 1954 based on an initial plan, which provision for four main activity zones, namely

- The residential zone comprising seven communities that would accommodate estimated populations of 10,000 to 12,000 inhabitants each;
- A major industrial zone, including an area designated for the construction of the aluminium smelter;
- The town centre; and
- The harbour area.

By 1961, two of the seven proposed residential communities accommodating some 22,000 people had been developed. To plan for future expansion and development of the Town, the government initiated another landmark Master Plan project for the town. The plan, formulated by Doxiadis Associates, proposed a general Master Plan to cover the whole 8000-acre acquisition areas of Tema as well as a 25-year programme for the development of the town and its industrial enclaves. It was estimated that over the 25-year period from 1961, the Town would accommodate an estimated population of 235,000 to 250,000 inhabitants.

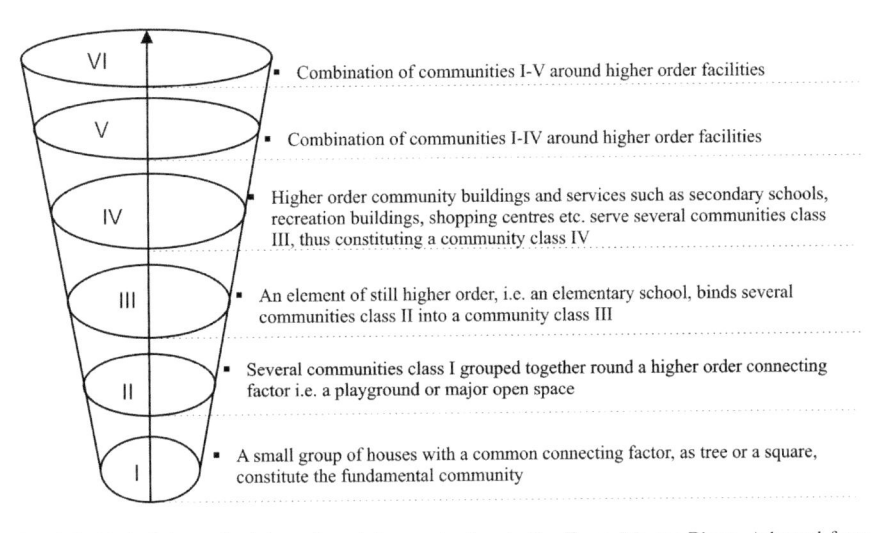

- Combination of communities I-V around higher order facilities
- Combination of communities I-IV around higher order facilities
- Higher order community buildings and services such as secondary schools, recreation buildings, shopping centres etc. serve several communities class III, thus constituting a community class IV
- An element of still higher order, i.e. an elementary school, binds several communities class II into a community class III
- Several communities class I grouped together round a higher order connecting factor i.e. a playground or major open space
- A small group of houses with a common connecting factor, as tree or a square, constitute the fundamental community

Fig. 3.5 Underlying principles of spatial organization in the Tema Master Plans. Adapted from Doxiadis Associates (1962)

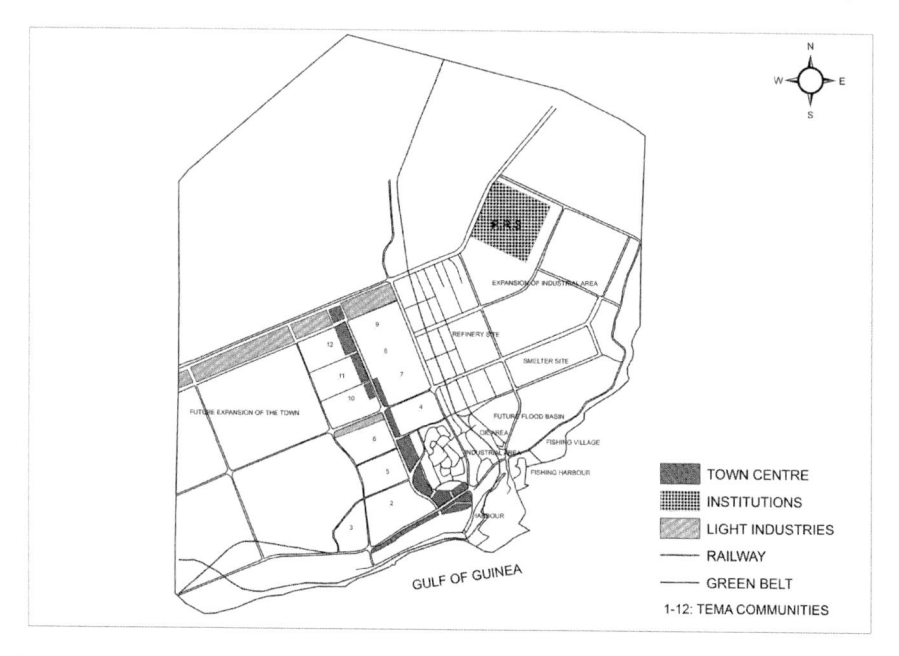

Fig. 3.6 Proposed road system in the Tema Master Plan. *Source* Reproduced digitally from analogue map obtained from the TCPD, Accra

Furthermore, the Tema Master Plan allocated about 2800 acres for industrial development and 15 acres for development of the town centre. The Master Plan concept followed a bottom-up organization of communities/neighbourhoods around a hierarchy of facilities and services. The underlying philosophy and the plan in its totality was stated as follows:

> The whole Master Plan reflects the overall conception of the Metropolitan Area so far as the part referring especially to Tema areas is concerned. The main grid of roads is formed by the principal arteries which define the limits of the various communities in both the residential and industrial areas and connect them with the port…[it] provides for the creation of the Centre of the town, the zone of light industry and handicrafts along the Accra-Tema Freeway, and green and recreation areas particularly towards the Sakumo Lagoon. (Doxiadis Associates 1962, p. 161)

The resulting spatial organization of the seven communities envisaged is illustrated in Fig. 3.5. Following this concept, each community on its own was to become relatively self-sufficient, while the Town in its entirety (i.e. aggregation of all seven communities) was to be organized around the town centre where higher order facilities and services would be provided. Moreover, the Town was planned to be built as a mixed community, housing households of different income groups (see Figs. 3.6 and 3.7) from the lowest to the highest living in different housing types from isolated dwellings, apartment buildings, small single-storey houses and multi-storey towers (Doxiadis Associates 1962).

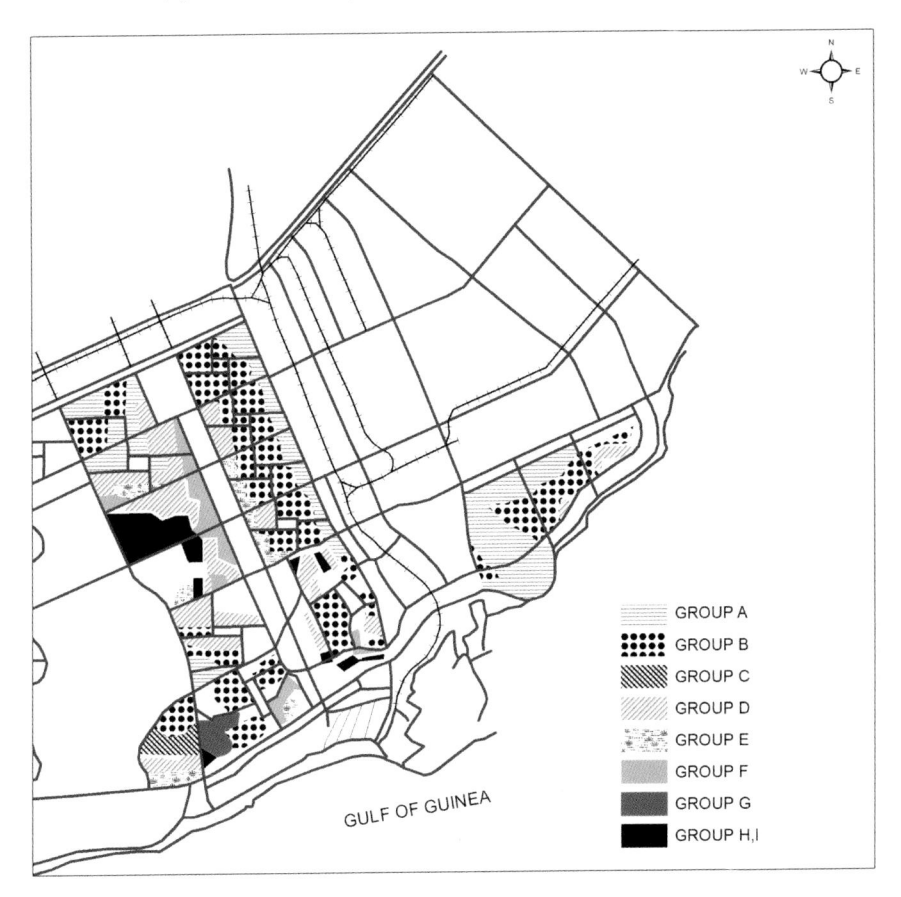

3.3 Landmark (Spatial) Plan-Making Projects in the Era of Independence

Fig. 3.7 Distribution of communities based on income groups in the Tema Master Plan. *Source* Reproduced digitally from analogue map obtained from TCPD, Accra

In addition to the land use zones proposed, sketches of recommended standardized housing units accompanied the final plan. The introduction of standardized types of houses, according to the planners was needed for easy application of mass production techniques and to achieve unity and rhythm by establishing a common measure of order required to realize the large-scale synthesis of the Master Plan (Doxiadis Associates 1962).

3.3.5 *Planning Scheme for Kumasi, 1963*

As one of the biggest cities in Ghana during the era of independence, Kumasi also benefited from the governments master planning interventions. In 1963, the Planning Scheme for the Kumasi metropolis was prepared to guide its development. The

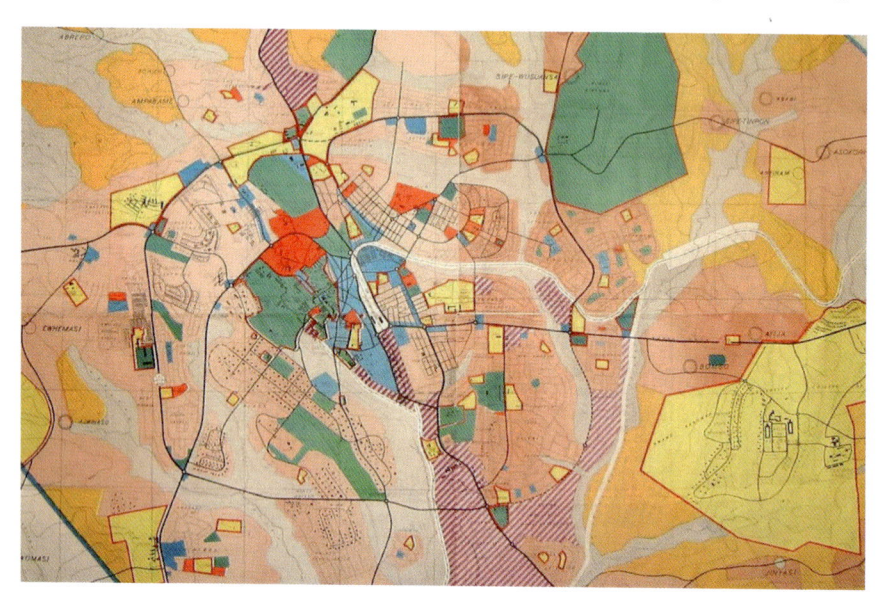

Fig. 3.8 Planning Scheme for Kumasi, 1963. *Source* TCPD, Kumasi

Plan (see Fig. 3.8), provided broad land use zones to with provisions for industrial, commercial and residential areas and became the principal document for land use regulation and development control. Prior to the preparation of this landmark Master Plan, the city of Kumasi had been noted for the urban overall urban greenery, the lush vegetation, colour shrubs and well-manicured lawns that characterized the landscape, earning it the accolade, 'Garden City of West Africa' (Korboe 2001). Thus, the 1963 Plan was initiated to among other things, preserve the Garden City character of the city and provide the basis for the development of modern infrastructure for the city.

The three plan-making projects outlined here is by no means an exhaustive list of the various schemes that were prepared in the period following independence Ghana. Indeed, there were other important spatial planning projects including the construction of 52 new towns, such as Adumasa, Akrade, Mpamu and Senchi around the Volta Lake as resettlement schemes for populations displaced by the construction of the Akosombo hydro-electric dam (Adarkwa 2012). Undoubtedly, these Master Plans covering the major Towns then (i.e. Accra, Tema and Kumasi) were significant in that they initiated and institutionalized the tradition of land use planning and development control at the level of towns and cities in Ghana.

Rather unfortunately, the relative peace and stability in the independence era that enabled these initial National Perspectives and Master Plans were short-lived. Dr. Kwame Nkrumah in 1966 was overthrown, and the period afterwards witnessed a series of military takeovers. A series of short-term plans characterized attempts at planning during this turbulent period, at least at the national level. The frequent changes in government in the latter half of the 1960s to the early 1980s and the political turmoil that ensued curtailed the implementation of most of the landmark spatial development plans, and the short-term plans that were intended to replace them. For most of these short-term plans, their overall achievement as far as spatial planning is concerned and impacts at the settlement level were rather minimal. For example, the Two-Year Development Plan initiated by The National Liberation Council (NLC), which overthrew Nkrumah's government, aimed to provide some 2,000 housing units across the country to accommodate the increasing urban population. However, only 50% of this target would be realized with the housing units provided in Accra (64%), Kumasi, 9% and Sekondi-Takoradi (8%) and Cape Coast (11%). Over this period of instability, the gains of socio-economic transformation pursued in the wake of independence diminished; a virtually dormant planning system could not keep up with the increasing population growth and urbanization pressures while infrastructure in most towns and cities deteriorated as a result of a lack of maintenance (Adarkwa 2012).

3.4 Decentralization and Contemporary Planning in Ghana

The long period of military takeovers which plunged the country into political turmoil and economic uncertainty also eroded the gains of planning as an activity all levels. Eventually, this period would give way for reforms in the late 1980s. In 1988, a decentralization programme was initiated to reform the governance structure of the country and increase local-level participation in the development process. The first major legislation on decentralization was the PNDC Law 207, which transferred greater decision-making functions to local governments. The law increased the number of local government authorities, called District Assemblies from 45 to 65. Later, the 1993 Local Government Act (Act 462) would replace the PNDC Law 207. The Act 462 delegated substantial administrative powers to district assemblies at the local level and mandated them to ensure overall development of local government areas. In 1994, the National Development Planning (System) Act (Act 480) was promulgated to provide the legal basis for planning at all levels in Ghana. Thus, the Local Government Act (Act 462) and the National Development Planning (System) Act (Act 480) became the two most important legislations that will set out the institutional framework for planning and define the nature and scope of the activity for close to three decades. The institutional arrangements and competences for contemporary planning in Ghana under the two legislations is summarized in Fig. 3.9.

From 1994, the new tradition of 'Development Planning' was instituted through the National Development Planning (System) Act (Act 480) and the Local Government Act (Act 462). The National Development Planning Commission (NDPC) was established in the same year. The NDPC was mandated to carry out socio-economic, environmental and physical planning as a single integrated task (Diaw et al. 2002). However, in practice, the spatial components of planning were largely de-emphasized. Instead, under this new tradition of development planning, the scope of planning undertaken by the NDPC at the national level was mainly concerned with

Administrative levels	1994 System Act (Act 480)	2016 Land Use and Spatial Planning Act (Act 925)
National	National Development Planning Commission (NDPC) & Sector Ministries	Land Use & Spatial Planning Authority (LUSPA)
Regional	Regional Coordinating Councils (RCCs)	Regional Spatial Planning Committee
Local (MMDAs)	District Planning Authority (Development Planning Units (DPUs))	District Planning Authority (Town and Country Planning Department)

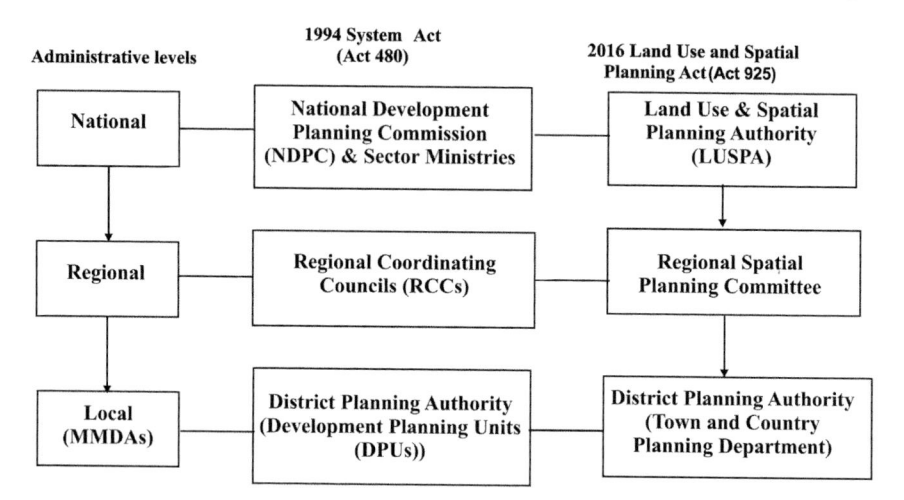

Fig. 3.9 Governance structure and institutional competences under the National Development Planning (System) Act (Act 480, 1994) and the Land Use and Spatial Planning Act (Act, 925, 2016). *Source* Acheampong and Alhassan (2016, p. 7)

addressing social economic issues, with a particular focus on issues such as economic restructuring, macroeconomic stability and poverty reduction. Through the NDPC, successive governments formulated National Medium-Term Development Plans to realize their visions the national level. At the regional level of political administration, Regional Coordinating Councils (RCCs) will monitor coordinate and evaluate the performance of local governments (i.e. Metropolitan, Municipal and District Assemblies, (MMDAs)). MMDAs, in turn, were responsible for the overall development of areas under their jurisdiction at the local level.

The tradition of 'Development Planning' initiated by the National Development Planning (System) Act (Act 480), therefore, had a profound impact on spatial planning at all levels, as well as on planning education in the country. Spatial planning was essentially neglected at the national and regional levels of political administration. At the level of MMDAs, a new department, the Development Planning Unit (DPU), was established as a decentralized department to translate national goals formulated by the NDPC into local Medium-Term Development Plans (MTDPs). MTDPs prepared under NDPC guidelines failed to incorporate spatial planning issues, in particular, the physical organization of activities in towns and cities within their respective districts (Diaw et al. 2002). The TCPD, another decentralized department at level of MMDAs, provided spatial planning functions, which was narrowly conceived as involving the designation of land uses and development control.

As a result of the institutional arrangements created by the decentralization programme and the National Development Planning Systems Act, two *types* of planning will from 1994 onwards, come to exist in Ghana. On the one hand was the *type* of planning called 'Development Planning' which in practice limited its scope to addressing prevailing socio-economic development issues, with particular emphasis on poverty alleviation. On the other hand was 'Land use Planning' which was also narrowly conceived as encompassing the preparation of layouts for neighbourhoods, towns and cities and development control (Acheampong and Alhassan 2016). Moreover, under this established notion of the 'spatial' being distinctively separate from the 'socio-economic', the activity of planning would become compartmentalized into the two separate decentralized departments at the local level (i.e. the DPU and TCDP with competence in development planning and land use or physical planning, respectively), with hardly any coordinated efforts leading to the identification of cross-cutting matters and dealing with them accordingly. Consequently, land use plans were prepared that had no bearing on the prevailing socio-economic realities of the day, except for assisting landowners to sell their land. Similarly, MTDPs were and continuous to be prepared without adequate consideration for the spatial implications of proposals.

The tradition of 'Development Planning', in its conception and in practices, would also change the focus, scope and content of planning education in Ghana. When the Development Planning (System) Act (Act 480) was promulgated to effectively entrench the doctrines of development planning, planning education at the KNUST also responded by introducing the B.Sc. Planning and later the B.Sc. Development Planning programmes, as well as postgraduate programmes in Spatial Planning for Regions in Growing Economies (SPRING) and Development Policy and Planning (DEPP). These new programmes were oriented towards training professionals to run the newly created DPUs at the level of MMDAs as well as other intuitions at the regional and national levels (i.e. NDPC). Their curricula were adapted to respond to the emerging context of decentralization and democratization and the need for a more inclusive approach to planning geared towards poverty reduction (Inkoom 2009; Diaw et al. 2002). In doing so, however, the design and land use organization aspects of the programmes were de-emphasized, effectively discontinuing the training of professional Town Planners in the country. This is one of the main reasons why for many years to come, the spatial planning system would weaken and become ineffective in delivering its objectives.

The justification for the prevailing notion then, that spatial planning was somehow irrelevant to the principles of bottom-up development and planning as a vehicle for poverty alleviation, hence de-emphasising it in planning education remains unclear to date. Perhaps, this was due to the technical nature of planning and its narrow focus on land use zoning, settlement design and development control then. Whatever could be the reason, what we know is that until 2004 when a new programme in human settlement planning was introduced at KNUST, the premier university for planning education in Ghana, spatial planning in both its traditional and contemporary doctrines had remained virtually dormant for over two decades as far as planning education in Ghana was concerned.

It is worth noting that the aforementioned shift in planning education from its traditional focus on the physical design and development of settlements to a new doctrine of planning that had at its heart, poverty alleviation and democratic governance, was not exclusive to Ghana alone. As Diaw and colleagues (2002) show, this shift was observed across sub-Saharan Africa from the early 1990s. The resulting impacts on planning education and planning practice dating are twofold. Firstly, with the traditional design and spatial organization focus of planning de-emphasized, the boundaries between spatial planning and new competing fields such as environmental management and local economic development blurred; universities were compelled to respond by providing more courses in these emerging fields, resulting in the decline of interest in urban and regional planning (Diaw et al. 2002). Secondly, in retrospect, this shift did not make planning education and the practices based on it very effective in dealing with the rapid urbanization levels that would be experienced from the beginning of the twentieth century. As Okpala, writing about urban education and planning practice in Anglophone, sub-Saharan African countries, argues:

> The earlier design attribute of planning education, instead of just being de-emphasized, seem to have largely been eliminated from several of the planning programmes, leaving its graduates with very little if any practical hands-on skills that identifies what they are expected to do as urban planners in the field. Many come out without any identifiable skills. This has tended to further befuddle what urban planning products of the evolving democratized planning curriculum should be. This situation tends to continue and sustain the dialectics and endemic turbulence to which planning, and planning education has been subject to for quite a while. This is a critical weak link in planning education in many Anglophone African countries. (Okpala 2009, p. 29)

Further analyses of the consequences of the reforms ushering in decentralized, development planning and the accompanying lack of emphasis on the 'spatial' in both planning education and practice will be provided in the subsequent chapters of this book. At this stage, we aim to point out the significance of the decentralization programme initiated in the late 1980s as well as the profound impacts the accompanying legal and intuitional arrangements have had spatial planning in contemporary times.

3.5 The Age of Reform: The Land Use Planning and Management Project and the New Spatial Planning System

Against the backdrop of the neglect of spatial planning at the national and regional levels under the development planning tradition and a weak land use planning system at the local level, reforms were initiated in 2007 by the government of Ghana under the Land Use Planning and Management Project (LUPMP). LUPMP was a three-year initiative (2007–2010), implemented under the broader umbrella of reforms initiated by the Land Administration Project (LAP), which started in 2003. Funded by the Government of Ghana and the Nordic Development Fund, the LUPMP's overall objective was to develop a coherent, streamlined and sustainable land use planning and management system, which is decentralized and based on consultative and participatory approaches in order to manage effectively human settlements development. LUPMP was effectively an ambitious attempt to reform and update Ghana's land use planning and management system. Under the project, the following four activities were implemented:

- Policy studies and the reform of the legal and institutional framework for land use planning and management;
- Development and testing of pilot decentralized land use models in selected high priority areas;
- Implementation of an information system for land use planning and management;
- Pilot plan-making and the implementation of plans at regional, district and local levels.

Four major developments that resulted from the reforms are:

i. Promulgation of the Land Use and Spatial Planning Act (Act 925), 2016;
ii. Restructuring of the institutional arrangements for spatial planning, leading to the establishment of a Land Use and Spatial Planning Authority (LUSPA);
iii. Institutionalization of the three-tier system of spatial planning instruments; and
iv. Establishment of Land Use and Spatial Planning and Development Fund.

In the sections that follow, the key developments under LUPMP outlined above and their significance for spatial planning in Ghana are discussed.

3.5.1 The Land Use and Spatial Planning Act (Act 925), 2016

Until the promulgation of Land Use and Spatial Planning Act (925) in 2016, the mandate for land use planning was derived from the Town and Country Planning Ordinance (CAP 84) that came into being 1945. Thus, for many years, the activity of planning as defined in CAP 84 was based on an obsolete piece of legislation. The fact that CAP 84 remained in force despite lacking the capacity to deal with the complexities of planning and development issues in twenty-first century further lends credence to the initial assertion that spatial planning missed the radar of various that preceded LUPMP. Later in Chapter 5, we will discover how the apparatus of local planning and development control in modern times have struggled to cope partly as a result of having to operate under an obsolete legal framework that was first introduced more than a decade before Ghana became an independent country. For now, we will focus on the new land use planning law (i.e. Act 925).

The intent and purpose of new spatial planning law as stated in the Act is that of:

> AN ACT to revise and consolidate the laws on land use and spatial planning, provide for sustainable development of land and human settlements through a decentralised planning system, ensure judicious use of land in order to improve quality of life, promote health and safety in respect of human settlements and to regulate national, regional, district and local spatial planning, and generally to provide for spatial aspects of socio economic development and for related matters (2016 Land use and Spatial Planning Act (Act 925)).

The Act itself marks a significant development in the history of spatial planning in Ghana. This is because, as mentioned previously, for close to 70 years, the activity of planning was conducted under an obsolete legal framework—the 1945 Town and Country Planning Ordinance (CAP 84). Its overall goal is ambitious and consistent with the challenges that confront the spatial planning system and planners in contributing to sustainable place-making at all levels in the twenty-first century. It delineates the scope and purpose of spatial planning by using the combined terms *land use* and *spatial planning* to suggest a deliberate attempt to institute a new tradition of integrated and multi-scale planning that delivers wider socio-economic and environmental development imperatives with the traditional design and regulatory function of town planning. Most importantly, it recognizes that effective planning is a precondition for judicious use of scarce resources such as land.

Another important feature of the new planning law is its definition of planning areas. As was previously discussed, the 1945 Town and Country Planning Ordinance (CAP 84) adopted a piecemeal and centralized procedure whereby only specific areas determined to qualify for planning through a legislative instrument received the benefits of planning. The new planning law, in contrast, abandons this procedure and instead, declares all areas in the country as potential planning areas, subject to the provisions of the Act. Specially, a planning area in the Land Use and Spatial Planning Act is defined as:

> The territory of Ghana as defined under the Constitution of the Republic of Ghana including the land mass, air space, sub-terrain territory, territorial waters and reclaimed lands shall be a planning area and subject to the planning system provided under this Act and other relevant laws (2016 Land use and Spatial Planning Act (Act 925)).

3.5.2 *Institutional Framework for Spatial Planning Under the Land Use and Spatial Planning Act (Act 925)*

Prior to reforms, the authority and competence for spatial planning were vested in the TCPD which has offices at the national, regional and district levels of public administration. This three-tier institutional set-up is maintained under the Land Use and Spatial Planning Act (Act 925), although the names and functions of the institutions, especially those at the national and regional levels have changed. Under the new spatial planning law, the Land Use and Spatial Planning Authority (LUSPA) has been established to replace the National Head Office of the TCPD. According to the Act, LUSPA is a body corporate with perpetual succession headed by a Chief Executive Officer appointed by the president.

Below the national structure, at the regional level, the Regional Spatial Planning Committee replaces the Old Regional TCPD while the authority and competence of spatial planning at the local level is now vested in the District Planning Authority, which replaces the former District TCPD. The functions of and competencies for the spatial planning institutions established under the new spatial planning law are

3.5 The Age of Reform ...

Administrative levels	Institutions	Functions and Competences
National	Land Use & Spatial Planning Authority	• Perform the spatial, land use and human settlements planning functions of the NDPC • Prepare and provide for the technical human settlements planning component as may be required by the NDPC for inclusion in the national development plans or infrastructure plan • Prescribe the format and content of the spatial development framework, structure plans and local plans • Provide directions, guidelines and manuals for spatial planning • Develop the capacities of the district assemblies and other Institutions in relation to spatial planning
Regional	Regional Spatial Planning Committee	• Develop a RSDF for the region • Adjudicate on appeals or complaints resulting from decisions, actions or inactions of the District Spatial Planning Committee of the MMDAs • Prepare sub-regional or multi-district SDFs for two or more districts within the region • Perform any other function to give effect to this Act within the region
Local Government	District Spatial Planning Authority	• Functions as the spatial, human settlement and planning authority for its area of authority (i.e. MMDAs).

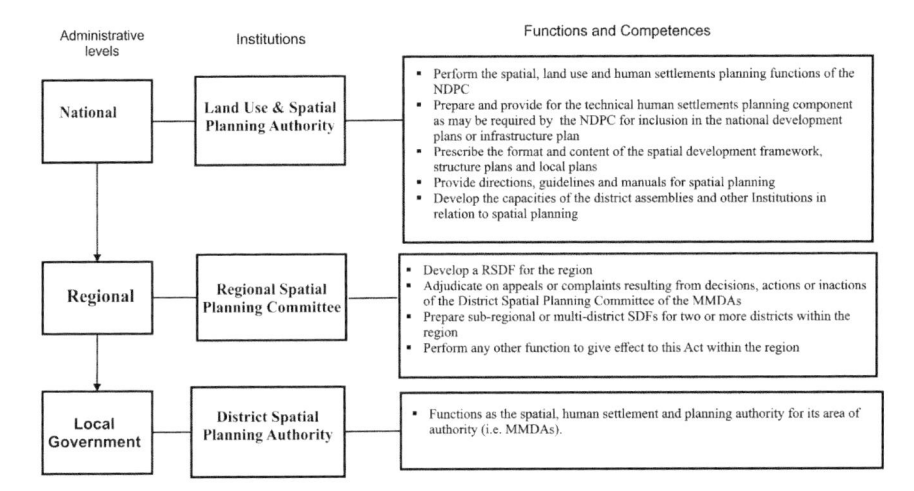

Fig. 3.10 Governance structure and institutional competences for spatial planning under the 2016 Land Use and Spatial Planning Act (Act 925)

illustrated in Fig. 3.10. These institutions will be revisited in the subsequent chapters where we deal in detail with spatial planning at various spatial scales (i.e. national, regional, sub-regional and local levels).

3.5.3 The Three-Tier, Hierarchical System of Spatial Planning Instruments

Another prominent development under the reforms introduced under LUPMP and enshrined in the new spatial planning law is the establishment of a decentralized spatial planning system based on a three-tier model of spatial planning instruments, which correspond with the three-tier national governance structure. These instruments are National Spatial Development Framework, Regional Spatial Development Framework and District Spatial Development Frameworks. Within this three-tier framework, derivative development framework, such as Sub-Regional Spatial Development Frameworks may be prepared where necessary (see Fig. 3.11).

The concept of Spatial Development Frameworks (SDFs) is defined in the National Spatial Planning System (NSPS) Module Guidelines as the following:

> ..an indicative plan, showing the expected development over a fifteen to twenty-year period, which will include the location of key components of the strategy aimed at achieving the desired development'. (MESTI, NSPS Module Guideline 2011: p. 9)

The three-tier system of spatial planning instruments is intended to provide a direct connection between national development strategies and their spatial realization and local development policies through a 'chain of conformity' (MESTI 2011). It is,

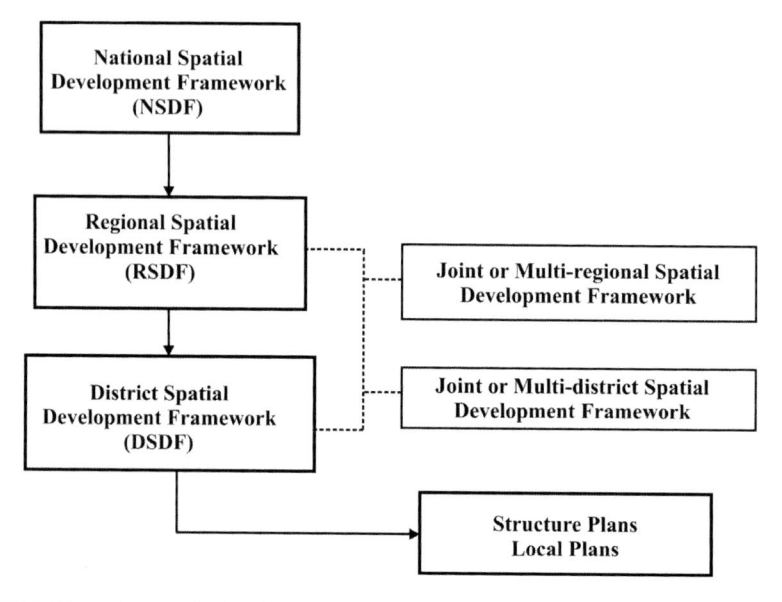

Fig. 3.11 Three-tier model of spatial development frameworks and derivative plans established by the new decentralized spatial planning system

therefore, a legal requirement under the new spatial planning system for an SDF to be prepared at the national level and for all the administrative regions and MMDAs in the country. Besides each of the administrative units above, which overlaps with the national governance structure having a spatial plan, the new system recognizes the need for multi-regional and multi-district or sub-regional SDFs, where there need arises to do so. Spatial plans prepared for the *region* must follow the parameters established in the National Spatial Development Framework. Similarly, frameworks prepared by MMDAs for their respective local areas must conform with their corresponding regional frameworks.

3.5.4 The Land Use and Spatial Planning Development Fund

The institutions mandated to perform spatial planning activities at all levels like many other public-sector institutions require adequate funding to carry out their mandates effectively. The activities of spatial planning, including planning scheme initiation, plan preparation, stakeholder consultations, development control and day-to-day administrative tasks require substantial amount of financial resources. Although central government do allocate funds to its Ministries, which in turn, allocate to their respective agencies and departments, including the TCPD, these funds have always not been adequate.

3.5 The Age of Reform …

Indeed, the new planning model experimentation processes under LUPMP, which saw the preparation of several spatial plans across the country relied heavily on financial support in the form of loans and grants from development partners and players in the emerging oil and gas industry (Acheampong and Alhassan 2016). For example, the preparation of the Western Regional Spatial Development Framework (WRSDF) was financed by a grant from the Norwegian government under the 'Oil for Development (OfD)' programme. Tullow Oil and its partners under the '*Town Planning: an Imperative for Sustainable Oil economy in Western Region*' programme provided substantial funding for the preparation of several lower-tier spatial development plans for selected urban centres in the oil and gas enclave of the Western region of Ghana. Similarly, the Greater Kumasi Spatial Development Framework was formulated with financial support from the Japan International Cooperation Agency (JICA).

While these external sources of funding have played significant role in institutionalizing the new spatial planning system, building on the gains realized so far would require mobilizing substantial funding domestically to undertake the activities of spatial planning. In response to this imperative, the new spatial planning law has proposed for the establishment of the Land Use and Spatial Planning Development Fund. The purpose of the fund as stated by the law is to provide financial resources to:

- Finance research into planning issues and capacity building;
- Defray costs of plan preparation, reports and dissemination of information;
- Fund promotional activities and public education; and
- Assist planning entities in the performance of their activities

According to the Land use and Spatial Planning Act (Act 925), sources of money for the fund would include central government allocation provided through the Ministry of Finance and Economic Planning with the approval of parliament, proceeds received by the newly established LUSPA from investments, moneys ceded to the Fund from the District Assemblies Common Fund (DACF),[1] and funding from development partners including grants and donations. As of the time of writing this book, the establishment of the fund remained a proposal with legal backing from the new planning law. That said, in many respects, it constitutes one of the innovative and significant achievements initiated by the reforms under LUPMP. Once up and running, the fund would, without any doubts, strengthen the spatial planning system to deliver its functions.

3.6 Conclusions

This chapter has traced the history of spatial planning from the mid-nineteenth century colonial days to the present day. The discussion has shown that from the early

---

[1] DACF is a system for the transfer and administration of funds transferred to MMDAs from central government.

Table 3.1 Periodizing significant developments in the history of spatial planning in Ghana

Year	Event
2017	Land Use and Spatial Planning Authority established
2016	Land Use and Spatial Planning Act (Act 925) promulgated
2015	Ghana National Spatial Development Framework Formulated
2013	Greater Kumasi Sub-Region Spatial Development Framework Formulated
2012	Western Region Spatial Development Framework Formulated
2012	National Urban Policy Framework Formulated
2010	Reforms initiated: Land Use Planning and Management Project
1994	National Development Planning System Act 480 promulgated; National Development Planning Commission established
1993	Decentralization, Local Government Act 462 promulgated
1963	Planning Scheme for Kumasi Metropolis prepared
1963	National Physical Development Plan prepared
1963	Seven-Year National Development Plan
1961	Tema Master Plan
1961	National Planning Commission established
1958	Planning programme established at Kwame Nkrumah University of Science and Technology (KNUST)
1958	Accra Master Plan
1951	Four-Year National Development Plan
1945	Town and Country Planning Department established
1945	Town and Country Planning Act (CAP 84) promulgated
1942	Alan Burns Four-Year National Development Plan
1919	Guggisberg 10-Year Development Plan

1920s when National Development Plans which were essentially spatial in scope and purpose were initiated, the planning system and the activity of spatial planning have evolved. A summary of the most significant development in the history of spatial planning in Ghana is provided in Table 3.1.

The introduction of the Town and Country Planning Act (CAP 84) in 1945 and the establishment of the TCPD around the same period marked the birth of formal spatial planning in the country. The 1950s ushered in a period of comprehensive planning at the national level with an ambitious vision of modernization and socio-economic transformation. From this period, up to the mid-1960s, a series of landmark spatial planning projects were initiated. Perhaps, the most significant event in the history of spatial planning was the formulation of the National Physical Development Plan in 1963, the first of its kind in Africa. This together with the National Perspectives it accompanied would inspire other landmark spatial planning activities at the level of towns and cities in Ghana. Significant among these were the 1958 Accra Master Plan; the 1961 Master Plan for Tema; the 1963 Master Plan for Kumasi; as well as

3.6 Conclusions

the planning and construction of several new towns across the country. These initial efforts institutionalized the tradition of comprehensive national spatial planning, and land use planning and development control at the level of towns and cities in Ghana.

After years of political turmoil and economic uncertainty, resulting from the series of military takeovers between the mid-1960s and late-1980s, which eroded much of the gains made earlier, the inception of decentralization marked another significant turning point in the history of spatial planning. The 1993 Local Government Act (Act 462) and the National Development Planning (System) Act (Act 480) not only initiated a new dawn of decentralized planning but also introduced institutional structures and planning tradition that will have profound impact on the scope of planning and planning education in the country. In particular, the discussion highlighted how the new tradition of 'Development Planning' in its narrow conception of planning, the new institutions it deployed as well as the impact it had on planning education put spatial planning at crossroads. Spatial planning would not be given the same priority it received in the 1960 s, instead, at the national and regional levels, the activity was neglected. During this period, spatial planning, conceived narrowly as concerned with land use plan preparation and development control together with an obsolete legal framework that had been in place for over seven decades, became weak and ineffective in dealing the prevailing challenges of the twentieth century.

Reforms initiated in 2007 by the Government of Ghana under the Land Use Planning and Management Project (LUPMP), marked another turning point in the history of spatial planning in Ghana. Aimed at resurrecting a weak and ineffective land use planning system, LUPMP introduced a number of significant reforms. This included the promulgation of a new planning law—The Land Use and Spatial Planning Act (Act 925) in 2016—to replace the rather obsolete 1945 Town and Country Planning Act. Under the new planning law, new institutions including LUSPA and the Regional Spatial Planning Committee have been established to carry out spatial planning activities at the national and regional levels, respectively. The law has also entrenched a new three-tier system of spatial planning instruments by which the preparation of spatial development frameworks at the national, regional and district levels have become a legal requirement.

Having traced the history of spatial planning in this chapter, the next three chapters will focus on contemporary spatial planning at the national, regional and local levels. Among other things, these chapters will focus on various plans that have been formulated under the new spatial planning system and examine critically the interface between the new spatial planning system and the existing development planning system established by the 2016 Land Use and Spatial Planning Act and the 1994 National Development Planning (System) Act, respectively.

References

Acheampong RA., Ibrahim A (2016) One nation, two planning systems? Spatial planning and multi-level policy integration in Ghana: mechanisms, challenges and the way forward. Urban Forum 27(1): 1–18

Adarkwa KK (2012) The changing face of Ghanaian towns. Afr Rev Econ Finan 4(1): 1–29

Diaw K, Nnkya T, Watson V (2002) Planning education in Sub-Saharan Africa: responding to the demands of a changing context. Plan Pract Res 17(3): 337–348

Doxiadis Associates (1962) The Town of Tema, Ghana: Plans for two communities. Doxiadis Associates. Athens Centre of Ekistics. https://www.jstor.org/stable/43615973?seq=1#page_scan_tab_contents. Accessed 24 July 2018

Fuseini I, Kemp J (2015) A review of spatial planning in Ghana's socio-economic development trajectory: a sustainable development perspective. Land Use Policy, 47: 309–320

Greed C, Johnson D (2014) Planning in the UK: an introduction. Palgrave Macmillan

Inkoom DKB (2009) Planning education in Ghana. Available at: http://www.unhabitat.org/grhs/2009. Accessed 30 July 2015

Korboe D (2001) Historical development and present structure of Kumasi. In: Adarkwa KK, Post J (eds) The Fate of the Tree: planning and managing the development of Kumasi, Ghana. Woeli Publishing, Accra

Mabogunje AL (1990) Urban planning and the post-colonial state in Africa: A Research Overview 1. Afr Stud Rev 33(2): 121–203

Ministry of Environment, Science, Technology and Innovation (MESTI) (2011) The new spatial planning model guidelines. Town and Country Planning Department. November 2011.

Nkrumah K (1973) Revolutionary path, vol 172, International Publishers, New York

Okpala D (2009) Regional overview of the status of urban planning and planning practice in Anglophone (Sub-Saharan) African countries. United Nations Human Settlement Programme, Regional study for Global Report on Human Settlement, Nairobi

Town and Country Planning Act (1945) CAP 84. http://www.epa.gov.gh/ghanalex/acts/Acts/TOWN%20AND%20COUNTRY%20PLANNING%20ACT,1945.pdf

Town and Country Planning Division (1958) Accra. A plan for the town. The Report for the Minister of Housing. Town and Country Planning Division of the Ministry of Housing. Url: http://mci.ei.columbia.edu/mci/files/2013/03/Accra-Town-Plan-1958.pdf. Accessed 24 July 2018

Town and Country Planning Act (1958) Act 30 The thirtieth act of the parliament of Ghana

Town and Country Planning (Amendment) Act (1960) Act 33 The thirty-third act of the parliament of the Republic of Ghana

Part II
Modern-Day Planning Systems and Scales of Spatial Planning

Chapter 4
Contemporary Traditions of Planning and Spatial Planning at the National Level

Abstract Contemporary planning at the national level in Ghana is realized under two legal instruments—the National Development Planning (System) Act (Act 480, 1994) and the Land Use and Spatial Planning Act (Act 925, 2016). These instruments have established two seemingly different planning systems—the development planning system and the new spatial planning system—all of which draws on the governance structures established by the decentralization law (i.e. 1933 Local Government Act (Act 462)) to deliver their objectives. This chapter firstly examines the focus and scope of three decades of national-level planning under the established tradition of development planning. Next, it presents a detailed discussion of the newly established spatial planning system and identifies the institutional arrangements for spatial planning at the national level. The ensuing institutional conflicts and duplication of competences, after the inception of the new spatial planning system that operates alongside the entrenched tradition of development planning are also examined. The two main national-level spatial policy instruments (i.e. the National Spatial Development Framework and National Urban Policy Framework) are briefly discussed with the aim to illustrate the functioning of the new spatial planning system in practice at the national level. Following from this, an analysis of the practical challenges of and pathways towards realizing long-term national spatial development imperatives, within a culture of medium-term national development planning, which is essentially dictated by the constitutionally guaranteed tenure of governments, is provided.

Keywords Strategic spatial planning · National policy framework
National spatial development framework · Development planning system
Spatial planning system · National urban policy framework · Medium-term plan
Urban policy · Ghana

4.1 Introduction

The purpose of this chapter is to provide a detailed discussion on modern-day planning at the national level in Ghana. In Chap. 3, we discovered that the decentralization Act, Act 462 of 1993, and the National Development Planning (System) Act, Act 480 of 1994, instituted a tradition of development planning. These legislations would set out the institutional arrangements and define the scope and purpose of planning in general from the mid-1990s to the present day. It was also mentioned that the renewed interest in spatial planning almost a decade ago and the reforms that followed between 2007 and 2010 under LUPMP have culminated in the establishment of a new spatial planning system, which derives legal backing from the Land Use and Spatial Planning Act (Act 925) of 2016. These two legislations (i.e. the National Development Planning (System) Act of 1994 and the Land Use and Spatial Planning Act of 2016) currently operate side by side, deploying different institutional arrangements to accomplish the activities of 'development planning' and 'spatial planning', essentially conceived as two different *types* of planning in Ghana.

Building on the historical context provided in the previous chapter, we will begin the discussion in this chapter by examining the focus and scope of national planning instruments that have been formulated under the National Development Planning (System) Act (Act 480) and its development planning tradition. Next, critical analyses of the interface between the conception of planning and the associated practices under the System Act (Act 480) and that of the new spatial planning system under the 2016 spatial planning law (i.e. Act 925) will be presented. Here, the question as to whether the seemingly two different planning systems would be a recipe for institutional conflict and duplication of efforts will be addressed. We will also discuss progress in national-level spatial planning, looking at the newly formulated National Spatial Development Framework and the National Urban Policy framework. The challenges of achieving long-term national spatial development goals under a culture of medium-term national development planning will be examined. The chapter will end with concluding thoughts and reflections on the future of national-level spatial planning in Ghana.

4.2 Three Decades of National Medium-Term Development Planning

In 1983 following years of political instability and economic recession, Ghana adopted the International Monetary Fund (IMF)/World Bank programme of structural adjustment. Under the Structural Adjustment Programme (SAP), governments sought to rebuild the economy through devaluation, increase in domestic currency prices for export, budget deficit reduction, price liberalization and increased participation of the private sector (Kraus 1991; Loxley 1990). Under SAP, the preparation of Public Investment Programmes (PIP) was the main instrument of national-level planning. The first PIP was prepared in 1986 to be implemented over a nine-year period until 1995, on a three-year rolling basis.

PIPs had several shortcomings. As the name suggests, they focused exclusively on government spending to boost the economy. A PIP, therefore, lacked synergy between sectoral programmes, was characterized by an arbitrary allocation of ministerial areas of responsibility that was not based on any well-defined sectors of economic activity and relied exclusively on economic criteria, in particular the projected economic rate of return (ERR) on government investment projects (Government of Ghana 1995). In fact, PIPs were more of a list of government investments and cost-benefit analyses designed to produce quantifiable economic benefits rather than an elaborate national development plan. Consequently, they lacked any detailed elaboration of the spatial distribution and timing of government projects and investments.

With a new decentralization law (i.e. Local Government Act, Act 462) and the National Development Planning (System) Act, Act 480 in place, coupled with the aforementioned weaknesses of PIPs, the first attempt at National Development Planning was initiated. This culminated in the formulation of first National Development Policy Framework (1995–2020) following the transition to a constitutional democracy in 1992. Vision 2020, as the National Development Policy Framework would be known, was intended to be a long-term national development plan spanning a period of 25 years, phased into five, five-year rolling programmes. Vision 2020 was initiated under the National Democratic Congress (NDC), which governed from 1992 to 2000. A change in government in 2001 will see the New Patriotic Party (NPP) in power for the first time as a democratically elected government. The NPP government would, however, abandon Vision 2020 and initiate a new tradition of national medium-term development planning.

Since 2001, successive governments have followed the tradition of national medium-term development planning. Each government formulates a national policy framework (NPF), which articulates their goals and priorities as well as the programmes and strategies to realize them. The most recent of NPFs include the Growth and Poverty Reduction Strategy-GPRS I (2003–2005) and GPRS II (2005–2008)—see Government of Ghana, 2005, and Ghana Shared Growth and Development Agenda-GSGDA I (2010–2013) and GSGDA II (2014–2017)—see Government of Ghana (2010, 2014). Below, the focus of medium-term NPFs and their treatment of spatial issues are discussed, acknowledging the broader context of priorities at the supra-national and national levels that shape national development goals.

Firstly, the time horizon of NPFs overlaps with the constitutionally defined tenure of government. The national constitution, which has been the basis for democratic governance in Ghana since 1992, allows for a four-year mandate for a government. Thus, successive governments have used NPFs to articulate their visions and priorities for this four-year period. For the sixteen-year period between 2001 and 2017, for example, four NPFs were prepared and implemented between the two main political parties, the NPP and the NDC. The NPP government initiated GPRS I (2003–2005) and GPRS II (2003–2005) while the NDC government initiated GSGDA I (2010–2013) and GSGDA II (2014–2017). These national plans are prepared at the beginning of the four-year term of a government. When a government receives a second consecutive mandate, it would introduce a new medium-term NPF, which is often a sequel to the first framework, such as the case with GSGDA I and GSGDA II. Thus, since 2001, the longest NPF period has been eight years.

Secondly, NPFs have been profoundly affected by supra-national development goals of economic restructuring and macroeconomic stability set by international bodies. In the days of SAP, national frameworks derived their goals directly from agenda heavily influenced and funded by the IMF and the World Bank. After SAP came the poverty reduction agenda of the Millennium Development Goals (MDGs) and Sustainable Development Goals (SDGs) spearheaded by the United Nations and other partners including the World Bank. For example, under the influence of the MDGs, the first NPF, (i.e. GPRS I) formulated in 2003 was deployed as 'a comprehensive development policy framework in support of poverty reduction and growth'. Around the same time, Ghana was participating in the Heavily Indebted Poor Countries (HIPC) Initiative. HIPC was an initiative under which the IMF and International Development Association (IDA) would support a debt reduction package subject to conditions, including the implementation of a poverty reduction programme and structural reforms aimed at bringing about macroeconomic stability (IMF 2004). In 2013, just a decade after being declared HIPC and recieving debt relief, Ghana would once again enter another IMF programme to control its rapidly rising public debt. These international arrangements and the accompanying conditions, to a larger extent, have set the agenda for national development and defined the scope and purpose of national-level planning over the past thirty years.

In view of the aforementioned, NPFs themselves lacked detailed elaboration of the spatial manifestations of national development goals. Also, in sharp contrast to earlier attempts at national-level spatial planning initiated under the National Physical Development Plan in 1963, none of the NPFs have so far been accompanied by an elaborate Spatial Development Framework detailing the geographical distribution and timing of national development priorities. Instead, spatial issues were given broad thematic references in these national plans such as 'Spatial Organization' in Vision 2020, 'Urban Development, Housing and Slum Upgrading/Urban Regeneration' in GPRS II and Infrastructure and Human Settlements in GSGDA I and II (see Table 4.1).

For about thirty years starting from 1993, area-based issues were therefore treated as separate from other wider socio-economic development issues at the national level. In all cases, the synergies between the different thematic issues and their spatial dimensions and distributional implications, in particular, were ignored. In addition to the narrow thematic spatial references in NPFs, the associated goals were quite vague, lacking any carefully designed programme of action to implement them at the level of towns and cities. Effectively, one could argue that spatial planning at the national level was non-existent for the 30-year period of medium-term national development planning.

From the foregoing, we discover the forces that have shaped the national development agenda. Undoubtedly, supra-national development priorities and conditions have had profound influences on the development agenda set and pursued at the national level. The tradition of development planning has also helped to translate national development agenda into medium-term NPFs, which as we have discov-

4.2 Three Decades of National Medium-Term Development Planning

Table 4.1 Spatial planning-related issues captured in national policy frameworks under the tradition of development planning

National-level instruments	Thematic areas and spatial references	Policy goals
Ghana Shared Growth and Development Agenda (GSGDA II 2014–2017)	**Infrastructure and Human Settlements** • Spatial/Land Use Planning and Management • Urban Development and Management • Rural Development and Management • Housing/Shelter • Slum Prevention and Regeneration	• Facilitate ongoing institutional, technological and legal reforms in support of land use planning, streamlining spatial and land use planning system, and strengthening the human and institutional capacities for effective land use planning and management • Promote redistribution of urban population and spatially integrated hierarchy of urban settlements, as well as promoting urban infrastructure development, maintenance and provision of basic services • Improve access to adequate, safe, secure and affordable shelter, as well as improving and accelerating housing delivery in the rural areas • Upgrading existing slums and prevent the occurrence of new ones
Ghana Shared Growth and Development Agenda (GSGDA I 2010–2013)	**Infrastructure and Human Settlements** • Human Settlements Development • Infrastructure Development	• Ensure that all organized human activities are well planned and spatially integrated into line with the strategic direction of national development efforts in the medium to long term • Expand existing social and economic production infrastructure to ensure that services provided are reliable, affordable and efficient

(continued)

Table 4.1 (continued)

National-level instruments	Thematic areas and spatial references	Policy goals
Growth and Poverty Reduction Strategy (GPRS I 2006–2009)	**Urban Development, Housing and Slum Upgrading/Urban** • Regeneration • Housing • Slum Upgrading/Urban Regeneration	• Promote urban infrastructure development and provision of basic services • Increase access to safe and affordable shelter
Growth and Poverty Reduction Strategy (GPRS II 2003–2005)	**Production and Employment**	• Initiate and implement effective planning of human settlements in the interest of efficiency and amenity
Vision 2020-National Development Policy Framework (1995–2020)	**Spatial Organization** Regional population distribution Hierarchy of settlements Planned Human settlements	• Achieve an optimal geographical pattern of resource investment designed to strengthen and geographically extend national and, particularly, local linkages between settlements of varying functions and size and their hinterlands and between settlements with complementary functions
Public Investment Programmes (PIP 1986–1995)	No explicit spatial references	• No explicit spatial development goals

ered, failed to address the spatial aspects of national development programmes. The third force that has shaped the scope and content of national-level planning is planning education. Indeed, as we discovered in Chap. 3, the orientation of planning education not only determines the scope and content of what is taught but also the knowledge and skills of the professionals who are expected to perform planning functions at all levels.

Thus, the shift in focus of planning education from its initial spatial orientation to 'Development Planning' in line with the prevailing emphasis on poverty alleviation could also account for the lack of elaborate National Spatial Development Frameworks either as stand-alone frameworks or adjunct frameworks to NPFs. After all, the orientation of the cadre of planners trained over this period, who would go ahead to handle the task of planning at the NDPC, would also reflect in the type of plans they prepare. On the contrary, the cadre of town planners whose training predated the shift in focus of planning education to 'Development Planning' and hence were more spatially oriented continued their practice in public service with the Town and Country Planning Departments of the District Assemblies. Their expertise has so far been largely confined to preparing planning schemes and carrying out development control at the level of towns and cities. Consequently, the town planning profession for over thirty years in the history of contemporary planning in Ghana had no profound impact on national-level planning under the tradition of medium-term national development planning.

4.3 The Systems Act and the New Land Use and Spatial Planning Act: A Recipe for Institutional Conflicts and Duplications?

Having examined the scope and purpose of national-level planning over the past three decades in the previous sections, this section focuses on the legal and institutional arrangements for contemporary planning at the national level. The key focus of the discussion here is on the two main legislative instruments that have established two different *types* of planning and vested the authority and competence for planning in different institutions. Although the main concern in this chapter is on the impacts of these arrangements on national-level planning, we will also examine their implications for institutional arrangements and competences for sub-national-level planning.

Figure 4.1 illustrates the institutional arrangements under what has come to be known as the 'development planning system' established under the 1994 National Development Planning (System) Act (Act 480, 1994) and the 'spatial planning system' institutionalized by the newly promulgated Land Use and Spatial Planning Act (Act 925, 2016). The common attribute of the two planning systems is that they both draw on the decentralized governance structure established by the Local Government Act (Act 462). Consequently, both planning systems advocate for planning at the national, regional and local levels.

Administrative levels	1994 Systems Act (Act 480)	2016 Land Use and Spatial Planning Act (Act 904)
National	National Development Planning Commission (NDPC) & Sector Ministries	Land Use & Spatial Planning Authority (LUSPA)
Regional	Regional Coordinating Councils (RCCs)	Regional Spatial Planning Committee
Local (MMDAs)	District Planning Authority (Development Planning Units (DPUs))	District Planning Authority (Town and Country Planning Department)

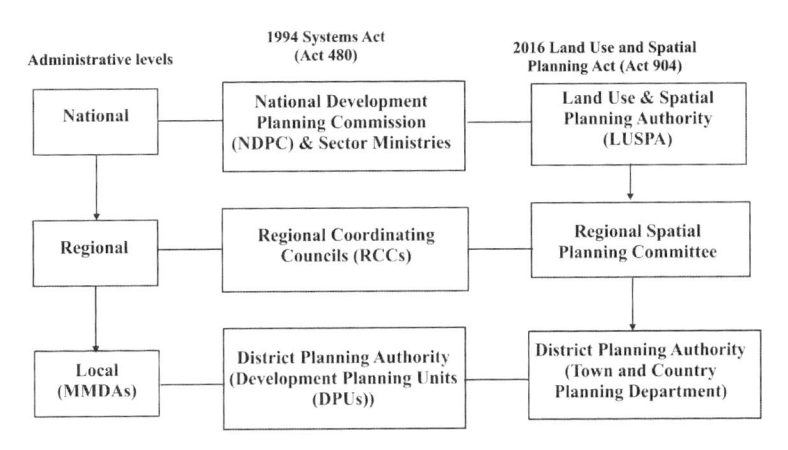

Fig. 4.1 Institutional arrangements for planning under two main planning laws in Ghana

Except for the levels of planning, the institutional arrangements that determine the authority and competences for planning under the two systems differ significantly. Under the System Act (Act 480) from which the tradition of 'development planning' has become entrenched, the authority and competence for development planning at the national level are vested in the NDPC and the sector Ministries. In practice, the 'development planning' system's design is premised on the practical realities that Government Ministries prepare sector plans and policies covering different areas, including transportation, education, health, agriculture, among others. The NDPC at the national level then co-ordinates the sectoral plans to derive a composite national plan or policy framework. At the regional level of administration, the Regional Coordinating Councils (RCCs) are mandated by the decentralization law to monitor, co-ordinate and evaluate the performance of local governments (i.e. MMDAs). It is worth noting, however, that RCCs are administrative entities and do not have planning competences. We will revisit the implications of the RCCs' limited planning powers later in Chap. 5 where we focus on planning at the regional level.

Moreover, under the development planning system, MMDAs are the planning and development authorities at the local level. At this level, the NDPC issues policy guidelines directly to MMDAs based on which they prepare Medium-Term Development Plans (MTDPs). MTDPs have the same time horizon as their respective NPFs. Local governments prepare their MTDPs through their Development Planning Units (DPUs), a decentralized department which is the equivalent of the NDPC at the local level. Thus, through the issuance of guidelines by NDPC at the national level to DPUs at the local level, the development planning system is meant in principle, to achieve synergy and consistency between the goals of NPFs formulated and implemented at the national level and those of MTDPs prepared by MMDAs at the local level.

On the contrary, under the spatial planning system instituted by the new land use and spatial planning law promulgated in 2016 (i.e. Act 925), the authority and competence for spatial planning at the national level are vested in a newly established Authority—the Land Use and Spatial Planning Authority (LUSPA). The governing body of LUSPA is a Board comprising a Chief Executive Officer, a Chairperson and representations from sector ministries, sector agencies and other private sector and professional bodies (see Fig. 4.2). At the regional and local levels, respectively, planning competences are vested in Regional Spatial Planning Committees and the TCPDs of MMDAs, respectively.

Acheampong and Alhassan (2016), in their initial analysis of the aforementioned arrangements, coined the phrase 'one nation, two planning systems' to refer to the conflicting institutional arrangements under the development planning system and the spatial planning system respectively. They argued that having two distinctly separate planning systems which are grounded in the established notion of the 'spatial' being distinctively separate from the 'socio-economic' in planning hampered effective policy integration and could potentially result in duplication of efforts and waste of resources. In the paragraphs that follow, their initial analysis of the contemporary planning systems are taken further. In doing so, we will examine in detail, the overlapping institutional mandates granted under the two planning systems as well

4.3 The Systems Act and the New Land Use and Spatial Planning Act ...

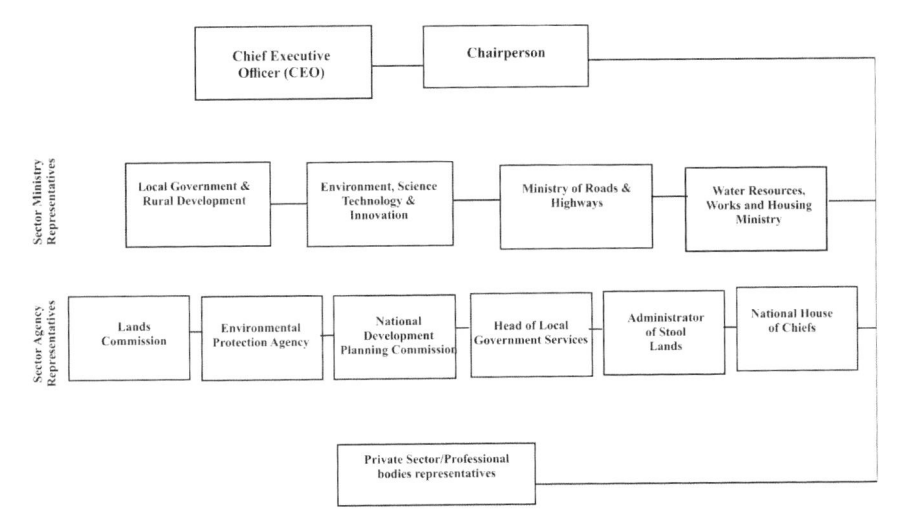

Fig. 4.2 Composition of the governing body of the Land Use and Spatial Planning Authority (LUSPA)

as the scope and content of the various planning instruments that these institutions are required to formulate and implement.

At the national level, the Land Use and Spatial Planning Act (Act 925) mandates LUSPA to perform the spatial, land use and human settlements planning functions of the NDPC. The assumptions underpinning this type of arrangement and their practical implications are worth examining. The notion that development is multidimensional and complex, and that its planning requires multiple institutions with specific competences certainly underpins this arrangement. By this notion, spatial planning is conceived as dealing with one of the several dimensions of development. Indeed, the multidimensionality notion of development is a valid one but as to whether this justifies the current arrangement where two autonomous institutions are deployed to carry out the tasks of planning at the national level is open to debate. As we shall soon discover, the emergent institutional arrangements for planning at the national level could potentially result in unhealthy competition, duplication of efforts and a waste of public resources.

The multidimensionality of development is reflected by the many sector ministries and agencies at the national level that are mandated to handle different aspects of government policy in the areas of health, education, transportation, employment and housing among others. Invariably, every plan or policy formulated by these ministries and agencies covers socio-economic issues that could be translated spatially. Therefore, an institutional arrangement that offers a common platform at the national level to bring together these sectoral programmes, policies and plans into a single coherent framework would be much more beneficial than one that further compartmentalizes these developmental issues into separate planning systems. In many respects, the current arrangements whereby two autonomous planning bodies undertake planning achieve the latter. If a single planning body at the national level were to be established, it would have a number of units/division housing professionals with specific technical competences deemed relevant for the task of integrated planning at the national level. Under this alternative arrangement, the challenge of co-ordinating sector plans would be minimized while the competences of all relevant units could be drawn upon to formulate integrated and comprehensive national development plans.

Whether seamless or sequential, the time horizon of the two instruments would remain one of the major challenges for realizing the attempts to link the plan-making activities of the NDPC mandated to perform national development planning on the one hand, and LUSPA, mandated to perform national spatial development planning on the other hand. This is because, as pointed out earlier, on the one hand, there is a prevailing tradition of medium-term planning at the national level whereby governments through the NDPC have historically prepared four-year NPFs to coincide with their constitutionally enshrined turner. The NSDF on the other hand, as stipulated by the new spatial planning law, should take a long-term perspective of about 25 years.

Last but not least, the current arrangement could potentially result in duplication of efforts and waste of resources at all levels. As initially observed by Acheampong and Alhassan (2016), the new spatial planning system has widened the scope of spatial planning from the traditional preoccupation with land use designation and development control to a more strategic and integrated focus that is meant to bring together all the dimensions of development into a single coherent framework at all spatial scales. NPFs and MTDPs under the development planning system are also supposed to address all these dimensions of development at the national and local levels. Thus, effectively, the separate institutions under each of the two systems would be formulating different instruments which cover essentially the same development issues, but only different in what they choose to call them. After all, is it not the primary purpose of a truly integrated National Development Plan to address all aspects of development so as to give direction to planning and development at the sub-national levels?

4.4 National-Level Spatial Planning Under the New Spatial Planning System

In Chap. 2, we identified that one of the significant success stories of the reforms implemented under LUPMP was the introduction of the three-tier system of spatial planning instruments: Under this system, we learnt that SDFs are to be prepared at the national, regional and district levels in Ghana. In keeping with this requirement and as part of the plan-making experimentation projects undertaken in the immediate aftermath of reforms, the first SDF covering the entire country—The Ghana National Spatial Development Framework—has been prepared to guide spatial development over a period of 20 years, beginning 2015. In this section, an overview of the NSDF will be provided focusing on the core spatial strategies proposed by the framework.

4.4 National-Level Spatial Planning Under the New Spatial Planning ...   69

But first, it is important to understand the purpose of SDFs in general as put forward in the New Spatial Planning (NSPS) Model Guidelines and the scope and content of an NSDF as outlined in the Land Use and Spatial Planning law.

The NSPS guidelines define an SDF as follows:

> ... an indicative plan, showing the expected development over a fifteen to twenty-year period, which will include the location of key components of the strategy aimed at achieving the desired development' (MESTI 2011, p. 9).

Moreover, the scope and content of a NSDF as outlined in the Land Use and Spatial Planning Act (Act 925) are the following:

- a statement from the Authority covering the spatial dimensions of general trends, prospects, opportunities and challenges of the country;
- the objectives that are to guide the Authority (i.e. LUSPA) in coping with the challenges and enable the Authority to contribute to the improvement of quality of life and sustainable management of land use and human settlements;
- the strategies designed or to be employed by the Authority to cope with and guide management on land use to meet the identified challenges; and
- the means to be employed in monitoring the efficiency of the strategies adopted by the Authority;
- the designation of the proposed hierarchy of human settlements, anticipated population growth and distribution;
- the location of major potential projects, development corridors and other areas of national importance;
- the designation of infrastructure, services and development corridors of national importance;
- the allocation of development centres of national importance and their likely development within the planning period;
- a Strategic Environmental Assessment of the Spatial Development Framework; and
- a statement of the consultative procedures undertaken in the preparation of the framework; and
- any other matter considered relevant for the purpose of the National Spatial Development Framework.

4.4.1 The Ghana National Spatial Development Framework (2015–2035)

The NSDF (Government of Ghana 2015) is the first comprehensive, long-term national spatial development to be formulated since the first national physical plan was prepared in 1963. The preparation of the framework begun in 2013 and took a period of two years to complete. The framework has been adopted and by the new planning law become a legally binding policy framework to guide spatial development in Ghana in the next 25 years. The framework is comprehensive and integrated, providing a spatial strategy to address a range of social, economic and environmental issues across the country. Below, an overview of the NSDF is provided.

4.4.1.1 Background to the NSDF

The historically weak spatial planning system, the absence of long-term national development planning, including spatial planning as well as the current realities of socio-economic and environmental development challenges have necessitated the formulation of the NSDF. A lot has changed with respect to the underlying processes and the emergent structure and of the national space-economy since the first National Physical Plan, that would remain the only one for nearly 60 years, was formulated. A snapshot analysis of population dynamics, one of the most important drivers of socio-economic and environmental change at all levels, affirms this. With a total population of nearly 7 million in 1960, the country's population has nearly quadrupled to about 25 million over the 60-year period between 1960 and 2010 (see Table 4.2). According to the latest National Population and Housing Census data, Ghana's population has been growing at average rate of 2.69% per annum (see Table 4.2). Over the same 60-year period, the proportion of the country's population living in urban areas (i.e. settlements with population 5000 and above) has nearly doubled from 23% in 1960 to 51% in 2010 (see Fig. 4.3).

Moreover, population growth in the ten regions confirms the national trend (Table 4.3): All ten regions have been growing at 1% or more over the period between 2000 and 2010 (see Fig. 3.4). The Greater Accra, Ashanti, Western and Northern regions have been the fastest growing areas with average annual growth rate of 3.27, 2.84, 2.13 and 3.24 %, respectively (Fig. 4.4). These regions also have the biggest urban agglomerations including Accra, Kumasi and Tamale metropolises, which accommodate 57, 43 and 23% share of their regional populations, respectively. The metropolitan areas of Accra and Kumasi alone are increasing their populations at an average rate of 4.2 and 5.6%, respectively.

Table 4.2 Historical population size and growth rate

Year	Population size	Growth rate
1960	6,726,815	–
1970	8,559,313	2.40
1984	12,296,081	2.62
2000	18,912,079	2.73
2010	24,658,823	2.69

Source Population and Housing Census data, Ghana Statistical Services (For more information, see http://www.statsghana.gov.gh/statistics.html)

4.4 National-Level Spatial Planning Under the New Spatial Planning ...

	1960	1970	1984	2000	2010
Total Population	6,726,815	8,559,313	12,296,081	18,912,079	24,658,823
Urban	1551000	2472000	3934000	8274000	12551000
Rural	5,175,815	6,087,313	8,362,081	10,638,079	12,107,823

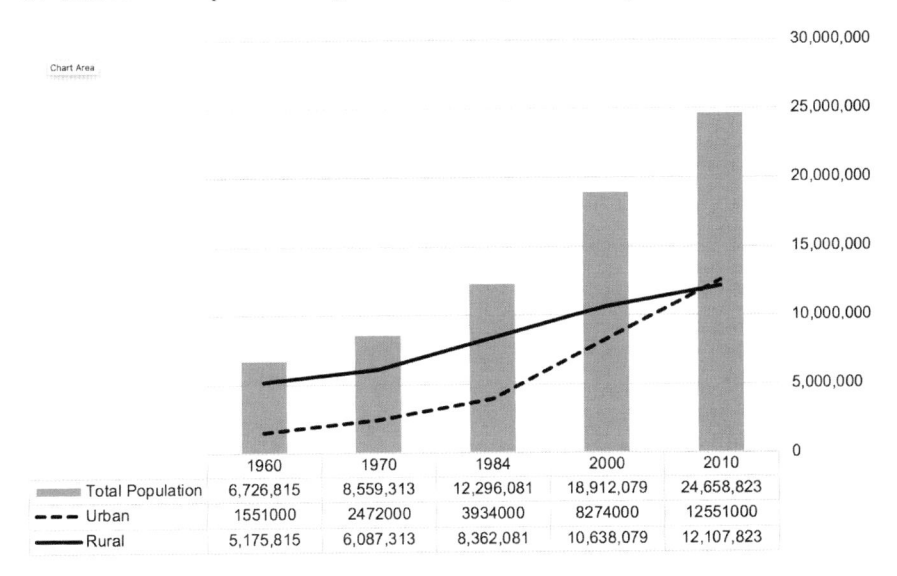

Fig. 4.3 Urbanization trends in Ghana. *Source* Based on Population and Housing Census data, Ghana Statistical Services (For more information see: http://www.statsghana.gov.gh/statistics.html)

Table 4.3 Historical population distribution in the ten administrative regions of Ghana

	1960	1970	1984	2000	2010
Ashanti	1,109,133	1,481,698	2,090,100	3,612,950	4,780,380
Greater Accra	541,933	903,447	1,431,099	2,905,726	4,010,054
Eastern	1,044,080	1,209,828	1,680,890	2,106,696	2,633,154
Northern	531,573	727,618	1,164,583	1,820,806	2,479,461
Western	626,155	770,087	1,157,807	1,924,577	2,376,021
Brong-Ahafo	587,920	766,509	1,206,608	1,815,408	2,310,983
Volta	777,285	947,268	1,211,907	1,635,421	2,118,252
Central	751,392	890,135	1,142,335	1,593,823	2,201,863
Upper East	468,638	542,858	772,744	920,089	1,046,545
Upper West	288,706	319,865	438,008	576,583	702,110
Ghana	**6,726,815**	**8,559,313**	**12,296,081**	**18,912,079**	**24,658,823**

Source Population and Housing Census data, Ghana Statistical Services

The rapid population growth and the increasing levels of urbanization present both opportunities and challenges. Opportunities exist in that bigger urban agglomerations are vital in supporting and driving economic growth. However, the growing population poses challenges in terms of adequately supplying infrastructure and social services, ensuring sustainable use of resources while protecting the natural environment to ensure the continuous supply of essential ecosystem services. In order to exploit the opportunities and manage the challenges, a deliberate and proactive plan

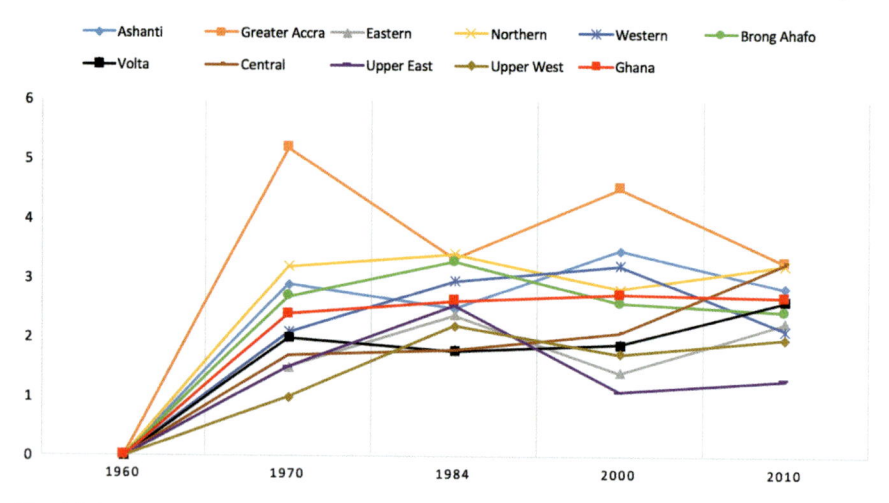

Fig. 4.4 Regional population growth rate (1960–2010). *Source* Based on Population and Housing Census data, Ghana Statistical Services

that identifies the needs of the population, interventions needed to address them and elaborates how public investments could be anchored to specific regions, districts, cities, towns and villages for equitable development outcomes is crucial. An SDF at the national level therefore becomes a prerequisite for sustainable development outcomes across the country.

4.4.1.2 Spatial Strategy of the NSDF: Settlement Hierarchy, City-Regions and Urban Networks

Theoretically, the role of towns in the national development process and the strategic imperative of their designation for well-defined function(s) have been expressed in regional spatial organization models and concepts such as the core–periphery model (Friedmann 1967); growth pole theory (Perroux 1955); central place theory (Christaller 1933; Lösch 1940); and new economic geography (Krugman 1998). The overall spatial strategy of the NSDF is anchored on the concept of settlement hierarchy—a hierarchical configuration of towns and cities according to some criteria such as population size, physical size, services and the functions they are to play now and in the future within the overall national systems of settlements.

Four key principles underpin the overall spatial strategy put forward by the NSDF. Referred to as its 'pillars', they are the following:

i. Emphasis on balanced polycentric development
ii. Improve regional, national and international connectivity
iii. Strengthen the metropolitan city-regions of Accra and Kumasi
iv. Promote development in secondary cities
v. Ensure sustainable development and protect ecological assets.

4.4 National-Level Spatial Planning Under the New Spatial Planning ...

Table 4.4 Proposed hierarchy of settlement in the NSDF

Grade	Settlements
Grade-1	*Metropolitan areas*: Accra, Kumasi, Sekondi-Takoradi and Tamale
Grade-2	*Remaining six regional capitals + five large towns*: Cape Coast, Sunyani, Koforidua, Ho, Wa and Bawku + Obuasi, Techiman, Yendi, Nkawkaw and Tarkwa
Grade-3	*Administrative centres and market towns*: They vary widely in population, from 58,000 to just over 5000 people. About 60 settlements have been identified and classified as Grade-3
Grade-4	*Rural service centres*: Their populations vary between 5000 and 10,000. About 92 settlements have been identified and classified as Grade-4

In Ghana, previous research has sought to demonstrate how an urban system can be produced to meet spatial development objectives. For example, Yeboah et al. (2013) identified a typology of 14 towns according to the services and functions they provide (see Yeboah et al. 2013). Building on this initial research and insights from the regional spatial organization models, the NSDF proposes a four-tier settlement system to reflect its core principles and to realize the overall spatial strategy. As shown in Table 4.4, the four largest urban agglomerations have been identified as *Grade-1* settlements. Being the only metropolitan areas and the administrative capitals of the four fastest growing regions in the country, these settlements have the population, infrastructure and services that could be further exploited to drive the socio-economic development process of the country. At the same time, some of the major development challenges, including lack of access to shelter, inadequate infrastructure and services, rapid urban expansion and environmental degradation are localized in their severest magnitudes in these larger urban centres. The remaining six regional capitals and additional five large towns have been identified as *Grade-2 settlements*. Following this population-and-function criterion, *Grade-3* and *Grade-4* settlements are also identified and designated as administrative and market towns, and rural service areas, respectively.

Some of the strategic imperatives of socio-spatial, economic and environmental development outcomes are best realized by thinking and acting beyond the limitations imposed by formal administrative boundaries. Indeed, it is widely acknowledged that in most cases, the solutions to problems of metropolitan areas as defined by their respective administrative boundaries lie beyond those boundaries. Consequently, the concept of 'functional regions' or 'special regions' has become useful as way of thinking about the interlinkages between places, especially between metropolitan units and the areas immediately surrounding them. 'Functional regions' or 'special regions' therefore cover contiguous spatial zones that are usually under separate local authority administrations and are of particular strategic importance to the overall national spatial economy.

In line with strategic imperatives of thinking about places as interconnected webs of interactions and functional interdependence, the NSDF delineates what it calls '*city-regions*'. Two such regions are identified and delineated, namely Accra city-region and the Greater Kumasi Sub-Region (GKSR). Later in Chap. 4, where the

Table 4.5 City-regions and urban networks identified and delineated in the NSDF

City-regions and urban networks	Description
Accra city-region	Defined as the Greater Accra Metropolitan Area, the contiguous built-up area as 2010 and the districts sharing part of the built-up area and within the present and future daily commuting zone of the metropolis
Greater Kumasi Sub-Region (GKSR)	Comprises one metropolitan area (i.e. Kumasi Metropolitan Area) and seven immediate surrounding districts, namely Afigya-Kwabre District, Kwabre East District, Ejisu-Juaben Municipality, Bosomtwe District, Atwima Kwanwoma District and Atwima Nwabiagya
Sekondi-Takoradi Urban Network (STUN)	Includes six districts: Sekondi-Takoradi (STMA), Ahanta West, Shama, Wassa East, Mpohor and Komenda/Edina/Eguafo/Abirem
Cape Coast Urban Network (CCUN)	Includes Cape Coast Metropolis, the central regional capital and six surrounding districts. Mfantsiman, Ekumfi, Komenda Edina Eguafo Abirem, Abura/Asebu/Kwamankese, Lower Twifo Denkyira and Ajumako/Enyan/Essiam
Tamale Urban Network (TUN)	Includes Tamale North Sub-metro and Sagnarigu districts and parts of eight neighbouring districts, namely Savelugu/Nanton, Kumbugu, Tolon, East Gonja, Gonja Central, Mion, Karaga and Gushegu
Sunyani Urban Network (SUN)	Sunyani Municipal Area and surrounding 60 urban and rural settlements. Major towns in the cluster include Techiman, Berekum, Dormaa Ahenkro, Wenchi, Mim, Kintampo
North-East Urban Network (NEUN)	Covers the whole of Upper East Region, and the North-Eastern part of Northern region; major towns in the network include Bolgatanga, Bawku, Walewale and Gambaga
Aflao Urban Network (AUN)	Comprises 22 urban settlements, including Afloa, the largest town Aflao in the network. Another major town in the cluster are Dzodze, Analoga and Akatsi
Ho-Hohoe Urban Network (HHUN)	Comprises 18 urban settlements, with Ho, the Volta regional capital being the largest town in the network. Other major towns in the cluster include Kpandu and Hohoe
Wa Urban Network (WUN)	Consists of six urban settlements, namely Wa, Jirapa, Tumu, Nandom, Lawra and Hamale in the Upper West Region of Ghana

focus will be on regional spatial planning, we will revisit these city-regions. In addition to the two city-regions, '*urban networks*' conceptualized as clusters of towns and their immediately surrounding rural settlements are also designated in the NSDF. Eight of such clusters are identified and delineated in the NSDF. A summary description of the city-regions and urban networks proposed in the NSDF is provided in Table 4.5.

The overall spatial development concept of the NSDF, which combines proposals for major infrastructure development such as trunk roads and railway networks with the proposed hierarchy of settlements, city-regions and urban-networks, is depicted in Fig. 4.5.

4.4 National-Level Spatial Planning Under the New Spatial Planning … 75

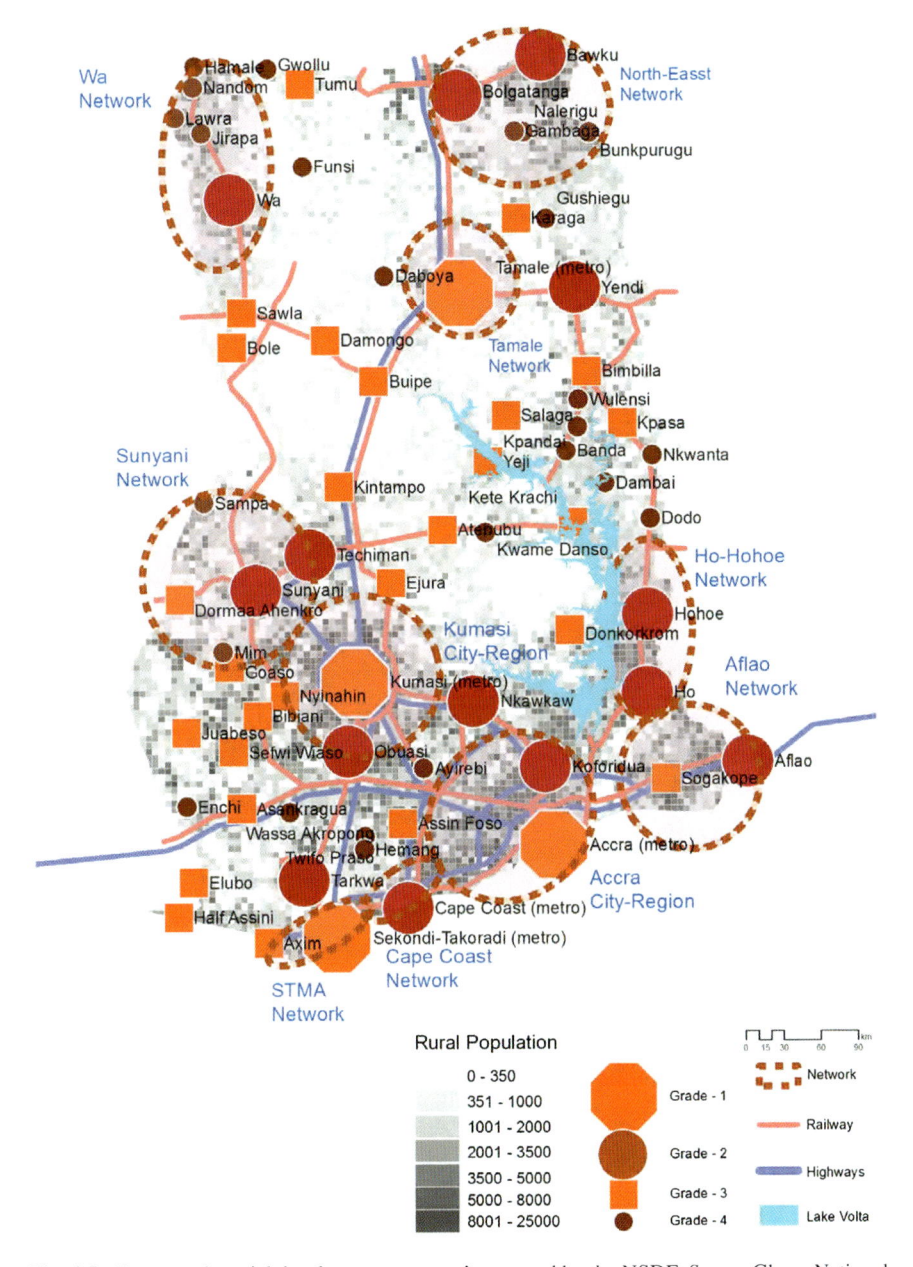

Fig. 4.5 'Integrated spatial development concept' proposed by the NSDF. *Source* Ghana National Spatial Development Framework

4.4.2 The National Urban Policy Framework

Ahead of the publication of the NSDF in 2016, the first comprehensive urban policy, the National Urban Policy Framework (NUP), was published in 2012. While the NSDF addresses spatial development issues of all settlements and how they are linked to the long-term national social, economic and environmental development priorities, the NUP, as the name suggests, focuses mainly on urban areas—settlements defined as having a population of 5000 or more in Ghana. Like the NSDF, the NUP, the first of its kind in Ghana, was necessitated by the rapid rate of urbanization, particularly in the beginning of the twenty-first century and the attendant socio-spatial, economic and environment challenges. The intent of the urban policy therefore was to provide a comprehensions framework that will guide decision-making at multiple scales and across sectors that affect the pattern of urban growth and management of urban areas in the country. The overall goal of the national policy as stated in the NUP document is to:

> … promote a sustainable, spatially integrated and orderly development of urban settlements with adequate housing, infrastructure and services, efficient institutions, and a sound living and working environment for all people to support the rapid socio- economic development of Ghana (Government of Ghana 2012, p. 12).

The NUP goals and objectives are founded on the seven core principles of:

- Promoting urban centres as engines of growth;
- Promoting development through an integrated settlement system;
- Facilitating socio-economic development of rural and lagging regions;
- Mainstreaming environmental concerns into urban development;
- Enhancing participatory and accountable urban governance;
- Employing information, education and communication strategy; and
- Emphasizing the roles of central and local governments.

Moreover, the overarching goal of the NUP is translated into twelve key objectives outlined as follows:

- To facilitate balanced redistribution of urban population.
- To promote a spatially integrated hierarchy of urban centres.
- To promote urban economic development
- To improve environmental quality of urban life.
- To ensure effective planning and management of urban growth and sprawl, especially of the primate cities and other large urban centres.
- To ensure efficient urban infrastructure and service delivery.
- To improve access to adequate and affordable low-income housing.
- To promote urban safety and security.
- To strengthen urban governance.
- To promote climate change adaptation and mitigation mechanisms.
- To strengthen applied research in urban and regional development.
- To expand sources of funding for urban development and strengthen urban financial management.

Under each of the twelve objectives, specific strategies to realize them are specified. A detailed programme of action, the Ghana National Urban Policy Action Plan, accompanied the NUP. The Action Plan outlines the key activities and initiatives that need undertaking in order to achieve the objectives of the NUP over a five-year time period. Specific institutions are identified and assigned responsibilities by the Action Plan.

4.5 Realizing Long-Term Spatial Development Imperatives Within a Culture of Medium-Term Development Planning

For the first time in nearly six decades, Ghana has adopted an integrated, long-range and legally binding National Spatial Development Framework and a comprehensive National Urban Policy. The NSDF, unlike the Physical Development Plan of 1963, is not just adjunct to a bigger National Development Plan, but a stand-alone comprehensive plan to realize social, economic and environmental development objectives over the next 20 years.

As significant as these developments are in the history of planning in the country, a number of practical questions need addressing. As was highlighted earlier, over the past three decades, the culture of preparing medium-term national-level plans that coincide with the constitutionally guaranteed tenure of governments has become an established norm. Within this culture of four-year development planning horizon at the national level, how could the strategic goals and objectives of the long-term NSDF be realized? Most importantly, how will a legally binding NSDF sit with political party priorities and promises that are based on election campaign manifestos? Below, these questions are reflected upon, suggesting ways in which the overall objectives of the NSDF could be achieved under the prevailing tradition of medium-term national development planning.

To begin with, it is important to highlight that although as of the time of formulating the NSDF, no long-term National Development Plans existed, it was still the intent as articulated in the new spatial planning law that the framework would accompany a long-term national development plan. In order words, it was anticipated, at least in principle, that a long-term national development plan would exist to which the NSDF would constitute a spatial elaboration. In reality, however, we have the reverse of the process, where a long-term NSDF has been formulated without any long-term National Development Plan in place. Over the 20-year horizon of the NSDF, three successive governments are expected. It is likely that at least the first two governments, over the next eight-year period, will follow the path of issuing medium-term NPFs. Should this trend continue, the major challenge would be how we achieve consistency between the NPFs and the NSDF. In principle, this should be straightforward. Given that the NSDF is a legally binding document with a presidential assent, successive governments in their NPFs would only need to identify and workout how their broad visions could be achieved through the NSDF. For example, political parties will have to identify aspects of the NSDF, especially the proposed physical projects that they would want to pursue and integrate them into the NPF they would eventually formulate on being elected to govern.

In reality, however, expecting politicians to align their priorities and visions to the existing framework would be difficult. Indeed, the divergence between long-term plan proposals and what politicians consider expedient partly reflect the highly political nature of the activity of planning. It also highlights the practical limitations of any form of long-range comprehensive planning in a process as dynamic as national development planning where socio-economic conditions and realities are bound to change over time. It may well be that the NSDF is legally binding, but the extent to which successive governments could be compelled to implement part or all the proposals of the framework remains to be seen.

The run-up to the 2016 general elections perhaps exemplifies the aforementioned challenges very well. In the wake of the political campaigns, the NSDF had already been formulated and formally adopted. At the same time, the NPP, the political party which eventually won the elections, outlined several priorities in their manifesto, which in their words, emphasized their philosophy of a 'place-based, bottom-up' approach to development. Under this overarching theme, several spatially oriented initiatives were outlined, including

- the 'one-district-one-factory' policy by which every district in the country was to get a manufacturing industry;
- establishment of the Infrastructure for Poverty Eradication Project (IPEP);
- the 'one-village-one-dam' policy by which a dam will be constructed in every village in the northern part of the country to provide irrigation for farming;
- establishment of the Zongo Development Fund to support the provision of critical infrastructure and services in deprived inner-city areas;
- and implementation of national digital addressing system to provide unique addresses for all properties in Ghana.

These above policy proposals have already reflected in the Budget Statement and Economic Policy of Government of Ghana for the 2017 Fiscal Year, clearly signalling a commitment to implement them. Although these proposals are not explicitly captured in the long-term NSDF and are the result of priorities defined from the perspective of a political party with respect to what is expedient, they also do not in principle contradict the long-term framework. In fact, the possibility exists for the 'one-district-one-factory' policy to be anchored on the national spatial strategy advanced in the NSDF.

Furthermore, in January 2010, the Constitution Review Commission was set up to consult with the people of Ghana on the operation of the 1992 Constitution and on any changes that need to be made to the Constitution. Among the major national issues over which there was public consensus was the need for a long-term National Development Plan. The Commission has recommended for this to be enshrined in the national Constitution, and as of the time of writing this book, a consultative process was underway by the NDPC towards the formulation of the first long-term National Development Plan. The proposed plan, which is expected to cover a 40-year horizon

4.5 Realizing Long-Term Spatial Development Imperatives … 79

(i.e. 2018–2057), could in principle absorb the NSDF proposals for the first 20 years in a reverse order as stipulated in the planning laws of the country. The next NSDF after 2030 would then become the spatial elaboration of the National Development Plan's last 20 years.

The above said, the proposed long-term National Development Plan, according to the NDPC, would be indicative rather than comprehensive, setting out broad development goals and allowing successive governments the flexibility to fill in the details. Aligning a comprehensive, legally binding NSDF with an indicative plan that outlines broadly, long-range national development goals will also present many practical challenges. This is because, while the former is detailed, the latter would lack in the level of detail needed to match to content of the two. Also, in its implementation, the National Development Plan is expected to go by the four-year medium-term planning horizon, effectively yielding ten NPFs over the 40-year period. Like the NSDF, the proposed 40-year plan is expected to be legally binding on successive governments. As a remedy, the successive medium-term NPFs could be made much more comprehensive to an extent that they could be aligned with the NSDF preceding them.

Summarising, with the entrenched norm of medium-term national development planning, outcomes of which reflect strongly political party manifestos and promises, a pragmatic approach of incrementally seeking a balance between priorities as defined by politicians and the technically grounded strategic imperatives of the NSDF will have to be adopted. The success or otherwise of this process will depend on the two major institutions which have the technical competence and legal mandate for planning at the national level—the National Development Planning Commission and the Land Use and Spatial Planning Authority. While relying on formal and legally established structures such as invoking the legality of the NSDF could be a major asset in this process, these institutions would largely have to rely on their technical competence as well as bring to bear their negotiation skills as players in the public policy arena to attempt to achieve reasonable consistency between the NSDF strategy and the goals of politicians. Should a long-term National Development Plan replace the norm of short to medium-term planning in the future, the ability to overcome the compartmentalization of planning between the NDPC, handling what they call 'Development Planning' on the one hand and LUSPA, dealing with what they conceive to be 'Spatial Planning' would be crucial to ensuring that spatial planning fulfils its potential and bring about sustainable development.

4.6 Conclusions

In the immediate aftermath of independence in 1957, prominence was given to an integrated National Development Plan that was accompanied by a comprehensive spatial development plan. This would no longer be embraced under the contemporary tradition of medium-term National Development Planning that was entrenched by the promulgation of the National Development Planning (System) Act (Act 480) and the Local Government Act (Act 462) at the beginning of the twentieth century. Under this tradition of medium-term national development planning, comprehensive spatial elaboration of NPFs, as was done in the first National Physical Development Plan in 1963, was ignored. Instead, over a period of 30 years, NPFs attempted to capture spatial development issues as just one of the thematic areas of wider national development priorities and as separate from other cross-cutting development issues. Even so, the associated goals were vague, lacking any carefully designed programme of action indicating the distribution and timing of programmes, projects and investments of NPFs at the level of regions, districts, cities and towns. National-level spatial planning effectively died during this 30-year period.

We also discovered that the promulgation of the Land Use and Spatial Planning Act (Act 925) in 2016 would resuscitate spatial planning at all levels. Operating alongside the existing Systems Act (Act 462), however, two separate planning systems—the development planning system and the new spatial planning system—would emerge, each deploying separate institutional arrangements for planning at the national and local levels. While acknowledging the significance of the reforms leading to re-establishment of spatial planning, we saw that the two separate and seemingly competing planning systems have compartmentalized the authority and competence for planning at the national between the NDPC, handling the narrowly conceived task of development planning and the newly established LUSPA, handling the task of spatial planning. The discussion highlighted the tendency for institutional conflict, duplication of effort and waste of resources under the emergent situation of 'one nation, two planning systems'. It was also highlighted that the impacts of the existing arrangements for planning transcend the national to the local levels.

The formulation of the Ghana National Spatial Development Framework (2015–2035) and the National Urban Policy Framework has so far been the two most important demonstration of spatial planning at the national level in Ghana. For the first time in more than five decades, a long-term, comprehensive and legally binding NSDF has been adopted. Despite this being a remarkable development in the history of spatial planning in Ghana, it was argued that the established culture of medium-term national development planning, the divergence between the strategic imperatives of the NSDF and the priorities of politicians and political party manifestoes, pose fundamental challenges to realizing the objectives of the long-term NSDF. It was argued that these challenges reflect the highly political nature of spatial planning and that a pragmatic approach of incrementally seeking a balance between priorities as defined by politicians and the technically grounded strategic imperatives of the NSDF will have to be adopted. This pragmatic approach would require the institutions of spatial planning actively engaging with the political process. This should include planning professional as public servants dialoguing with successive governments in a bid to reach consensus as to how best to bridge the gap between priorities as perceived and pursued by politicians on the one hand and the strategic imperatives put forward in the NSDF on the other hand.

References

Acheampong RA, Ibrahim A (2016) one nation, two planning systems? Spatial planning and multi-level policy integration in Ghana: mechanisms, challenges and the way forward. Urban Forum 27(1): 1–18

Christaller W (1933) Central places in Southern Germany. Fischer, Jena

Government of Ghana (1995). Ghana—Vision 2020 (The First Step: 1996–2000). Presidential report on coordinated programme of economic and social development policies (Policies for the preparation of 1996–2000 Development Plan)

Government of Ghana (2005) Growth and Poverty Reduction Strategy (GPRS II), (2006–2009). Republic of Ghana. National Development Planning Commission. http://planipolis.iiep.unesco.org/sites/planipolis/files/ressources/ghana_prsp_june_2006.pdf. Accessed on 24 Jul 2018

Government of Ghana (2010) Medium-term national development policy framework: Ghana Shared Growth and Development Agenda (GSGDA), 2010–2013. In: Volume I: policy framework. Government of Ghana, National Development Planning Commission (NPDC), Dec 2010

Government of Ghana (2012) National Urban Policy Framework. Action Plan. Ministry of Local Government and Rural Development.http://www.washwatch.org/uploads/filer_public/38/d4/38d4a952-f123-479c-ae10-8623b91582d8/national_urban_policy_framework_ghana_2012.pdf. Accessed on 24 Jul 2018

Government of Ghana (2014) Ghana Shared Growth and Development Agenda (GSGDA) II, 2014–2017. Medium-term national development policy framework. Government of Ghana. National Development Planning Commission (NDPC). Dec 2014 http://www.un-page.org/files/public/gsgda.pdf. Accessed on 24 Jul 2018

Government of Ghana (2015) Ghana National Spatial Development Framework (2015–2035). Executive Summary. Government of Ghana. February 2015.http://www.ghanalap.gov.gh/files/NSDF-Final-Report-EXECSUM-Vol-III-Final-Edition-TAC.pdf. Accessed on 24 Jul 2018

International Monetary Fund and International Development Association (2004) Enhanced heavily indebted poor countries (HIPC) Initiative completion point document. 15 June 2004. https://www.imf.org/en/Publications/CR/Issues/2016/12/31/Ghana-Enhanced-Initiative-for-Heavily-Indebted-Poor-Countries-Completion-Point-Document-17548. Accessed on 24 Jul 2018

Kraus J (1991) The struggle over structural adjustment in Ghana. Afr Today 38(4): 19–37

Krugman P (1998) What's new about the new economic geography? Oxford Rev Econ Policy 14(2): 7–17

Land use and spatial planning Act (2016) Act 925 The nine hundred and twenty fifth act of the parliament of the Republic of Ghana

Local Government Act (1933) Act 462, the four hundred and sixty two act of the parliament of the Republic of Ghana

Loxley J (1990) Structural adjustment in Africa: reflections on Ghana and Zambia. Rev Afr Polit Econ 17(47): 8–27

Lösch A (1940) The economics of location. Fischer, Jena

Ministry of Environment, Science, Technology and Innovation (MESTI) (2011) The new spatial planning model guidelines. Town and Country Planning Department. November 2011

National Development Planning (System) Act (1994) Act 480. The four hundred and eightieth act of the parliament of the Republic of Ghana. http://urbanlex.unhabitat.org/sites/default/files/urbanlex//gh_national_development_planning_act_1994.pdf. Accessed on 24 Jul 2018

Perroux F (1955) Note sur la notion de pôle de croissance: l'economie du xxeme siecle. Presses Universitaires de France

Town and Country Planning Act (1945) CAP 84.http://www.epa.gov.gh/ghanalex/acts/Acts/TOWN%20AND%20COUNTRY%20PLANNING%20ACT,1945.pdf

Chapter 5
The Inception of Regional Spatial Planning

Abstract Regional spatial planning is necessary for realizing at the sub-national level goals articulated in national spatial development policy instruments. However, what exactly constitutes a *region* is a contested concept as different types of *regions* may be imagined and delineated for spatial planning. The focus of this chapter is on spatial planning at the scale of regions in Ghana. The chapter begins by exploring the different types of regions that could be delineated, and outlines the rationale for planning at these spatial scales globally. Zooming into the Ghana context, this chapter explores the story of regional planning in Ghana and critically examines the effect on contemporary regional spatial planning of what has previously been described as a situation of '*one nation, two planning systems*' in Ghana. Ongoing efforts at institutionalizing regional spatial planning are discussed, using recent Regional Spatial Development Framework experimentation projects to illustrate spatial planning in practice for different types of regions in Ghana. The chapter also reflects on the current structures for delivering the objectives of regional spatial planning and the experiences gained so far from the experimentation projects, and suggests ways for an effective regional spatial planning system.

Keywords Regional development · Regional spatial planning
Regional Spatial Development Framework · City-regions · Urban networks
Special Development Zones · Development authority · Ghana

5.1 Introduction

Spatial planning involves taking decisions and addressing social, economic and environmental problems at multiple scales. As we discovered in Chap. 4, national-level planning instruments broadly set out government priorities and goals to be realized over a period of time. Unless specifically conceived and formulated as National Spatial Development Frameworks, they tend to focus mainly on the economy and are therefore generally limited in terms of their spatial expressions. These higher-level, national planning instruments therefore rely on lower-tier instruments to further translate strategic policies into sub-national policy frameworks. In most countries, the spatial planning system deploys regional plans or development frameworks as intermediate instruments between the national and local levels of decision-making. Regional Spatial Development Frameworks or plans anchor national policies to relatively larger spatial units that transcend the boundary of a single local government authority but smaller than the state.

Although the rationale for spatial planning at the regional scale is recognized, the concept of a '*region*' is contested. Practically speaking, it is far less complex when dealing with a single demonstrative unit (i.e. administrative region or district) as the spatial unit of a plan. This is partly because of the clearly delineated geographical boundaries, a well-defined administrative structures and institutional mandates, which are clearly specified in the relevant legislations. However, for strategic reasons, it becomes necessary to delineate other types of regions as spatial units for planning. The boundaries of such *regions* tend to be fuzzy, covering heterogeneous, yet contiguous landscape. Consequently, spatial planning for these areas may not be readily supported by existing institutional structures and legislative frameworks. This invariably calls for some form of institutional readjustments and a bringing together of several, sometimes conflicting pieces of legislations either through formally established channels or informal, ad hock procedures to accomplish the task of spatial planning.

This chapter focuses on spatial planning at the *regional* level in Ghana. We will begin the discussion by exploring what we mean when we use the word *region or regional*. We will also identify the different types of *regions* and outline the core arguments for spatial planning at this spatial scale. Next the experience of Ghana with respect to regional (spatial) planning will be discussed followed by a critical examination of the institutional and legal arrangements for contemporary regional (spatial) planning. Here, as was done in the discussion of national-level spatial planning in the previous chapter, we will examine the interface between the three main legislative frameworks for planning, focusing on how they affect regional planning (i.e. the 1994 National Development Planning (System) Act (Act 480), the 1993 Local Government Act (Act 462) and the 2016 Land Use and Spatial Planning Act (Act 925)). In the fourth section of this chapter, we will focus on the practices and experiences with respect to regional spatial planning under the new spatial planning system. Here, efforts to institutionalize regional spatial planning for different types of regions will be discussed. The penultimate section will reflect on the regional planning experiences so far and suggest ways for effective regional spatial planning in the future. The discussion will then conclude with a summary of the key issues and core arguments advanced in this chapter.

5.2 The Concept of Regions and Rationale for Regional Spatial Planning

Sovereign nations have clearly defined boundaries which are in turn subdivided into several geographical entities. These entities, in different countries, are given various names such as *district*, *municipality*, *province* and *region*. The names often imply the relative position of the spatial unit in question within a clearly defined, hierarchical system such as the units of political administration in a decentralized governance structure. This form of hierarchical arrangement of the units of spatio-political and administrative governance may vary according to the types of national governance structures adopted by a country, such as a decentralized unitary state, regionalized state and federal state.

Whereas local government areas (i.e. districts, municipalities and provinces) are easy to delineate for planning purposes because their boundaries are established by legislations backing the various tiers of governance in a country, the *region* as a spatial entity is quite difficult to define in any precise terms. This is because, geographically, a *region* is bigger than a single local government area but smaller than the state. It may encompass, for example, two or more local government areas or the whole of one local government area plus parts of neighbouring local government areas. Below, we explore the meanings of a *region* and identify some types of *regions*.

Powell (2012), advances one of the all-encompassing definitions of the concept of a *region*. According to him, a *region* is:

> a sizable division of territory separated from other areas by a mixture of tangible characteristics which simultaneously sets it apart from neighbouring areas and declares a degree of commonality or shared identity, among the physical features and/or the inhabitants of that divisions. (Quoted in Collits 2012, 180)

Stilwell (1992) also conceptualizes a region simply as a contiguous set of places which have something in common. In terms of spatial extent, regions are bigger than local authorities but smaller than the state (Murphy 1984).

From the above definitions, it becomes clear that a *region* covers a relatively bigger geographical area within a country, which may or may not share similar characteristics in, for example, the ethnicity and cultures of the population who live in the region. Collits (2012) distinguishes between three types of regions, namely

i. Homogenous regions, as one way of conceptualizing regions that emphasizes the commonality between the places that come together to constitute this geographical unit. The delineation of a region around a group of people based on a common language and culture could be considered as an example of a homogenous region;

ii. Functional regions, as another type of region based on places that share some functional linkages. Places may be linked functionally by, for example, the distribution of major residential and economic activities between them and infrastructure and commuting patterns. Functional regions are therefore not necessarily homogenous but are borne out of the strategic imperatives of strengthening and exploiting functional linkages between places with unique characteristics and comparative advantage for national development; and

iii. Policy or programming regions, as regions that have formal political dimensions that involves boundaries and governance. These types of regions are administrative divisions below the central government, which comprising several local authorities with clearly defined boundaries. In Ghana, for example, the ten administrative divisions in the country fall under this category of regions.

Collits (2012) further distinguishes between the *noun* and *verb* forms of the word region—that is 'regionalism' and 'regionalization'. *Regionalism* encompasses the theory and practice of bottom-up decision-making which stresses that policies are best formulated and implemented in and for regions rather than at the national level by central government. In spatial planning terms, it reflects the view that the region is one of the appropriate units for planning where people within it and their governance institutions can have the opportunity to shape decisions that affect them. *Regionalization*, on the other hand, refers to a deliberate, often imposed policy of reorganizing places and their governance structures by central governments to achieve efficiency objectives. Through the process of regionalization, several formally existing local authorities may be merged to form a single administrative unit.

Regional spatial planning can therefore be defined as a conscious attempt to develop policies, plans and strategies to influence the location and timing of social, economic and environmental development priorities for contiguous areas larger than local authorities but smaller than the state. It may cover established administrative regions, or a contiguous area delineated based on the functional relationships between them. The regional spatial plan, therefore, becomes a mid-tier instrument within the hierarchy of spatial planning instruments to anchor national visions to the geographical units they cover while at the same time providing the critical linkage for achieving conformity between national and local policy goals and objectives.

The tradition of regional planning dates back to the 1960s when the first regional plans were formulated in the USA and across Europe (Lichfield 1967). Over the years, regional planning doctrines and rationale have not only become established but also evolved in line with changes in the social, economic, political and environmental conditions that necessitate the need for meso-level spatial development planning. Several authors have written extensively about regional planning and the rationale for this approach to planning (see, e.g., Collits 2012; Haughton and Counsell 2004; Kuklinski 1970; Murphy 1984; Friedmann and Weaver 1979). The central claim of their arguments in favour of regional planning is summarized as follows.

Firstly, evidence show that problems that define national development goals and priorities do not necessarily manifest the same way across all areas. This means that magnitude and severity of development problems are not always the same across different areas in a country. Indeed, disparities in regional development can be observed in many countries, where certain areas experience prosperity while other areas experience deprivation and decline. The disparities between the historically deprived north and relatively prosperous south in Ghana are a case in point. In situations such as this, a regional approach to planning becomes crucial in not only formulating tailor-made

5.2 The Concept of Regions and Rationale for Regional Spatial Planning

policies for specific areas with unique challenges, but also as a means to achieve equitable allocation of resources to reduce the prevailing imbalances. Invariably, problems such as disparities in regional development require co-ordinated efforts in terms of the planning and implementation of programmes and strategies to bring about spatial convergence. As Collits (2012) argue, a regional approach to planning become necessary in instances where the problems that need addressing are considered too big for a single local government but too small to warrant direct intervention from central national government.

Another justification for regional planning is related to the need to harness the comparative advantage of different regions within a country. Within a country, different regions may possess unique opportunities and potentials from which they assert strategic importance within the overall national spatial economy. Identifying and strategically planning for these regions therefore become useful in strengthening their role as hubs of innovation and economic growth at the national level, while establishing their competitiveness at the global scale. A typical example of this would be the Western Region of Ghana, which furnishes most of the natural resources of the country.

Furthermore, regional planning implies devolution of decision-making powers to administrative structures below central government. As such, the activity itself constitutes an important part of the process of democratic governance. Planning at the level of regions, in the context of regionalism, affords people in the region the opportunity to influence decisions that affect them. This is true both under decentralized unitary states and under federal and quasi-federal arrangements, such as in the UK, Germany and France, where regional governments have the autonomy to negotiate and chart a unique path for their respective regions based on existing potentials, prevailing challenges and priorities for a given period. In addition, regional spatial planning and the instructional and legal arrangements that facilitates it have the potential to address political and administrative fragmentation that is inherent in decentralization programmes that often results in the establishment of several local government authorities (Haughton and Counsell 2004).

Finally, a regional approach to planning can provide the platform to co-ordinate the activities of several local governments. Collaboration and co-ordination are particularly useful in, for example, resolving land use conflicts, planning for and delivering major infrastructure projects and services, achieving environmental conservation objectives and assessing the implications of individual local authority's plans and policies on neighbouring local governments and the region as a whole. Later in Chap. 6, we will look at the idea of integration in spatial planning and how that is crucial in dealing with the inherently complex process of plan formulation, implementation and development management.

5.3 Spatial Planning for Regions in Ghana: A Previously Uncharted Territory

The importance of regional spatial planning in Ghana was recognized as far back as the 1960s when the first National Physical Development Plan was formulated. However, it was not until the beginning of 2010 that spatial planning at the regional level would first be experimented in the country. Prior to this period, the 1945 Town and Country Planning Ordinance (CAP 84) and its subsequent amendments—1958 Town and Country Planning Act (Act 30) and 1960 Town and Country Planning (Amendment) Act (Act 33)—did not mention regional planning at all. The scope of application of these initial legislations was mainly to guide planning and development at the level of towns and cities. Indeed, globally, the doctrines of regional planning would become prominent later in the 1960s; hence, these ordinances reflected the prevailing understanding of the nature and scope of planning then.

The National Physical Development Plan of 1963, which incidentally coincided with the period when the doctrines and practices of regional planning were gaining importance globally, was the first to explicitly state the need for planning at the regional level in Ghana. As we found out from the history of planning in Chap. 2, one of the core aims of the National Physical Development Plan was to guide the elaboration of regional development plans. These regional physical development plans, which would have possibly covered each of the eight administrative regions that existed at the time, were, however, never to be realized. In the post-independence era, governance below the central government was generally weak (Crawford 2004). Thus, public-sector activities, including spatial planning, took a top-down approach, where decisions affecting towns and cities were taken at the highest levels by the central government. Within this centralized governance structure, where decision-making powers had not been devolved to regions as administrative units, regional spatial planning would have been difficult to execute. Indeed, as we will later find out in this chapter, regions have remained administrative entities to date, without any clearly defined powers to formulate and implement development plans. Besides, the size of the population for the whole country back then was rather small, not exceeding seven million people. This might have necessitated focusing on major towns and cities such as Accra, Tema and Kumasi rather than venturing into planning at the regional scale.

In 1988, at the beginning of the decentralization and democratization programme, the regional tier of governance, overseen by RCCs, would be established by the PNDC Law 270. RCCs would become the main administrative bodies for the ten administrative regions that currently exist in Ghana. RCCs derive their mandate from the Local Government Act of 1993 (Act 462). The 1994 National Development Planning (System) Act (Act 480) formally instituted development planning at the regional level. The Systems Act, which as we have previously established, instituted the development planning system, required regional development plans to be formulated for different types of regions, including the administrative regions, and what it referred to as '*special development areas*' and '*joint development planning areas*'. According to the Act, special and joint development areas are to be designated if special

5.3 Spatial Planning for Regions in Ghana: A Previously Uncharted Territory

physical and socio-economic characteristics necessitate bringing together a number of areas as a single unit for the purposes of development planning.

Despite the legal framework supporting the formulation of regional development plans, not a single regional plan was prepared under the development planning system (see Acheampong and Alhassn 2016; Ofori 2017). As observed earlier by Acheampong and Alhassan (2016),

> the concept [and practice] of regional planning, either at the level of the administrative regions or functional regions [was] never pursued in Ghana within the established development planning system ... the National Development Planning (System) Act (Act 480), introduced the concepts of 'Joint-Development Planning Areas' where contiguous areas could be designated for planning purposes. However, the idea of jointly planning development has only existed as a concept in statutes without any experimental cases indicating how this could work in practice, in terms of the institutional, legal and financial arrangements needed to support such an endeavour .... (Acheampong and Alhassan 2016, 9)

Reasons for the lack of regional planning within the development planning system are outlined as follows: firstly, it appears that the institutional structures created by the decentralization law did not support the activity of regional development planning. Although RCCs were established as regional governance bodies, their powers and mandate, as stipulated in the local government law, did not extend into undertaking the task of planning themselves. Instead, their mandate, to date, is restricted to co-ordinating, monitoring and evaluating the activities of DPUs of local governments. They perform these functions through their Regional Planning Co-ordinating Units (RPCUs). That RCCs do not have the authority and by extension, competences for regional planning are evidenced by the functions assigned to their RPCUs under the development planning system. According to the National Development Planning (System) Act (Act 480), RPCUs shall

i. provide the District Planning Authorities (DPA) with such information and data as is necessary to assist them in the formulation of district development plans;
ii. co-ordinate the plans and programmes of the District Planning Authorities and harmonize the plans and programmes with national development policies and priorities for consideration and approval by the NDPC;
iii. monitor and evaluate the implementation of the programmes and projects of the District Planning Authorities within the region;
iv. act on behalf of the NDPC with respect to such national programmes and projects in the region as the NDPC may direct; and
v. perform such other planning functions as may be assigned to it by the NDPC.

Similarly, the regional offices of the TCPD, although staffed with professional town planners, never ventured into regional spatial planning until the reforms initiated under LUPMP, first introduced the concept of Regional Spatial Development Frameworks and enshrined it in the 2016 Land use and Spatial Development Act (Act 925). Instead, the TCPD at the regional level also confined their role mainly to supervising the activities of TCPD offices at the local government level. As has been mentioned elsewhere, the TCPD was operating under the CAP 84 of 1945 when the National Development Planning (System) Act was promulgated in 1994 and continued to do so despite the legislation becoming obsolete, until in 2016 when the Land

Use and Spatial Planning Act (Ac 925) was promulgated. Thus, the old planning ordinance did not make provisions for regional spatial planning.

Moreover, the National Development Planning (System) Act and the ethos of development planning that accompanied it would introduce new institutional arrangements, which instead of explicitly recognizing the TCPD as being part of it, sidelined it. This is notwithstanding the fact that until the tradition of development planning would become established from the mid-1990s, TCPDs had long been performing the functions of planning—as far back as the 1940s when they were first established. In fact, not a single mention of the TCPD as an institution can be found in the National Development Planning (System) Act of 1994. This further supports the argument advanced elsewhere that the development planning system, instituted by the aforementioned legislative instrument, was conceived to be different and separate from the spatial planning system that had been established earlier by the 1945 Town and Country Planning Ordinance (CAP 84). It is therefore fair to assert that the TCPD at all levels, including the regions, would have found the new institutional arrangements of the Systems Act not only an attempt to diminish their importance but also an invasion of their core competences. With its existence ostensibly threatened by the new-found ethos of development planning, the TCPD retreated into carrying out its traditional land use planning and development control functions mainly at the level of towns and cities. The opportunity for the formulation and implementation of regional development plans either by the regional TCDPs or in collaboration with the RCPUs of RCCs would be missed for more than six decades, until reforms initiated at the beginning of twenty-first century would culminate in the formulation of the first-ever Regional Spatial Development Framework in 2012.

5.4 Contemporary Planning Systems and Authority and Competences for Regional Spatial Planning

As has been elaborated previously, the 1994 National Development Planning (System) Act (Act 480) establishes the development planning system, while the 2016 Land Use and Spatial Planning Act (Act 925) establishes the new spatial planning system. Both legislations draw on the structures established by the decentralization law, Local Government Act (Act 462), to set out separate institutional authorities and competences for contemporary planning in Ghana.

Furthermore, in the previous section, we discovered that the National Development Planning (System) Act (Act 480) introduced the concept of regional development planning for different types of regions including administrative regions, special development areas and joint development areas. Alongside these initial requirements, the new Land Use and Spatial Planning law also provides the legal requirement for the formulation of Regional Spatial Development Frameworks (RSDFs) as intermediate frameworks between the national SDF and District Spatial Development Frameworks (DSDFs) at the local government level. As outlined in the Land Use and Spatial Planning Act (Act 925), RSPCs shall perform the following functions on behalf of RPCUs:

5.4 Contemporary Planning Systems and Authority and Competences for …

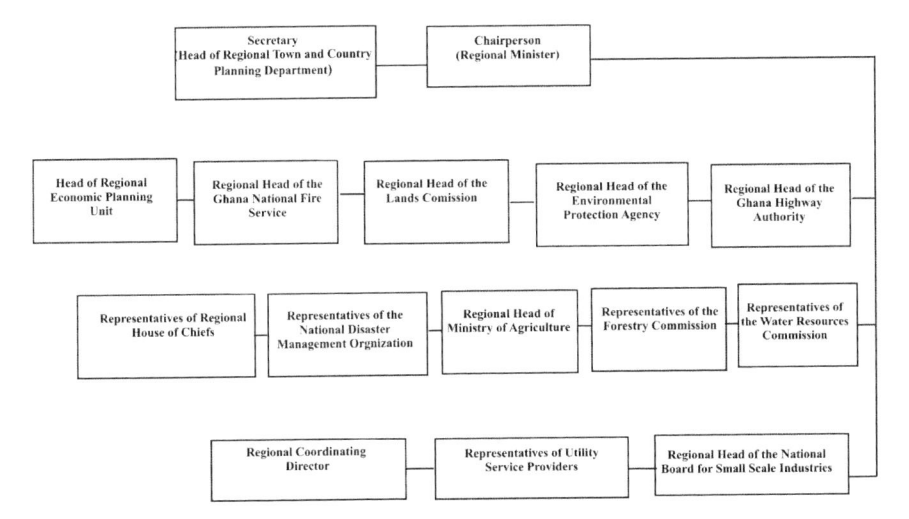

Fig. 5.1 Composition of the Regional Spatial Planning Committee established by the Land Use and Spatial Planning Act (Act 925)

i. develop a Regional Spatial Development Framework for the region in consultation with the district assemblies as part of the spatial development component of the Regional Development Plan;
ii. adjudicate on appeals or complaints resulting from decisions, actions or inactions of the District Spatial Planning Committee of the district assemblies;
iii. where required, prepare sub-regional or multi-district spatial development framework for two or more districts;
iv. within the region; and perform any other function to give effect to this Act within the region.

A critical examination of the legal provisions for regional planning is necessary to highlight their points of convergence and departure. Firstly, the National Development Planning (System) Act vests the authority and competence for regional development planning in RCCs. As we have already established, the decentralisation law does not explicitly grant RCCs the powers to perform the task of regional development planning. In addition, The Systems Act designates RPCUs as the technical wing of RCCs to accomplish their largely administrative and oversight functions. On the side of spatial planning, the Land Use and Spatial Planning law establishes RSPCs to deliver regional spatial planning. Unlike the Systems Act, the new spatial planning law outlines all the institutions that should constitute the RSPC to undertake the task of regional spatial planning (see Fig. 5.1). Thus, one of the major differences between the two legislations, which spell out the provisions for modern-day regional planning in Ghana, can be found in the separate institutional arrangements they deploy.

Secondly, some attempts have been made to harmonize the rather unnecessary compartmentalization of the functions of regional planning and competing roles of the institution involved that has emerged as a result of the two legislations which have instituted the two separate planning systems. The efforts at harmonization, articulated in the new spatial planning law, are intended, in principle, to be realized through the two mechanisms outlined as follows:

i. RSDFs prepared under the Land Use and Spatial Planning law are intended to be the spatial elaboration of regional development plans prepared under the National Development Planning Systems law. Thus, implicitly, it is assumed that an RSDF would be formulated under the spatial planning system alongside a Regional Development Plan formulated under the provisions of the development planning system. The two regional planning documents would then be issued together. Both regional plans would, in principle, cover the same time horizon.

ii. The Land Use and Spatial Planning law for purposes of harmonization also admonishes RSPCs to maintain a close working relationship with MMDAs and other key stakeholders in the formulation of RSDFs and that the procedure for the preparation and approval of plans shall be in accordance with the provisions for joint development planning areas under Act 480 and the manuals for the preparation of spatial development framework.

Thus, we find the spatial planning system reaching out to the development planning system to imitate some modalities to resolve a problem which could have been avoided in the first place. In principle, the aforementioned attempt to overcome the inherent challenges of the two separate and competing planning systems might be considered a step in the right direction. In practice, however, the two main mechanisms intended to resolve the problem would be fraught with several challenges. One reason is that, as previously highlighted, the RCCs expected to formulate Regional Development Plans to which RSDFs prepared by RSPCs would accompany do not have a clearly stated legal mandate to do so. This could probably explain why they have never endeavoured ventured into this territory in the first place since coming into existence nearly three decades ago.

In addition, while the spatial planning law clearly specifies RSDFs as long-term plans covering a period of 20 years, the Systems Act does not contain similar provisions with respect to Regional Development Plans. That said, it has been established that the culture of four-year medium-term development planning typifies planning horizons under the 1994 National Development Planning (System) Act (Act 480). Should this culture continue at the regional level, the practical challenge of aligning long-term spatial development frameworks with medium-term development plans that exists at the national level will also arise at the regional level.

The other important challenge to harmonizing development planning and spatial planning at the regional level arises from the scope and content of these plans. It is now recognized that spatial planning should take an integrative view and strive to engage with the wider issues affecting all aspects of development instead of focusing narrowly on settlement design and development control. Consequently, the new spatial planning law requires RSDFs to consider and integrate economic, social and environmental development objectives. It therefore not quite clear how different a Regional Development Plan, should one be formulated, would be from an RSDF. Again, it is becoming quite clear that at the regional level of planning, the conditions created for unhealthy institutional competition would lead to unnecessary conflicts and duplication of efforts. This would not only frustrate the institutions and other stakeholders involved in the process but also have a negative impact on the ability of a system that lacks clarity and consistency to effectively deliver regional development objectives.

5.5 Regional Spatial Planning in Practice Under the New Spatial Planning System

Since 2007, a series of spatial development frameworks and plans are being prepared to experiment with the three-tier system of spatial planning instruments defined under the land use and spatial planning law, and as part of a much broader programme to institutionalize the new spatial planning system. The Ghana National Spatial Development Framework (2015–2035), which was presented in Chap. 3, has been one of the major outputs of the experimentation process. This section provides a brief overview of two other major frameworks that have been formulated for two different types of regions in Ghana. These are the Western Region Spatial Development Framework (WRSDF 2012–2022), which covers an administrative region, and the Greater Kumasi Sub-Regional Spatial Development Framework, a multi-district framework covering eight local authorities which have been designated as a functional region. The overview of these two frameworks will cover the overall vision as stated in the frameworks, the underlying spatial development concepts, the core spatial strategies and how these are linked to wider socio-economic and environmental development objectives of the areas they cover.

5.5.1 Western Region Spatial Development Framework (2012–2022)

5.5.1.1 Background to the WRSDF

The Western Region Spatial Development Framework (WRSDF) is the first long-term regional framework to be prepared as part of the experimental plan-making process to entrench the new spatial planning system. The framework covers the whole of the Western Region of Ghana, one of the ten administrative regions in the country. The size of the region is approximately 24,000 km^2, representing 10% of the total land area of Ghana. The population of the region as of 2010 was nearly 2.5 million, representing about 9.3% of the country's total population. As of 2017, 22 local governments (i.e. MMDAs) constituted the Western Region, with Sekondi-Takoradi Metropolis being the largest urban agglomeration within it. The region is the richest of all the ten regions in terms of natural resources. It therefore functions as a major hub for economic activities in agriculture, mining, manufacturing and forestry in the national spatial economy. The offshore oil and gas exploitation, which has been underway in the region since 2007, has further boosted the areas economic potential and importance within the national economy. The economic boom anticipated by the oil and gas industry and the concomitant increase in economic activities, population and pressure on land and infrastructure therefore necessitated the formulation of the WRSDF. The framework preparation was financed by a grant from the Norwegian government under the 'Oil for Development (OfD)' programme.

The overall aim of WRSDF is to articulate long-term policies that would guide the spatial development of the regions over a period of 20 years. In view of this, the framework prescribes an orderly, sustainable, and cohesive land use plan as a platform for integrating social, economic, infrastructure and environmental development strategies of the regions. The overall vision of as stated in the framework is that:

> The Western Region of Ghana will be a spatially balanced, diversified and environmentally friendly economy that brings sufficient employment and social services to its people and the nation, based on sustainable use of its natural resource endowment. (WRSDF 2012–2022, iv)

Moreover, the framers of the Regional Spatial Development Framework aim for the WRSDF to provide the platform to co-ordinate and harmonize the priorities and plans of key stakeholders including the public-sector, the private-sector, traditional authorities and civil society organizations to bring about development in the region. As a regional framework, the WRSDF would also guide the formulation of lower-level frameworks including SDFs covering the districts in the region and Structure Plans and local sub-visions plans for cities, towns and neighbourhoods within these districts.

5.5.1.2 Spatial Strategy of the WRSDF

The core spatial development strategy of the WRSDF focused on proposals to ensure

- Balanced spatial development,
- Equal access to employment,
- Equitable distribution of services,
- Control of sprawling development and
- Affordable housing supply
- Biodiversity conservation and environmental protection

The goals of ensuring balanced spatial development would be realized through a proposed hierarchy of settlements, with each designated to provide specific strategic functions and services. The framework proposes a five-tier configuration of settlements (see Table 5.1). Following the growth-pole strategy, major settlements, within the regional system of settlements, are positioned to become key service centres, serving as catchment areas of growth for their surrounding peri-urban and rural areas.

5.5 Regional Spatial Planning in Practice Under the New Spatial ...

Table 5.1 Proposed hierarchy of settlement in the WRSDF

Grade	Settlements
Grade-1	*National centres*: Sekondi-Takoradi and Kumasi
Grade-2	*Regional capital + Four major towns*: Sekondi-Takoradi + Asankragwa, Tarkwa, Aiyinase
Grade-3	*District capital + Major towns*. About 46 settlements have been identified and classified as Grade-3
Grade-4	Other urban settlements with population 5000 or above. About 31 settlements have been identified and classified as Grade-4
Grade-5	*Rural settlements*: having population greater than 2500 but not exceeding 5000 inhabitants. About 57 settlements have been identified and classified as Grade-5

Linkages between the proposed hierarchy of settlements and wider socio-economic development goals are explicitly established. The framework designates four main economic activity areas, namely areas designated solely for agricultural activities; mixed agriculture and mining areas; areas designated exclusively for mining; and the coastal districts specializing in fishing as well as activities in the emerging oil and gas industry (see Fig. 5.2). To promote agriculture and agro-processing, the WRSDF identifies sites within the different grades of settlements to be developed for the storage, processing and packaging of agricultural produce.

Moreover, to ensure that the population has access to employment and benefit from the emerging oil and gas economy, the framework proposes suitable locations for various downstream oil and gas activities outside of the coastal areas where the oil drilling is taking place, linked by modern transport systems. Strategies to promote low-cost housing including the provision of suitable housing development sites to accommodate the expected increase in population as well as plans to covert construction camps as potential future housing areas are proposed in the WRSDF. The framework also identities and designates ecologically sensitive areas and biodiversity hotspots for conservation with the aim to ensuring continuous supply of essential ecosystem services.

5.5.2 Greater Kumasi Sub-Regional Spatial Development Framework (2013–2033)

Whereas the WRSDF covers an administrative region, the Greater Kumasi Sub-Regional Spatial Development Framework (GKSR-SDF) is the first spatial development framework formulated for a functional regional in Ghana. The Greater Kumasi sub-region (GKSR) was designated in 2010 by the Town and Country Planning Department as a functional planning area for purposes of strategic regional planning and sustainable growth management. The sub-region comprises one metropolitan area, the Kumasi Metropolis, which is the second largest metropolitan area in Ghana and seven immediate surrounding districts, namely Afigya-Kwabre District, Kwabre

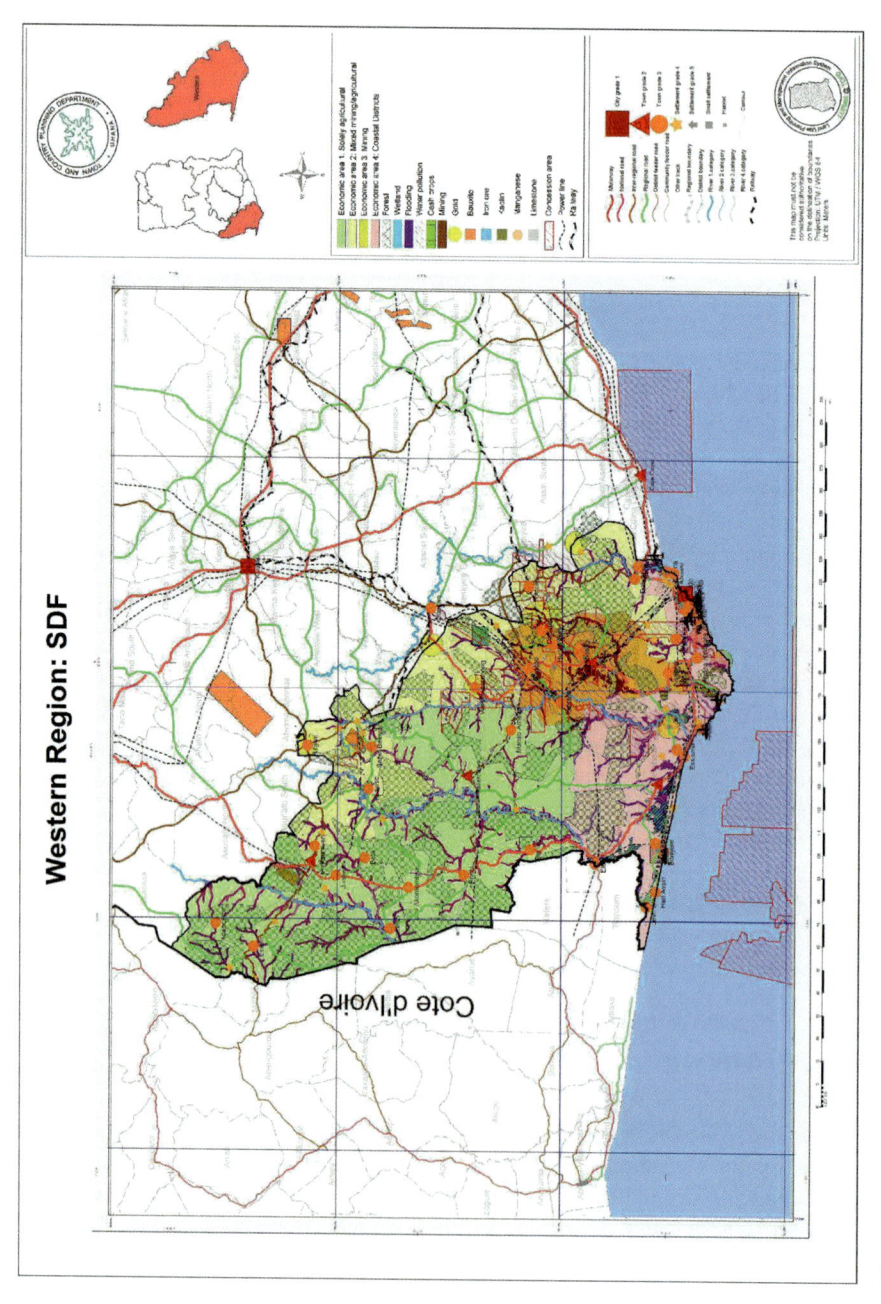

96 5 The Inception of Regional Spatial Planning

Fig. 5.2 Diagram of the Western Region spatial development framework *Source* Town and Country Planning Department, Accra Head Office

5.5 Regional Spatial Planning in Practice Under the New Spatial …

East District, Ejisu-Juaben Municipality, Bosomtwe District, Atwima-Kwanwoma District and Atwima-Nwabiagya.

It covers a contiguous area of approximately 2850 km^2 of urban, peri-urban and rural land, with an estimated population of about 2,764,091 in 2010. The sub-region's population represents about 58% of the total population of the Ashanti Region and 11% of Ghana's population. Growing at an average rate of 4.62% between 2000 and 2010 compared to the average annual growth rates of 2.84 and 2.69% for the Ashanti Region and Ghana, respectively (Acheampong et al. 2017), the sub-region's population is expected to rise to 5,761,463 by the end of the framework horizon in 2033. The Kumasi Metropolitan Area (KMA), being the most populous area within the sub-region, has a population of nearly two-million residents, which accounts for 74% of the sub-region's total population. The KMA is the fastest growing urban area in Ghana, increasing its population at an average annual rate of 5.69 since 2000 and 2010.

The rapid rate of urbanization in the sub-region and the need to provide a clear spatial strategy linked to the future socio-economic development of all areas within the sub-region necessitated the formulation of the GKSR-SDF. As a multi-district, sub-regional development framework, it would also provide the basis for the preparation of various local planning instruments to guide growth and development in areas within the sub-region. The framework envisages that:

> The Greater Kumasi Sub-Region will become a pioneer to transform the current economy into a vibrant, modernized and diversified economy including commerce, logistics, manufacturing and knowledge-based industries, by creating a liveable and efficient urban space, while maintaining the historical and cultural aspirations of the Ashanti Region. (GKSR-SDF 2013)

5.5.2.1 Spatial Strategy of the GKSR-SDF

Polycentricism—the principle of configuring urban spatial structure as a network of multiple activity centres performing designated functions—underpins the overall spatial strategy of the GKSR-SDF. The framework therefore proposes a major shift from the current monocentric spatial structure that characterizes the sub-region, where population and major functions are concentrated in the KMA, the sub-regional core (we will elaborate on the sub-region's spatial structure later in Chap. 10 where we focus on the relationship between spatial development and transportation). Instead, it puts forward a decentralized urban spatial structure by promoting development in designated sub-centres around the KMA as well as urban corridors which connect the KMA to the surrounding districts by means of modern transit systems such as a Bus Rapid Transit (BRT).

According to the framework, sub-centres would be developed as hubs of commercial, industrial, business and public administration functions supported by large-scale estates development for residential. As shown in the framework diagram (Fig. 5.3), the multi-nuclei spatial configuration follows a hierarchical ordering starting with a 'Primary' centre, being the KMA which performs commercial and

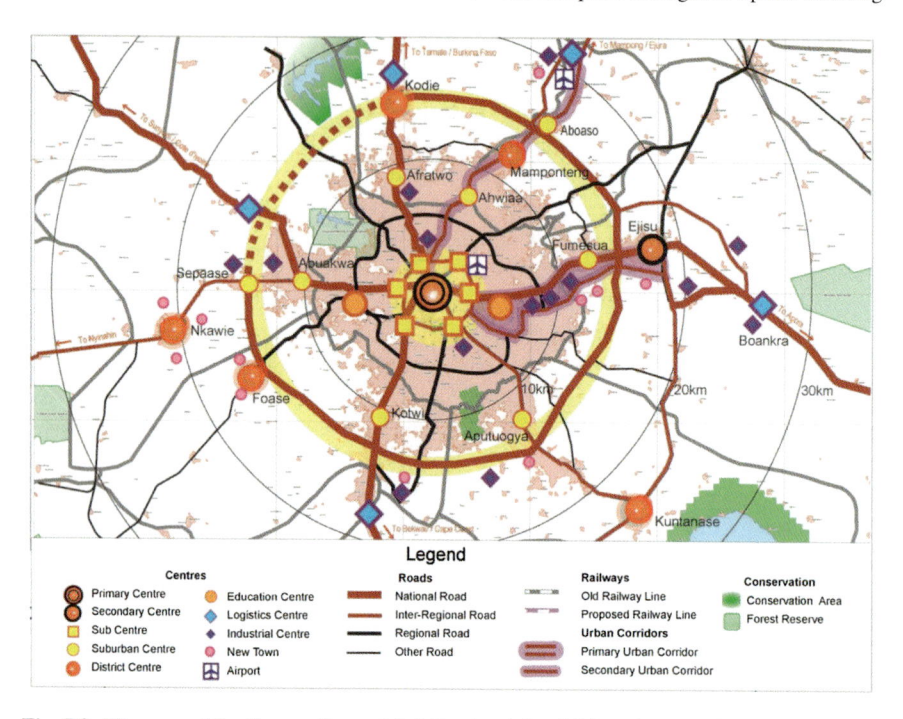

Fig. 5.3 Diagram of the Greater Kumasi Sub-Regional Spatial Development Framework *Source* Town and Country Planning Department, Kumasi Ghana

administrative functions. The Ejisu-Juaben Municipality located to the east of the KMA is designated as the 'Secondary' centre with the main functions being residential and administration. There are also 'District' centres which are the administrative capitals of the seven districts that constitute the sub-region. The eight sub-metropolitan units within the KMA have been designated as sub-centres followed by major peri-urban towns designated as sub-urban areas for residential development. Along the main road arteries, urban corridors linking surrounding districts to the KMA as well as areas designated for specific functions including education, logistics, industry and the development of new towns have been designated. To realize the overall spatial structure, the framework delineates areas in which lower-tier plans would be prepared, namely the Central Business District in KMA, and Kuntanase, Boaman, Juaben, Nobewam and Barekese Townships.

5.6 The Emerging Paradigm of Special Development Zones and Development Authorities

In addition to designating and planning for administrative and functional regions, a new paradigm of spatially oriented programming, whereby large geographical areas is designated by governments to be overseen by independent and autonomous bodies, commonly referred to as Development Authorities, is becoming common in recent years. Development Authorities implement area-based programmes that cut across several local government areas and sometimes across administrative regions. They operate directly under the office of the Presidency and derive their mandates from specific legislative instruments, often passed in the form of Act of parliament. Given their geographical scope, Special Development Zones in Ghana could be categorized under '*special regions*'.

> The designation of special development zones by various governments in Ghana do refer to the provisions of the Fourth Republican Constitution with respect to the directive principles of state policy. Article 36 (2) (d) of the national constitution states that:

The state shall, in particular, take all necessary steps to establish a sound and healthy economy whose underlying principles shall include undertaking even and balanced development of all regions and every part of each region of Ghana, and, in particular, improving the conditions of life in the rural areas, and generally, redressing any imbalance in development between the rural and the urban areas.

Successive governments over the past decade have invoked this constitutional provision to establish Special Development Zones and to pass Acts of parliaments to establish Development Authorities to oversee the formulation and implementation of programmes, plans and strategies in these special regions. The aim has been to prioritize development and investments in specific geographical areas with peculiar conditions and to address development disparities that characterize the national space economy. Table 5.1 outlines Special Development Zones that have been established or proposed since 2006, their geographical scope and their respective Development Authorities.

The designation of Special Development Zones and their respective Development Authorities has come to serve both strategic and political objectives. Strategically, Special Development Zones have become useful vehicles for channelling public investments to deprived regions in the country. For example, addressing the uneven development that has historically existed between the southern and northern regions of Ghana was the main justification for the establishment the Northern Savannah Ecological Zone. Historical bias in government policy towards the resource-rich southern and coastal regions that goes as far back as the era of colonization, coupled with natural factors such harsh weather conditions, intense heat, creeping desertification and natural disasters in the north (Songsore 2003), have been responsible for the widening north–south divide. In response to the relative underdevelopment of the Northern part of Ghana, the Savannah Accelerated Development Authority (SADA), which oversees the Northern Savannah Ecological Zone, was established in 2007 as a vehicle to prioritize investments to these economically depressed areas. In

2015, a long-term spatial development framework for the zone—The Spatial Development Framework for the Northern Savannah Ecological Zone (2015–2035)—was formulated by SADA with technical assistance from the TCPD.

Since the establishment of SADA, other Special Development Zones have either been established or proposed by successive governments. Designating such zones has ostensibly become linked to the election-winning strategies of political parties. This works in two ways. From the point of view of political parties, the promise of a Development Authority for a region is often aimed at signalling intention to prioritize development in that region, if elected. The population of the region respond or at least are expected to respond by electing the political party proposing the establishment of the Special Development Zone and development authority.

As a result of the aforementioned political expedience, Development Authorities and Special Development Zones have proliferated in recent years. Within the past decade alone, eight Development Authorities and their corresponding Special Development Zones have come up in the regional development discourse of the country. Most of them remain at the conceptual stages with no clear indication of their geographical coverage as well as the programmes that will be implemented.

Moreover, consensus as to the number of Special Development Zones that is feasible and practical in a relatively small country like Ghana, and the geographical extent of these zones are lacking across the political divide. Consequently, the creation of new Special Development Zones or reorganization of existing zones has become subject to the discretionary powers exercised by successive governments. For example, prior to introducing the Western Corridor Development Initiative, which now covers the Western and Central Regions, and its Development Authority, the Central Region Development Commission (CEDECOM) had already been established. The former therefore subsumed the latter as a result of the reorganization. Similarly, the 5th Government of the 4th Republic has proposed to establish four Special Development Zones with their respective Development Authorities across the country (see Table 5.2). Meanwhile, the proposed extent of these Development Zones, as could be guessed from their names and jurisdictions of their would-be Development Authorities, is already covered by Authorities established by the previous administration.

What the foregoing shows is that, while special regions such as those designated as Special Development Zones in Ghana have become necessary in addressing perceived development needs of various areas, especially areas lagging in the national socio-economic development process, the emphasis placed on political expedience over strategic imperatives in their designation has meant that the process has been overpoliticized, chaotic and uncoordinated.

5.7 Towards Effective Regional Spatial Planning

The contemporary legal frameworks and institutional structures which have created two seemingly different planning systems that are in competition with one another poses challenges for spatial planning at all levels, including regional planning. In

Table 5.2 Spatial Development Zones

Special Development Zone/Development Authority	Scope	Status
The Millennium Development Authority (MiDA)	Nationwide—oversee, manage and implement the Programmes under the Millennium Challenge Account for poverty reduction through economic growth as set out in each agreement between the Government of Ghana and the Millennium Challenge Corporation acting for and on behalf of the Government of the United States of America and for any other national development programme of similar nature funded by the Government of Ghana, a Development Partner or both and to provide for related matters	Established by an Act of Parliament (Act 702, 709 and 897 as amended), 2006
Northern Savannah Ecological Zone and The Savannah Accelerated Development Authority (SADA)	The upper-east, upper-west and northern regions, and districts of the Brong Ahafo and Volta regions that share similar ecological and social conditions as the three northern regions	Established by an Act of Parliament (Act 805, 2010)
Bui Power Authority	Mandate to plan, execute and manage the Bui Hydroelectric Project which includes: generation of electrical power for general industrial and domestic use; construction of a transmission system linked to the national grid; supply of electrical power to certified and licensed utility companies; promotion of activities consistent with the provision of facilities for multipurpose uses such as agro businesses, fisheries and tourism.	Established by the Act of Parliament (BPA Act 740, 2007)
Western Corridor Development Initiative and Western Corridor Development Authority	Western and central regions, and subsume the Central Region Development Commission (CEDECOM)	Projects ongoing, proposed Authority not established yet
Eastern Corridor Development Initiative and Eastern Corridor Development Authority	The southern and middle belts of the Volta region and the Accra Plains and Afram Plains. This initiative focuses on road upgrading from Tema, Asikuma, Hohoe, Jasikan, Yendi, and Nalerigu to Kulungugu	Projects ongoing, proposed Authority not established yet

(continued)

Table 5.2 (continued)

Special Development Zone/Development Authority	Scope	Status
Capital City Development Initiative and Capital City Development Authority	Special Development Zone covering the metropolitan, municipal and district assemblies within the Greater Accra Metropolitan Area including Accra and Tema and their surrounding districts	Proposed
Forest Belt Development Initiative and Forest Belt Development Authority	Cover the forest zones in the eastern, Ashanti and Brong Ahafo regions	Proposed*
Northern Development Authority,	The three northern regions	Proposed*
Middle-Belt Development Authority	Brong Ahafo, Ashanti, and Eastern Regions	Proposed*
Coastal Development Authority	The four coastal regions	Proposed*

*New Development Authorities Proposed by the 5th Government of the 4th Republic whose tenure runs from 07 January 2017 to 07 January 2020. It is possible that these new Authority, should they come in force will result in reorganization of existing Authorities

a similar vein, the increasingly politicalized motives that determine the designation and/or reorganization of Special Development Zones raises several implications for effective spatial planning at the regional level. One of the major highlights of the discussions in this chapter is how through a combination of past initiatives and contemporary reforms, the National Development Planning (System) Act (Act 480, 1994) and the Land Use and Spatial Planning Act (Act 925, 2016), both drawing on the decentralized governance structure established by the Local Government Act (Act 462, 1993), have created competing institutional arrangements for regional development planning. Also highlighted is the resultant compartmentalization of the task of regional planning between what appears, under the new spatial planning system, to be a well-laid out cross-agency arrangement constituting RSPCs to handle the task of regional spatial planning on the one hand and RCCs whose mandate to carry out the task of regional development planning, are less clearly defined by the decentralization law on the other hand.

Quite clearly, regional planning risks are being ineffective under the current institutional arrangements. Resolving the emergent institutional conflicts, while addressing the attendant problems with duplication of functions and resource wastage in the making and implementation of regional plans, is by no means an easy and straight-

5.7 Towards Effective Regional Spatial Planning

forward task. That said, in the short to medium term, effective co-ordination between the institutions with competence for regional planning as defined in the two systems (i.e. development planning system and spatial planning system) would be crucial in overcoming some of these challenges.

By indicating the intent that RSPCs would work closely with RPCUs, the Land Use and Spatial Planning Act has laid the foundation for intuitional co-ordination that may facilitate a more integrative approach to regional development planning—an approach which brings together social, economic and environmental objectives in a coherent framework to address the developmental challenges at the scale of different types of regions in Ghana. However, formal arrangements compelling institutions to co-ordinate do not always work in practice. This would be particularly so in the context of Ghana where the existing institutional arrangements, by default, are designed to promote compartmentalization of tasks and unhealthy competition, rather than ensuring integration and collaboration. Beyond this formally established mechanism, it may be possible to rely on informal, ad hoc arrangements to build consensus and to harmonize any potential conflicts and duplications of efforts that may arise in the regional development planning process.

In the interest of building strong and effective institutional arrangements to deliver the development needs of the citizenry, a sober reflection, resulting in a dispassionate debate on the current arrangement, is needed. Hopefully, this would lead to an overhaul of the overall state machinery planning at all levels, including the regions. A new law which establishes a unified planning system by merging the two currently competing planning systems might be needed in the future.

If regional spatial planning would deliver its goals effectively, conscious efforts at achieving some form of balance between political expedience on the one hand and technical, strategic imperatives on the other hand in the designation of Special Development Zones must be pursued. The current situation whereby Special Development Zones and their respective development authorities are delineated, and/or the boundaries of existing zones are redemarcated by successive governments, would further complicate the institutional structures beyond what might be reasonable and necessary to bring about efficient outcomes. This ensuing chaos is not far-fetched, as the practice creates a complex layer of overlapping institutions, jurisdictions and spatial units that could obfuscate regional spatial planning endeavours. For example, there are the jurisdictions of development authorities that oversee Special Development Zones; jurisdictions of already established administrative structures at the regional and sub-regional levels where multiple districts are involved, and the physical boundaries of the city-regions and urban networks proposed in the NSDF (see Chap. 3). Superimposing all the above layers reveals the complexity of institutional arrangements, involving a plethora of agencies, departments and authorities, as well as overlaps among spatial units that regional spatial development planning must grapple with.

Avoiding the negative consequences of the aforementioned institutional arrangements will depend largely on dialogue and consensus building. The institutions with competence and authority in planning at the national level (i.e. NDPC and LUSPA) should facilitate a consultative process that would bring together multiple stakeholders, most importantly, the various political parties. The aim of this process should be on assessing the current situation of regional development disparities and debating the adequacy or otherwise of existing regional institutions and their capacity to deliver regional development objectives as recognized in the national constitution, the decentralization law and the two major legislations establishing the contemporary planning systems. This would lead to answering the question of whether Special Development Zones and Development Authorities are needed as part of the long-term approach to regional development. At the end of the consultative process, the number of such special regions and the basis for their designation would be agreed upon and incorporated in existing national level spatial planning instruments to guide regional development.

Furthermore, although regional spatial planning has effectively been institutionalized with the promulgation of the Land Use and Spatial Planning Act, the various regional bodies mandated that would emerge to assume the regional spatial planning functions would be new and quite inexperienced. This is because, although initial pilot plan-making projects played a significant role in establishing the new spatial planning system and regional spatial planning, the process relied mainly on the expertise of consultants. That is not to say that the administrative and functional regions that have so far benefited from the process were not actively involved in the process. Rather, one of the reasons for this assertion is that the regional planning instruments have been prepared even before the new planning law that would establish the formal institutions for regional spatial planning came into force. In the absence of the planning law and the formal institutions, the experimentation process relied on ad hoc arrangements such as the formation of Regional Oversight Committees to assist consultants to formulate the plans (Acheampong and Alhassan 2016). These ad hoc committees were, however, dissolved after the plan formulation, creating a vacuum for the implementation, monitoring and evaluation activities of the planning process.

Lastly, in the course of piloting the new model of regional spatial planning, only a handful of regions and districts had the experience of working through ad hoc committees and consultants to formulate their regional spatial development plans. This implies that most of the administrative regions and districts are yet to practically experience the process of regional spatial planning. In view of this, one of the major challenges envisaged for these regions and district as they set out to prepare their own regional and multi-district spatial development frameworks would be how to effectively handle the enormity of co-ordinating tasks so vital to the process. In this regard, the newly created LUSPA which implemented the plan experimentation process while under the old arrangement as the head office of the TCPD would have to assist inexperienced regions to formulate their own plans. Of course, the option also exists for these regions to rely on consultants to assist them in the formulation of these plans. In the end, however, conscious attempt at instituting a culture of co-operation between multiple districts in the event of functional regional spatial planning, and multiple administrative regions in the event of multi-region spatial planning or planning for Special Development Zones, would be crucial for regional spatial planning to effectively deliver its objective.

5.8 Conclusions

The role of spatial planning in achieving regional development objectives, though recognized as early as 1963 when the first National Physical Development Plan was formulated, was never to be realized until the beginning of the twenty-first century. Besides the recognition of planning at regional level in the 1960s, this discussion in this chapter showed that despite the National Development Systems Act (Act 480, 1994), establishing the Development Planning System, requiring the formulation of different types of regional development plans not a single regional development plan conceptualized by this system has been formulated to date. In recent years, experimentation with regional spatial planning as part of measures to entrench the new spatial planning system and the promulgation of the Land Use and Spatial Planning Act (Act 925, 2016) have institutionalized spatial planning for administrative regions and functional regions comprising multiple districts in Ghana.

It has also been argued that while reforms leading to the establishment of the new spatial planning system and experimentations with regional spatial planning are commendable, the emergent situation of competing institutional structures taken in their entirety does not come quite close to the ideal arrangements that could facilitate effective regional planning. Besides planning for functional and administrative regions, the discussion has shown that area-based programmes whose geographies extend over areas that could be considered as special regions have become common in the recent decade. It was argued that the politicization of the process leading to the designation of Special Development Zones and their respective Development Authorities does not lend itself for strategic issues to be considered critically. Consequently, the entire activity of spatial planning at the regional level risk is being ineffective in delivering expected economic, social and environmental development objectives.

Against the backdrop of the complex institutional arrangements and the inefficiencies that would result, various measures to strengthen regional spatial planning for to deliver its objectives have been recommended. It was argued that in the short to medium term, effective co-ordination between the institutions with competence for regional planning within the two systems (i.e. Development Planning System and Spatial Planning System) through formal and ad hoc informal mechanisms would be required to make regional spatial planning effective. In the long-term, genuine reflections on the current system and a dispassionate debate should result in the unification of the two competing planning systems, in order to eliminate the needless distinction between regional spatial planning and regional development planning, and the resultant compartmentalization of the authority and competence for these planning tasks into separate institutions at the regional level. This will certainly require a new planning law that harmonizes the two competing laws as well as some form of institutional restricting. Also, a multi-stakeholder platform is needed to build consensus as to the need for Special Development Zones beyond the established administrative regions, multi-district functional regions, city-regions and urban networks that exist currently in the country. The capacity of the majority institutions that are currently inexperience in the practices of regional spatial planning but are be required to formulate their spatial plans either alone or in joint arrangements with other districts and regions will have to be strengthened. Most importantly, the effectiveness of regional spatial planning being it for special regions, multi-district functional regions or city-regions would depend on the ability to promote a culture of co-operation between multiple districts and multiple administrative regions.

References

Acheampong RA, Ibrahim A (2016) One nation, two planning systems? Spatial planning and multi-level policy integration in Ghana: mechanisms, challenges and the way forward. Urban Forum 27(1): 1–18

Acheampong RA. Agyemang, FS, Abdul-Fatawu M (2017) Quantifying the spatio-temporal patterns of settlement growth in a metropolitan region of Ghana. GeoJournal, 82(4): 823–840

Collits P (2012) Planning for regions in Australia. In Thompson S, Maginn P (eds) Planning Australia: an overview of urban and regional planning. Cambridge University Press

Counsell D, Haughton G (2004) Regions, spatial strategies and sustainable development. Routledge

Crawford G (2004) Democratic decentralisation in Ghana: issues and prospects. POLIS working paper, 9: 1–13

Friedmann J, Weaver C (1979) Territory and function: the evolution of regional planning. University of California Press

Kuklinski AR (1970) Regional development, regional policies and regional planning: problems and issues. Reg Stud 4(3): 269–278

Land use and spatial planning Act (2016) Act 925 The nine hundred and twenty fifth act of the parliament of the Republic of Ghana

Lichfield N (1967) Scope of the regional plan. Reg Stud 1(1): 11–16

Local Government Act (1933) Act 462 The four hundred and sixty two act of the parliament of the Republic of Ghana

Murphy PA (1984) Regional planning: purpose, scope and approach in New South Wales. Australian Planner 22(4): 16–19

National Development Planning (System) Act (1994) Act 480 The four hundred and eightieth act of the parliament of the Republic of Ghana. http://urbanlex.unhabitat.org/sites/default/files/urbanlex//gh_national_development_planning_act_1994.pdf. Accessed 24 Jul 2018

Ofori S (2017) Regional policy and regional planning in Ghana: making things happen in the territorial community. Routledge

Powell DR (2012) Critical regionalism: connecting politics and culture in the American landscape. UNC Press Books

Stilwell F (1992) Understanding cities and regions. Spatial Political Economy Pluto Press, Leichhardt, Sydney

Songsore J (2003) Regional development in Ghana: the theory and the reality. Woeli Pub. Services

Town and Country Planning Act (1945) CAP 84. http://www.epa.gov.gh/ghanalex/acts/Acts/TOWN%20AND%20COUNTRY%20PLANNING%20ACT,1945.pdf

Town and Country Planning Act (1958) Act 30 The thirtieth act of the parliament of Ghana

Town and Country Planning (Amendment) Act (1960) Act 33 The thirty-third act of the parliament of the Republic of Ghana

Chapter 6
Local-Level Spatial Planning and Development Management

Abstract In Ghana, local governments (i.e. MMDAs) are responsible for the overall development of areas under their jurisdiction. As such, MMDAs are mandated to perform spatial planning functions. In Ghana's three-tier spatial planning system, MMDAs have competences for local spatial planning. This chapter deals with local-level spatial planning in Ghana. It outlines the various policy instruments intended for use under the new spatial planning system to translate national and regional development policies to district-wide frameworks and subsequently to cities, towns, neighbourhoods and rural areas within the district. The contextual issues influencing planning and development management at the local level are identified to provide the context for the discussion of local spatial planning practices and challenges. In particular, the chapter examines the impact of the historical disconnect between national development planning and local land use planning prior to the establishment of the new spatial planning system; the complex indigenous landownership systems that spatial planning must grapple with; and the challenges posed by the existing institutional and legal structures established formally to mediate the separation between landownership rights on the one hand and the determination of land use and development control on the other hand. The gap between the established development management systems in normative terms as opposed to its functioning in practice is also examined. Based on the challenges of local spatial planning and development management identified, ways in which planning at this level could be made effective are identified.

Keywords Local plan · Structure plan · District Spatial Development Framework
Development control · Zoning · Traditional authorities · Land tenure
Non-compliance · Ghana

6.1 Introduction

In Ghana, local governments (i.e. MMDAs) are mandated by law to undertake the task of spatial planning. The spatial planning powers exercised by MMDAs and the various types of instruments and tools they use to realize their mandate is what is referred in this chapter as local spatial planning. In principle, local planning should reflect visions, goals and objectives articulated by national and Regional Spatial Development Frameworks in various frameworks and plans formulated and implemented at the local level by MMDAs. The local plans would, in turn, shape the development of cities, towns and rural areas within their respective jurisdictions of local governments.

In corollary of the above, local governments in Ghana are mandated to formulate and implement spatial planning instruments in the form of District Spatial Development Frameworks to influence the distribution of activities at district-wide level, and Structure Plans and detailed subdivision schemes to determine the physical layout and land use configuration for cities, towns and villages within the district. Following the normative sequence of spatial planning preceding physical development, various tools and policies are to be deployed to manage development in conformity with the policies of existing plans. While the plan-led approach established by law is considered the ideal for the primary reason that proactive planning provides the potential benefits to mitigate the negative effects of what would otherwise become haphazard and uncoordinated development of settlements, it is important to bear in mind that in some instances, local spatial planning does include reactive interventions aimed at managing development that has not necessarily emerged according to a formal land use plan.

The purpose of this chapter is to examine local spatial planning in Ghana. Firstly, the various types of local spatial planning instruments that MMDAs are required to prepare and implement under the new spatial planning system are outlined. Next, various contextual issues that have shaped local planning and will continue to influence planning and development management in future are discussed. The issues discussed here are the historical disconnect between national-level planning and local land use planning, the complex land tenure system that local planning must grapple with, as well as the evolving, formal institutional and legal arrangements that determine local spatial planning practices in the country. With this context, the next section will focus on the development process and the role of development management, which has long been known as development control. Here, the key stakeholders will be identified, examining their roles in the development process. The development management system will be discussed by first identifying the various tools, standards, regulations and procedures that are applied, and moving on to examine the gap between the established development management system in normative terms as opposed to its functioning in practice. In the penultimate section, the challenges of local planning and development management as well as ways in which spatial planning at the local level could effectively deliver its objectives will be discussed.

6.2 Local-Level Spatial Planning Instruments

As explained in Chap. 3, the new spatial planning system which has formally been established by the Land Use and Spatial Planning Act (Act 925, 2016) instituted a three-tier system of spatial planning instruments from the national to the local levels in the following order: National Spatial Development Framework (NSDF), Regional Spatial Development Framework (RSDF) and District Spatial Development Framework (DSDF). Within this main hierarchy of SDFs, a similar hierarchy can be identified at the local level in the following order: DSDFs are the highest level of frameworks prepared by MMDAs from which *Structure Plans* are prepared. Detailed *Local Plans* then emerge from the *Structure Plans*. In 2012, the TCPD, under the Ministry of Science, Technology and Innovation (MESTI) published the New Spatial Planning Model Guidelines which sets out the geographical coverage and contents of all the different types of spatial planning instruments from the national to the local levels. A brief description of the local spatial planning instruments specified in the model guidelines (i.e. DSDF, Structure Plan and Local Plan) is provided as follows.

6.2.1 District Spatial Development Framework

DSDF, according to the New Spatial Planning Model Guidelines, is a long-term indicative plan having fifteen- to twenty-year time horizon that articulated the overall spatial strategy for achieving defined social, economic and environmental policy goals and objectives of local governments. To realize the chain of conformity between national and regional frameworks, a DSDF is intended to emerge logically from an RSDF and coherent with national and regional policies.

The framework should address the spatial development implications of economic development, employment, housing, and infrastructure and services such as education, health care, tourism and leisure, transportation communications and environmental protection at the district level. Under the new spatial planning law, DSDFs are to be prepared and adopted by all MMDAs in Ghana and to use them to influence the type, location, timing and intensity of development within their respective jurisdictions. They are to designated structure plan areas within the boundaries of local governments and spell out the broad policies for these areas as well as the parameters for the formulation of the *Structure Plans*. Once adopted, MMDAs are required by law to review their DSDFs every four years.

6.2.2 Structure Plan

Whereas a DSDF is indicative, a Structure Plan (SP) that emerges from it is a legally binding document. As statutory long-term instruments, SPs are formulated to guide future development or redevelopment of all areas within their boundaries. The main tool employed in SPs is zoning—the practice of defining broad land uses including residential, commercial, industrial, mixed-use areas, major open spaces, agricultural areas and areas for upgrading and regeneration. The alignment and corridors of major transportation routes, major water, sewerage and power networks and other key features for managing the effects of development are to be delineated in SPs. The final SP consists of maps showing the designated land uses and infrastructure networks, identifying which parts of the network will need upgrading, which are additional to the existing network and which areas need repair. According to the planning model guidelines, at the MMDA level, the basic zoning classification will be included in the plan, while a SP for a sector of a town or for smaller areas may provide an additional layer of zoning ordinances that provide further information on permissible types of development and densities, the height and form of the building, site lines and setbacks and even use of construction materials. Finally, a phasing plan that indicates the priorities for infrastructure supply as the basis for all other types of developments in the SP area should accompany the Plan.

6.2.3 Local Plan

From SPs, Local Plans are prepared. A local plan according to the planning model guidelines is a plan which proposes the disposition of land by function and purpose, or land to be preserved in its present state, to meet the present and future identified community needs within the time frame for which the plan is valid. Local plans should be prepared when needed, and the uses of land must be in conformity with permitted uses of the land in the designated zoning classification, as identified in the approved SP. In Ghana, local plans are detailed subdivision plans showing individual land parcels and their proposed uses. The local plan is also supposed to identify open areas unsuitable for development, including land where the slope exceeds that is permissible for construction and water bodies, including floodplains (though the latter may be designated for recreational areas), and areas where existing trees are to be preserved or new trees planted. Local plans are the primary instruments of development management at the level of towns, neighbourhoods and specific sites and are as such legally binding documents. This also implies that where local plans exist, planning authorities are required to grant development and building permits in strict adherence to the plan unless a proposal for rezoning or revision of the plan has been accepted and the existing plan revised accordingly to accommodate any new proposals. The role of permitting and rezoning in the development management process is addressed later in this chapter.

6.3 The Context of Local Spatial Planning in Ghana

Before going ahead to discuss local spatial planning and development management practices, it is essential to understand the major forces that continue to shape planning at this level in Ghana. These forces emanate from historical and contemporary institutional and legal arrangements under supports the activity of planning and determine how national goals are translated and realized at the local levels, as well as the complex and the complex land tenure system within which land use planning objectives are expected to be realized. In the sections that follow, these contextual issues are explained to provide the framework for the subsequent discussions in this chapter.

6.3.1 Four Decades of Severed Connection Between National Policy Frameworks and Local Land Use Plans

In Chaps. 3 and 4, the profound impacts on spatial planning at all levels of the decentralization programme that started in the late 1980s and the decentralized planning system that accompanied it were explained. It was established, among other things, that the tradition of development planning which became established by the National Development Planning (System) Act (Act 480, 1994) virtually eliminated spatial planning at the national level, while the concept of regional development planning it introduced has, to date, not been tested in practice. The dearth of spatial planning at the national and regional levels that prevailed from 1970 when the implementation period of the first National Physical Development Plan ended would also have profound implications for local spatial planning. Thus, examining the historical neglect of spatial planning at the national and regional levels is vital to understanding the forces that have shaped spatial planning and development practices and the state of human settlements in Ghana today.

As explained previously, the contemporary tradition of development planning instituted by the National Development Planning (System) Act (Act 480, 1994) established the NDPC at the national level, and drawing on the administrative structures of the Decentralization Act (Act 462, 1993) mandated MMDAs through their DPUs to formulate MTDPs to guide socio-economic development at the local level. Between the national and local, RCCs were mandated to co-ordinate and monitor the activities of DPUs and by extension, MMDAs. It was further explained that within this structure, national development policy goals found their way directly into MTDPs by means of guidelines issued by the NDPC to DPUs of local governments.

On the contrary, spatial planning at the local level did not have any such mechanism to directly reflect national spatial development goals. The primary reason for this is that, the lack of a framework that would elaborate the spatial dimensions of medium-term NPFs meant that although the TCPD head office existed at the national level, it could not issue guidelines directly to TCPD offices at the district level as was done by the NDPC. This coupled with the fact that the Systems Act did not mention the TCPD at all meant that there was no legal obligation for the national TCPD to issue national policy guidelines relating to spatial development strategies. Consequently, MTDPs prepared by DPUs under NDPC guidelines not only neglect spatial planning issues of the kind currently required in DSDFs under the new spatial planning law but also do not reflect how changing socio-economic conditions anticipated by their plan proposals would affect population growth and the physical structure of settlements, and the need to plan spatially for growth. The compartmentalization of the tasks of planning whereby DPUs were empowered legally to formulate MDTPs while TCPDs operating under the obsolete Town and Country Planning Act (CAP 84) of 1945 retreated into performing their traditional land use planning and development control functions as separate from district-wide socio-economic development objectives meant that no effective linkages between the two departments existed. The lack of linkages between national policies and local land use plans and the notion of the spatial being separate from wider socio-economic goals persist even to date, hence justifying the compartmentalization described above. This situation was summed up by Acheampong and Alhassan as follows:

> The dearth of a spatial vision at the national level coupled with the severed interaction between the NDPC and the Town and Country Planning Departments at the local levels have over the years, established and perpetuated the notion that the 'spatial' is separate from the 'socio-economic' when in fact, the two are integrated and are together, essential to socio-economic transformation. Consequently, at the level of MMDAs, two separate planning committees exist for the purposes of co-ordinating and approving plans. Whereas Development Planning Subcommittees co-ordinate the process of MTDP preparation, the Statutory Planning Committees exercise co-ordinating and approval responsibilities over land-use plans. (Acheampong and Alhassan 2016, p. 9)

The tradition of development planning that accompanied the decentralization programme coupled with the obsolete 1945 Town and Country Planning Ordinance (CAP 84), which despite being revised only twice in 1958 and 1960 remained in force as the legal framework for land use planning until it was replaced in 2016 by the Land Use and Spatial Planning Act (Act 925) weakened spatial planning and the development management system at the local level. For years, TCPD offices operated under severe resource constraints, lacking the institutional capacity and up-to-date legal backing to effectively undertake their mandate even at the local level where their activities were confined. The accumulated impacts of the arrangements that crippled town and country planning as is called locally are today reflected in the haphazard and uncoordinated development that characterizes many towns and cities in the country.

6.3.2 A Complex Land Tenure System and the Separation Between Ownership Rights and Determination of Use of Land

Spatial planning at the local level is fundamentally about land, a critical input in the national socio-economic development process. It is about the determination of its highest and best use and its management to guarantee availability for future use. Land is not only a scarce resource but also has several vested interests. In Ghana, the state, indigenous governance institutions, communities, families and individuals all have vested interests in this precious resource called land. Understanding the prevailing land tenure system and its complexities is therefore fundamental to appreciating the challenges spatial planning in Ghana must grapple with as it strives to achieve the social, economic and environmental development objectives expected of it.

The prevailing land tenure system in Ghana can be divided into two broad typologies, namely public land and customary land. There are two main mechanisms by which land becomes public. First, the state may compulsorily acquire land in the public interest under the State Lands Act, 1962 (Act 125), or other relevant statute. Second, under the Administration of Lands Act, 1962 (Act 123), land may become vested in the president in trust for a landholding community. The customary sectors, on the other hand, are rather complex. Having its origins in the traditional extended family system and community organization and governance around the chieftaincy institution, landholders under this system include individuals and families who hold a customary freehold title to land, and stools, skins and families that symbolize and represent communities and 'Tendamba' (i.e. the first settlers) or clans who hold Allodial title to land (Kasanga and Kotey 2001). About 90% of all land in Ghana falls under different customary ownership arrangements (Kasanga and Kotey 2001).

Whereas all vested and public lands are administered by the Lands Commission as provided in the 1992 Constitution and in the Lands Commission Act 1994 (Act 483), the management of customary land is based on indigenous law, which is founded on the principle stools, skins and families hold land in trust for their people and that such lands should be managed according to the fiduciary duty of the traditional authorities towards their people on the basis of customary law (Ubink and Quan 2008). The concept of trusteeship as it applies to landownership and management in the customary sector is recognized by the state. Article 36(8) of the 1992 Constitution states:

> the state shall recognise that ownership and possession of land carry a social obligation to serve the larger community and, in particular, the state shall recognise that the managers of public, stool, skin and family lands are fiduciaries charged with the obligation to discharge their functions for the benefit respectively of the people of Ghana of the stool, skin or family concerned, and are accountable as fiduciaries in this regard.

The land tenure system in Ghana therefore raises a number of implications for the activity of spatial planning at the local level. On the one hand, the state recognizes customary landownership and the associated management arrangements. On the other hand, the state, through MMDAs and their TCPDs, is mandated to determine the uses to which both lands vested in the state (i.e. public lands) and lands owned under the different customary arrangements are put. Spatial planning as a public-sector activity must thread the rather uneasy path of interfering in these prevailing interests and mediating the essential tensions that arise as a result of the separation between prevailing ownership rights, which is not always fully vested in the state, and the determination of various uses to which land should be put, an activity vested fully in state institutions. The effectiveness of spatial planning and development management in delivering sustainable outcomes therefore depends largely on the ability of mandated public-sector institutions to bring about compromise between powerful vested interests in land. The nature of the conflicts that arises as spatial planning attempts to deliver within the prevailing context of complex landownership arrangement, especially with respect to the customary land tenure system, will be examined in greater detail later in this chapter as we discuss the plan making and development management functions of the planning system.

6.3.3 Legal and Institutional Arrangements for Contemporary Local Spatial Planning and Development Management

MMDAs derive their legal mandate to perform planning functions primarily from the Local Government Act (Act 462, 1933) which establishes them as the planning authority for their respective jurisdictions. It appears, on critical interpretation of the Local Government Act, that the type of planning which MMDAs are mandated to undertake referred mainly to the formulation on MTDPs which would become established a year later in 1994 by the National Development Planning (System) Act (Act 480). As has been established at the beginning of this chapter, this tradition of development planning not only isolates TCPDs at the district level but with its narrow focus on socio-economic development issues fails to integrate the spatial dimensions of development.

One would argue the contrary by referring to the fact that Section 47 of Act 462 makes mention of physical development. However, the Act mentions physical development solely in relation to the permitting process but not explicitly in terms of the fundamental land use planning functions that must precede development management of which permitting only becomes one of the tools for ensuring that physical

6.3 The Context of Local Spatial Planning in Ghana

development conforms to an existing land use plan. Furthermore, even where Act 468 refers to a plan in relation to physical development, it mentions the district development plan, which ostensibly is the MTDP, but not a spatial plan of the kind prepared by the TCPD. The focus of the Act on permitting but not the wider land use planning functions of the TCPD is possibly due to the revenue side benefits of the application for development and building permits and not necessarily out of genuine concern to ensure orderly physical development. Thus, physical development, and the district TCPDs by the extension, was only important to MMDAs in so far as they contribute to the revenue mobilization priorities of the MMDAs. Under this arrangement, which persists to date, district TCPDs are essentially money-making departments for their local governments.

Moreover, Act 462 enjoins MMDAs to constitute a development planning subcommittee, which to date is the established committee responsible for approving MTDPs prepared by DPUs of MMDAs. Similar subcommittee arrangement for land use planning is not mentioned by the Act; instead, the TCPD, acting through CAP 84, would also constitute its own Statutory Planning Committee (SPC) to exercise co-ordinating and approval responsibilities over land use plans (Acheampong and Alhassan 2016). Indeed, the eighth schedule of Section 161 of the Local Government Act (Act 462) names the district TCPD as one of the departments or organizations ceasing to exist in districts, implying that although the TCPD has offices within MMDAs, the department is not recognized by the Act as one of the decentralized departments of the local government system. It is therefore not merely coincidental that the Decentralization Act did not repeal the 1945 Town and Country Planning Ordinance (Cap 84). Clearly, by keeping CAP 84, the intent was that TCPDs at the district level would continue to maintain relationship with their head office in Accra. This arrangement meant that TCPDs at the district level, although providing services for and under the Ministry of Local Government, belonged to the sector Ministry of its head office in Accra, which for nearly a decade has been the Ministry of Environment, Science, Technology and Innovation.

In 2016, the promulgation of the Land Use and Spatial Planning Act (Act 925) would replace the obsolete CAP 84 and strengthen the institutional arrangements for local spatial planning. Under the new spatial planning law, which operates alongside the Local Government Act (Act 462), spatial planning competences are vested in a District Planning Authority, in which the Act refers to as the MMDAs. The spatial planning functions of the Authority are to be realized through a District Spatial Planning Committee (DSPC) and its Technical Subcommittee whose functions are summarized in Box 6.1.

Box 6.1: Functions of the District Spatial Planning Committee and Technical Subcommittee under the Land Use and Spatial Planning Act (Act 925, 2016)

A District Spatial Planning Committee shall:	The Technical Subcommittee shall:
• Ensure that physical development is not carried out in the district unless that development is duly authorized in accordance with this Act; • Ensure that the preparation of the District Spatial Development Framework is in accordance with this Act; • Ensure that the preparation of the SP and local plan in the district is in accordance with this Act; • Deliberate on and approve the recommendation of the Technical Subcommittee or request further consideration by the Technical Subcommittee where necessary; • Consider and approve applications for permit; and • Perform other functions required to be performed by this Act within the district. • The District Spatial Planning Committee may impose conditions that it considers appropriate in giving approval to the recommendations of the Technical Subcommittee.	• Prepare or review the District Spatial Development Framework, Structure Plans, Local Plans and Rezoning Plans; • Review applications for physical development; • Recommend to the District Spatial Planning Committee applications for approval; • Provide the Authority with reports as required for the enforcement of this Act; • Make recommendations to the District Spatial Planning Committee to approve any of the items; documents or matters required to be approved under this Act; • Make input into the discussions of site advisory and site selection teams set up for public projects by the Site Advisory Committee established under the State Lands Regulations, 1962 (L.I. 230); • Provide technical services, establish conditions in relation to the various plans and monitor implementation of the plans; and • Perform any other function assigned to the committee by the District Spatial Planning Committee.

The Land Use and Spatial Planning Act (Act 925) also specifies the composition of the District Spatial Planning Committee and its Technical Subcommittee as shown in Box 6.2.

Box 6.2: Composition of District Spatial Planning Committee and Technical Subcommittee under the Land Use and Spatial Planning Act (Act 925, 2016)

A District Spatial Planning Committee consists of
  i. the District Chief Executive of the district who shall be the chairperson and in the absence of the District Chief Executive, the chairperson of the subcommittee on works shall act as chairperson, and in the absence of both, the District Coordinating Director shall act as the chairperson;

ii. the head of the Physical Planning Department of the district who shall be the secretary of the committee;
iii. the District Coordinating Director;
iv. the chairperson of the subcommittee on development planning of the District Assembly, and works of the District Assembly;
v. the District Development Planning Officer;
vi. the head of the Works Department;
vii. the head of the Urban Roads Unit of the District Assembly;
viii. a representative of the regional director of the Environmental Protection Agency;
ix. the head of the Disaster Prevention Department of the District Assembly;
x. one representative, of the Lands Commission in the district not below the rank of a Staff Surveyor appointed from the Survey and Mapping Division of the Lands Commission;
xi. one representative from the traditional council of the district and in districts where there are more than one traditional council, the person elected by the traditional councils within the district to represent them on a rotating basis;
xii. persons nominated by the elected members of the assembly from among their number to represent them as follows: (i) in the case of a Metropolitan Assembly, three representatives; (ii) in the case of a Municipal Assembly, two representatives; and (iii) in the case of a District Assembly, one representative except that in the nomination, preference shall be given to female elected members of the Assembly;
xiii. A District Spatial Planning Committee may co-opt or invite any other qualified person as a consultant, to attend a meeting of the District Spatial Planning Committee for the purpose of the specific subject matter being considered by the District Spatial Planning Committee.

Technical Subcommittee
A Technical Subcommittee consists of

i. the head of the Physical Planning Department of the district who shall be the secretary of the committee;
ii. the District Development Planning Officer;
iii. the head of the Works Department;
iv. the head of the Roads Unit or Urban Roads Department of the District Assembly;
v. the district head of the Disaster Prevention Department of the District Assembly;
vi. one representative of the Lands Commission in the District;
vii. one representative of the regional head of the Environmental Protection Agency;

> viii. the District Fire Officer;
> ix. the head of the District Health Department;
> x. two co-opted members at least, one of whom is the chairperson of a sub-metro or urban council as appropriate;
> xi. A representative from any of the utility agencies or other relevant agencies may be co-opted if required.

The institutional arrangements specified in the new spatial planning law clearly is an attempt to address the isolation of the TCPD at the level of MMDAs highlighted previously by making the district TCPD an integral part of the local governance system. The recognition of the problem and the attempt made to resolve it is commendable. However, the new arrangement appears to fail to at addressing this problem. The reasons are as follows; firstly, the Decentralization Act, which has not been amended, still maintains that TCPD is not one of its decentralized departments. Secondly, the old arrangement whereby district TCPD works for and under one sector Ministry while the head office is responsible to a different sector Ministry has not be addressed by recent reforms in spatial planning and the new spatial planning law. Instead, as of the time of writing, LUSPA, the new authority established in 2016 by the coming into force Act 925 to replace the TCPD national head office, was responsible to the Ministry of Science, Technology and Innovation (MESTI), while district TCPD offices within MMDAs are expected to work under the Ministry of Local Government and Rural Development. This arrangement is expected to continue into the foreseeable future. Thus, some of the original reasons for which district TCPDs have been marginalized in their respective MMDAs remain, despite the reforms. The ability of the new institutional arrangements to respond effectively to the almost overwhelming challenges of sustainable land use planning and management that confronts human settlements of varying sizes remains to be seen. In the interim, it is important to bear in mind the contemporary legal and institutional arrangements in order to appreciate the emergent physical development outcomes at the local level today and potentially in future.

6.4 Local Plans and Development Management in Practice

In the context of spatial planning, social, economic and environmental development objectives articulated in planning instruments at the various spatial scales are to be realized locally through the development of land. In Ghana, the legal meaning implied when the term physical development is used is contained in Section 162 of the Local Government Act (Act 462). Physical development, it states:

> the carrying out of building, engineering, mining and other operations on, in, under or over land or the material change in the existing use of land or building and includes subdivisions of land or disposal of waste on land including the discharge of effluent into a body of still

6.4 Local Plans and Development Management in Practice

or running water and the erection of advertisement or other hoarding. (Section 162, Local Government Act)

The physical development process is a dynamic and continuous one which involves various actors realizing their objectives and needs. The key actors and stakeholders in the development process therefore include:

i. Developers, which include private property developers, individuals, households, businesses, public-sector institutions;
ii. Landowners, including traditional authorities, government, individuals, families, communities and corporate entities;
iii. Public-sector institutions such as MMDAs and their TCPDs, other land sector agencies (e.g. Lands Commission, Administrator of Stool Lands, Survey Department), the Environmental Protection Agency, Utility Service Providers, Security Agencies;
iv. Civil society groups (CSGs) and non-governmental organizations (NGOs);
v. Utility service providers in the private sector.

These actors realized their objectives by carrying out transformations of various forms and intensity to land and landed properties at different locations over different time periods. Housing development tasks place through different channels (i.e. public sector, private sector, households, etc.) to provide shelter for the population. Commercial, industrial and civic buildings are the physical manifestations of the public and private sectors achieving various aims including establishing businesses, generating economic activities, providing employment and carrying out the task of civic governance. Road systems are engineered and built to enable access to social services and to facilitate movement between different activity zones to generate economic growth. Some actors may not directly undertake any form of physical development as legally defined but are regarded key stakeholders by virtue of the fact that they own land, probably the most vital input in the physical development process. The public sector, including central governments, local governments, sector ministries, departments and agencies, could both be developers and managers of the development process at same time or at different times.

6.4.1 Plan-Led Development versus Non-Plan Development-First and the Permitting Process

In normative terms, the legal definition of development presumes the existence of an up-to-date local plan to which all types of development should conform. The formal establishment of town planning, which in Ghana dates back to the introduction of the first planning ordinance in 1945, is a declaration of intent in favour of the normative, plan-led approach to physical development. Planned development has several benefits including the following:

i. An authoritative plan provides the vehicle for the provision of infrastructure and services. It enables local governments to programme their investments in infrastructure.
ii. An approved local plan signals to utility service providers that an area has been earmarked for future development and that it will require construction of building and supporting infrastructure and services. From the perspective of the providers of these services, the local plan is not only a legal document for development control but also an opportunity to conduct business.
iii. From the point of view of developers including property development companies, individuals and households, having an authoritative plan provides certainty: knowing that their developments conform to an existing local plan means that they authorized and therefore would not face penalties and possibly demolition in future, providing certainty that their investments are protected.
iv. The local plan, through its zoning regulations and development standards, eliminates negative externalities, thereby contributing to wider measures of protecting public health.
v. By prescribing standards to which development should conform and specifying permissible and non-permissible uses, a local plan may provide the guarantee that the cumulative effect of individual developments is positive on adjacent properties rather than having a devaluing effect.
vi. Potential conflicts relating to land and landed property could be averted on the basis of the local plan since it demarcates the boundaries of individual parcels of land and ensures that public spaces such as road reservations, parks and open spaces are clearly demarcated from private spaces.

However, the ideal sequence of having a local plan to dictate the type, pace and intensity of physical development is violated, curtailing some of the associated benefits. In Fig. 6.1, three different scenarios of the plan-development nexus are depicted. Scenario 1 depicts the ideal plan-led sequence, while Scenarios 2 and 3 show exceptions to the normative order that fundamentally underpins the spatial planning system. For purposes of keeping things simple at the start, the plan-led approach will be explained first and used to illustrate the development management process. Next, the two scenarios that constitute the exceptions to the established procedures will be explained.

6.4.1.1 Local Plan Preparation and Approval Processes

In Chap. 2, it was explained briefly that the plan-led approach first instituted under the 1945 ordinance followed the principle of declaration of statutory planning areas, which involved a piecemeal approach of deciding which areas should be planned for development under powers exercised by a Minister responsible for town planning. The incremental designation of areas for statutory planning would, however, be abandoned in the 1970s, implying that since then all areas in Ghana have become

6.4 Local Plans and Development Management in Practice

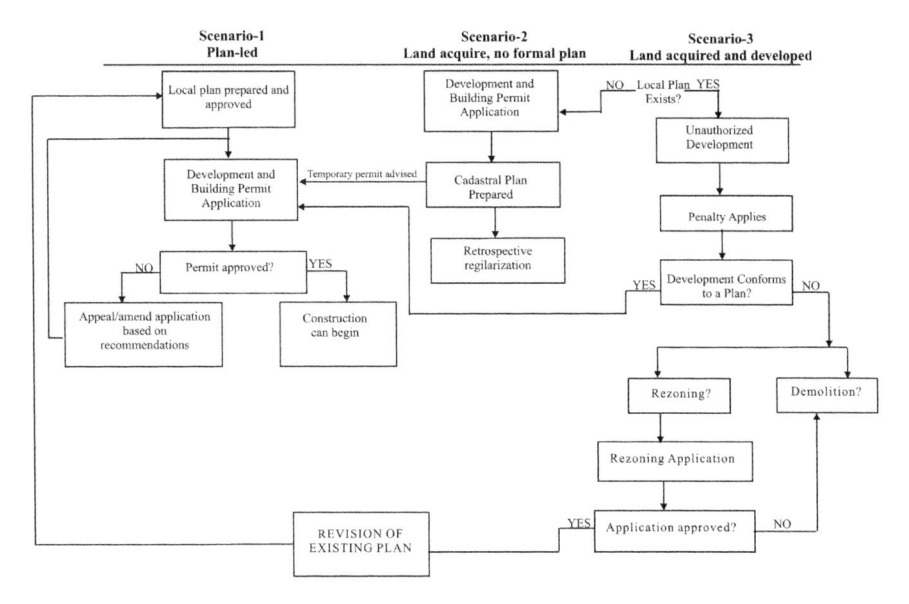

Fig. 6.1 Physical development and development control under normative plan-led versus non-plan scenarios

possible planning areas. This notion is firmly entrenched in the new Land Use and Spatial Planning Act (Act 925, 2016).

The types of local plans required by law have evolved over the years. For example, until the reforms leading to establishment of the new spatial planning system, the concept of SDFs at all levels, including local governments, did not exist. Thus, the requirement that MMDAs should formulate DSDFs is recent. This also implies that as of the time of writing this book, only a handful of MMDAs that were selected as part of the plan experimentation process that begun around 2010 to institutionalize the new model of three-tier spatial planning instruments have formulated and adopted DSDFs. Prior to this, SPs also called *Master Plans,* and *Town Planning Scheme*s in the form of detailed subdivision plans were to be prepared. In Chap. 3, some of the earliest examples of Master Plans prepared for major towns including Accra and Tema were discussed. It is worth mentioning that very few towns and cities have SPs even today. Thus, for years detailed subdivision plans commonly known as Sector Layouts have been prepared by district TCPDs to provide the basis for development control at the local level.

Local plans go through a series of stages beginning with plan initiation, preparation, approval and adoption. In principle, MMDAs through the TCPDs should initiate the preparation of a local plan for any area within their jurisdiction. In practice, other actors may initiate the process. For example, traditional authorities as custodians of land may approach the TCPD of MMDAs to formally request for a plan to be prepared to cover areas under their jurisdiction. Similarly, other actors in the development process such as private property developers planning to develop large tracts of land may also approach the TCPD to discuss the possibility of formulating a local plan. In all cases, the decision to go ahead with the preparation of the local plan must be sanctioned by MMDAs, relying on the technical advice of their respective TCPDs. A decision to go ahead initiates the local plan preparation process. The district TCPD in consultation with all stakeholders prepares the plan. At this stage of the process, the main activities undertaken include obtaining an official base map covering the plan area from the Survey Department, the government department authorized to undertake land surveying services. The TCPD undertakes all the relevant background studies that will inform the plan. Relevant stakeholders including landowners, traditional authorities and communities must be consulted.

The local plan is prepared using established zoning guidelines and planning standards. Zoning is done at both the SP and local subdivision plan levels. The zoning guidelines and planning standards published by the TCPD are applied to local plans. In zoning, broad development zones are designated in the plan. It prescribes the acceptable/permissible uses and form of development that should occur in the area covered by the plan. Within a local plan, each individual parcel of land is prescribed a permissible use. A standardized system of colour coding is used to communicate the zoning policies for all legally binding types of local plans. Zoning schedule published in the TCPD zoning guidelines and planning standards identifies 25 development zones that could be designated in a local plan. These are outlined in Box 6.3.

Box 6.3: Development Zones in the Zoning Guidelines and Planning Standards
　　i. Rural area (RU A–B);
　　ii. Residential zone (RZ A–E);
　　iii. Redevelopment/renewal/upgrading zone (ReZ);
　　iv. Education (E2 A–C);
　　v. Places of Worship Zone (PW);
　　vi. Health Zone (HZ A–F)
　　vii. Central Business District (CBD)
　　viii. Sub-regional Business Centres (SBC)
　　ix. Mixed Business Zone (MBZ)
　　x. Mixed Business Zone (BM);
　　xi. Informal business zone (BL);
　　xii. Government Business Zone (BG);
　　xiii. Recreation and Sports Zone (RS);
　　xiv. Public Open Space Zone (POS);
　　xv. Transportation and Warehousing Zone (TW);
　　xvi. Light Industrial Zone (LI);
　　xvii. Service Industry Zone (IS);
　　xviii. General Industrial Zone (IG);
　　xix. Noxious, Offensive, Hazardous Industrial Zone (IN);
　　xx. Special Use/Security/Military Zone (SM).

The final plan comprising a map of all proposed land uses and the accompanying written policies is approved by the District Spatial Planning Committee. Once the local plan is approved and adopted, it becomes legal basis for development control, which in the context of local governments in Ghana is achieved mainly through the permitting process.

6.4.1.2 Development Permitting under an Approved Local Plan

Prospective developers are supposed to obtain a development permit from relevant planning authorities (MMDAs). Once an application for a permit has been submitted to the MMDAs, it goes through a two-stage vetting process. The first stage of the vetting process ensures that proposed developments are consistent with the approved local plan. As explained previously, the plan prescribes what is known as permitted uses—all types of development that may be approved subject to the zoning policies specified in the plan. For certain types of permitted uses, the first stage of the vetting process also ensures that the proposed development conforms to EPA requirements. A proposal to site an industry in an industrial zone is one example of application that should demonstrate compliance with EPA requirements. In all cases, applicants must demonstrate proof of landownership. The requirements of proof of ownership depend on whether the subject is a public land or under customary ownership. In the case of the former, the Lands Commission certifies proof of landownership. In the latter case, the Administrator of Stool Lands certifies proof of landownership on the basis on an allocation certificate provided by the landowner when land is purchased. When all requirements are met under the first stage of the vetting process, a *development permit* is granted.

The second stage of the permit vetting process focuses on the proposed building by ensuring that it meets standards and regulations stipulated in the National Building Regulations 1996 (LI 1630). The rationale for these regulatory requirements is to ensure that minimum acceptable development standards are met. The items that should accompany the application include the structural drawings and architectural designs of the proposed building for simple residential building. In the case of multi-purpose and multi-usage buildings, additional requirements may apply including soil test report, Ghana National Fire Service report, EPA report, business registration and operating permit in the case of organizations among others. An application that meets all the requirements at the second stage of vetting process received a *building permit*. The final permit is a single permit granted by the Technical Committee of the District Spatial Planning Committee on the basis of the outcomes of the two-stage vetting process. It therefore grants the permission for the proposed development to commence.

6.4.1.3 The Exceptions to Normative Plan-Led Development

The ideal sequence of having an approved Local Plan first and managing physical development based on that Plan does not always happen in reality. In most cases, Local Plans are either non-existent or out of date. Developers would not necessarily wait for a local plan to be prepared before going ahead to build. Also, an out-of-date local plan is virtually the same as not having a Plan at all since such a plan lacks the legitimacy to be used as the basis for development control.

In Fig. 6.1, two scenarios that violate the plan-led approach were illustrated. The first considers a scenario where a prospective developer has acquired land in an area where a statutory local plan does not exist. In fact, this scenario is common in Ghana because landowners, mostly traditional authorities, do sell their lands to buyers who invest in land with the aim of benefiting from speculated rise in value in future, or would-be developers looking for cheap land in the urban periphery to build immediately. Under this scenario, two outcomes are plausible as far as the development-permitting nexus is concerned. Should the would-be developer approach the MMDA under whose jurisdiction the land falls to acquire a development permit, a temporary permit may be advised in the absence of a local plan. For this permit, the prospective developer may be required to produce a Cadastral Plan—a map produced by a licensed surveyor, delineating the boundaries of the land in question. The permit application will then be submitted to the MMDAs where the two-stage vetting process described previously would be applied. A successful application grants the developer the permission to build, with the view that the development would be incorporated in any future local plan prepared for the area and regularized retrospectively.

The second plausible outcome under scenario two, which happens in most cases, is that in the absence of a local plan, developers proceed to build without any permit at all. In such situations, physical development tends to occur in a haphazard, uncoordinated manner, and may not always meet the minimum established planning standards and building regulations. If a local plan is to be prepared later for the area, some of the development that meets minimum planning regulation may be regularized. In principle, those that do not meet the minimum standards and are regarded unfit for their respective uses become unauthorized development. The local authorities could, however, face legal action, compelling them to offer compensation should they decide to demolish such development. The case for compensation could be made legally on the basis of the planning law which states that all areas have been declared as planning areas, implying that the absence of a plan to a larger extent could be argued as failure on the part of the MMDAs to carry out their mandate. To avert legal action and the burden of having to pay compensation, development considered structurally unfit is allowed to remain until perhaps they approach the end of their technical life where redevelopment becomes necessary. During redevelopment, all regulations in the plan and permitting procedures will apply to the site and the proposed development.

6.4 Local Plans and Development Management in Practice

The third scenario (see Fig. 6.1) considers the situation where land has been acquired and developed. Here two outcomes of plausible depending on whether a local plan existed for the area or not. If a local plan does not exist, then the processes and outcomes discussed under Scenario 2 would apply. However, if a plan existed and land has been acquired and developed without a permit from the relevant MMDA, then such development is considered unauthorized. The MMDAs on identifying unauthorized development should serve the developer a notice to stop work. In most towns in Ghana, it is common to see the 'stop work, produce permit' notice in red paint scribbled on building under construction. This is how MMDAs serve their notice to developers who have violated the existing development control regulations. The effectiveness of this approach is another question that could be debated. Indeed, how could a developer produce a permit they do not have in the first place? Perhaps, the notice should read 'stop work, come for permit'? According to the planning law and building regulations, a penalty would automatically apply to such development. The exact amount to be paid as penalty is determined by the relevant MMDAs in their bye-laws that relate to the permitting process. Where a penalty is paid on the basis of the development proceeding in an area with a local plan without a permit, the planning authority may first have to decide as to whether the development, despite not having a permit, conforms to the zoning regulations of the plan. In other words, a residential development without a permit is an unauthorized development; however, if it is built on land zoned for residential purpose, then the local planning authority may initiate a process of regularizing the development after the appropriate penalty has been applied. Under such circumstances, the owner of the development would be required to formally apply for a building permit, where all the regulations and procedures described under the plan-led scenario will apply. On the contrary, if the development in question does not conform to the basic zoning of the land but meets minimum standards, then a request for rezoning, which will effectively alter the existing Plan, may be advised. The rezoning process will be explained shortly. If an unauthorized development does not meet the basic zoning regulations of the local plan as well as building standards, then by law, it must face demolishing.

Do unauthorized developments get demolished? In very few cases, properties get demolished for not complying with planning and building regulations. In most cases, however, planning and building laws are not applied due to corrupt government officials whom developers bribe to avoid facing demolition. In some other cases, such developments are owned by the rich and powerful in society including politicians, chiefs, technocrats and businessmen who by virtue of their connections to power are able to escape the law. In other cases, planning laws are relaxed or not applied at all for what some have argued as the need for planning to have a 'human face'. The 'human face' argument when it enters planning and development management discourse effectively follows the reasoning that under conditions of economic hardship, unemployment and severe housing shortages, no development should face demolition in so far as it provides shelter or a source of livelihood to people. Unfortunately, the immediate concern with shelter and livelihoods fails to consider the implications for wider public health and safety. The consequences of the lack of local plans and the lack of enforcement of planning and development management regulations are not far-fetched. Physical development across the country is chaotic and uncoordinated. Basic supporting facilities are lacking, especially in many residential neighbourhoods in the urban periphery. In some instances, buildings collapse, leading to loss of lives and investments because safety standards were not met.

6.4.2 Rezoning and Change of Use

The wider socio-economic conditions driving development is constantly changing. Local plans are prepared to influence development over time but at the same time the basic human limitation of not being able to foresee all possible future scenarios ought to be acknowledged in the planning and development management process. Over time, the wider socio-economic conditions that informed the plan and drive the development process change. The zoning policies of a plan may require revision to reflect market signals and to accommodate emerging needs and land use activities. Certain prescribed uses may over time become obsolete; for example, land could be zoned initially for industrial use, but its use will require changing when industry relocates. It is also possible that the local authority would not be able to attract industries as anticipated, implying that the industrial area may need rezoning to accommodate other types of uses. Agricultural land may also be rezoned for other uses such as residential and commercial in response to population growth and the attendant rise in demand for land to accommodate urban land use activities. Sometimes, the changes required may cover just a single plot or building in the planned area. At other times, large areas of the existing plan may be requiring revision in response to emerging trends. The process of varying existing zoning policies in an already approved local plan is called rezoning. Rezoning may become necessary at the SP level where the zoning regulations and policies that apply to large areas are altered or at the Local Plan level where a single plot or a number of sites may require revising to accommodate new uses. For example, an area zoned initially for commercial use may be rezoned later to accommodate residential functions. Rezoning may lead to changes in building heights and the densities of development in an existing Plan.

Change of use involves altering an existing plan in ways that is different from rezoning. According to the zoning regulation and planning standards published by the TCPD, in the Ghana context, change of use applies at the local plan level and is concerned only with altering the use of an individual plot or plots, but which remains within the list of permissible used for the land within a zoning schedule. Where the proposed change is not covered by the permissible uses, the zoning ordinance will need to be varied and this will require the land under consideration being rezoned. Change of use could also apply to alter an already approved building, for example changing a building from commercial use to residential and vice versa. The conditions

6.4 Local Plans and Development Management in Practice

for change of use as stipulated in the zoning guidelines and planning standards published by the TCPD are provided in Box 6.4.

> **Box 6.4: Conditions for Change of Use**
>
> An application for a 'change of use' will need to satisfy the following criteria:
>
> i. The type of development is within the permissible uses as given in the zoning guidelines.
> ii. Does not significantly alter the original intention of the plan or zone.
> iii. Does not cause disruption to the surrounding land uses by way of;
>
> - Significantly increasing traffic generation;
> - Significantly increasing noise and or odour;
> - Increasing the risk of fire or explosion;
> - Undermining the image of the area;
> - Being a risk to public health;
> - Intrusion of privacy.
>
> iv. Be of net benefit to the community in which the use is located.
> v. Has minimal impact on existing services and infrastructure.

MMDAs are required to review their SPs after a maximum of 15 years. The review may lead to the need to alter zoning ordinances. Thus, the review of the local plan could be one way of bringing about rezoning. In addition, at any time within the Plan implementation period, the zoning guidelines and planning standards document gives room for individuals, organizations or institutions proposing development of land that do not conform to an approved zoning ordinance of an area to apply for rezoning. Approval decision to alter the local plan or otherwise rests only with the relevant MMDAs. The preparation of the revised local plan and any revisions to the zoning ordinances affecting the land will be undertaken by the district TCPD office, or where it is outsourced to a consultant, the district TCPD must supervise the revision process. There may be instances where departures or waivers to the permitted land use under zoning schedules may be granted. In Box 6.5, the procedure for processing rezoning/change of use applications and the conditions under which of departures and waivers from the zoning schedule may be granted as specified in the zoning guidelines and planning standards published by the TCPD are provided.

Box 6.5: Procedure for Processing Rezoning and Change of Use Applications

The procedure for processing rezoning/change of use applications made by individuals or organizations is laid out in the document on permitting procedure as follows:

i. Applicant submits proposed request to Physical Planning Department, a comprehensive rezoning report prepared by a Professional Planning Consultant of good standing justifying the need for such rezoning/change of use.
ii. The report is studied and summarized, and recommendations for action prepared by the Physical Planning Department are presented before the Statutory Planning Committee, through the Technical Subcommittee.
iii. Physical Planning Department publishes the rezoning/change of use request in the Local Daily for two times at a two-week interval for public reaction within 28 working days of the first notice.
iv. The Statutory Planning Committee then considers the application for: (a) approval; (b) refusal/rejection or (c) deferred, with reasons within 90 days from the date of receipt of the application.
v. The applicant is informed by the secretariat (Physical Planning Department) of the Statutory Planning Committee to pay for a processing fee in respect of the approval.
vi. The requisite rezoning/change of use Plans are then prepared and distributed to all members of the Technical Subcommittee for study. In case of referral or refusal, the applicant is informed in writing.

Departures or Waivers to the Permitted Land Use Under Zoning Schedules—Discretion

From time to time, it will be necessary to consider departures or waivers to the zoning schedules. In granting such departures or waivers, the District Planning Authority should consider whether it would:

i. Not significantly alter the original intention of the plan or zone;
ii. Not cause disruption to the surrounding land uses by way of:

- Increasing traffic generation;
- Increasing noise and/or odour;
- Increasing the risk of fire or explosion;
- Undermining the image of the area;
- Being a risk to public health;
- Intrusion of privacy.

iii. Be of net benefit to the community in which the use is located;
iv. Have minimal impact on existing services and infrastructure.

6.5 Non-Compliance with Local Plans and Development Management Regulations

Perhaps, the biggest problem of spatial planning and development management at the local level in Ghana is non-compliance. Non-complaince takes many forms, including the construction of new buildings without a permit, changes made to approved development during construction, development in areas where it is prohibited, non-adherence to technical structural requirements and material specification during actual construction, among others (Korah et al. 2017; Arku et al. 2016). Non-compliance therefore leads to unplanned and chaotic development of large areas, risks to public health and lives of structurally deficient buildings and the negative consequences to economic growth. Development that complies to laid down regulations, on the other hand, provides several benefits to individuals and society at large. The question therefore is why do people not comply or what factors determine whether developers comply to local plans and building regulations? Below some of the factors of that explain (non-)compliance to planning regulations are examined.

As was explained under the normative plan-permit-development sequence, the existence of a Local Plan and the regulations and standards that accompany the local plan are basic requirements for compliance. However, in Ghana, it is either local plans are non-existent or that local plans exist but are out of date and therefore worthless as the basis to enforce development management regulations. In some major urban centres, statutory master plans and subdivision schemes some of which were prepared as far back as the 1960s without any revision are expected to guide development (Yeboah and Obeng-Odoom 2010). Outdated local plans contribute to non-compliance in two ways: these Plans tend to cover small areas previously planned for development, which means that as existing built-up areas expand without the scope of existing plans being expanded accordingly, new developments take place without any scheme to guide it. Thus, an out-of-date plan lacks authoritativeness required to ensure compliance. An out-of-date plan, to a larger extent, implies the absence of a plan. In addition to not being able to enforce development management regulations, outdated local plans lack the capability to address issues such as redevelopment of individual old buildings or several sites and modifications to existing buildings, for example through vertical and horizontal extensions or change of use.

Furthermore, town planning, since its inception has tended to focus on large towns to the neglect of small towns in rural and peri-urban areas. Settlements in these areas begin as small farming villages accommodating a few hundred people. Given their size and the dominantly agricultural uses that surround them, rural and peri-urban settlements are often allowed to grow organically without any formal planning intervention. At their early stages of existence, the absence of formal land use planning may be justified. However, over time these settlements become urbanized through a combination of in situ expansion and becoming engulfed by the outward expansion of bigger settlements. This process of rapid population growth and settlement expansion in hitherto rural areas, which has been termed as peri-urbanization, happens so quickly that it becomes extremely difficult, if not impossible, in the absence of advanced planning for local planning authorities to cope with. MMDAs attempt to cope by, for example, considering and approving individual development applications without reference to any plan. However pragmatic this case-by-case approach to approving development may appear to be, it has failed to ensure compliance to development and building regulations. This approach has also failed to bring about efficient utilization of land as well as coordinated physical development outcomes. Instead, in peri-urban areas, non-compliance as evidenced by building constructions without permits and the occasional collapse of structurally deficient buildings are commonplace. Besides the problems of non-compliance, these areas have poor accessibility and often lack access to basic facilities and services.

The prevailing landownership system in Ghana is also one of the underlying causes of non-compliance in the development process and the chaotic physical development that results thereof. As was explained at the beginning of this chapter, the customary land sector under which close to 90% of all lands in Ghana fall presents major challenges to land use planning. The separation between ownership rights vested in traditional authorities, families and individuals on the one hand and the powers to determine land use vested in state institutions, on the other hand, requires cooperation between the two sides to realize the benefits of land use planning and development management at the local level. In practice, however, such cooperation has been difficult to achieve. It is not uncommon for traditional authorities as trustees in the customary land sector to sell land without first preparing a land use plans either by contracting the services of a consultant or through the TCPD of MMDAs. This is particularly the case in many peri-urban areas where the high demand for land results in landowners sacrificing the short-term financial gains for the long-term public benefits that effective land use planning would generate. Evidence suggests that in some cases, traditional authorities prevent the preparation of land use planning schemes by withholding their cooperation (Ubink and Quan 2008). The reverence accorded the chieftaincy institution and chiefs in Ghana implies that the power they wield as community leaders in the traditional sense dwarfs any formal powers granted by the state to local governments and their technocrats, including town planners. In matters bordering on land, traditional authorities often tend to assert their powers leaving planners powerless to carry out their statutory functions.

The land management practices of traditional authorities have become a major source of land-related litigations which frustrates the planning system. As a result of the phenomenon that has come to be known as 'multiple sale of land', it is common for two or more people claim ownership to a piece of land after buying the same piece of land from the same landowner. Since proof of landownership is a key consideration in the development permit approval process, land disputes complicate the development-permitting process and often prolong the decision-making process. Consequently, in the major towns and cities, prospective developers in their attempt to protect their land from being sold and to avert possible land disputes under intense pressure to begin developing their land immediately without consideration for the statutory permitting (Arku et al. 2016).

6.5 Non-Compliance with Local Plans and Development Management Regulations

Even in instances where cooperation leads to the preparation and adoption of a local plan, traditional authorities can frustrate the implementation process as demonstrated in the quote below:

> Since the main aim of the chiefs is to maximise financial returns within the shortest possible time, important land uses such as open spaces, playgrounds, schools, markets, refuse dumps, roads, etc. are sacrificed, in order to augment the supply of building plots. This is a major cause of haphazard and unauthorised development in all statutory planning areas. (Ammissah et al. 1990; cited in Ubink and Quan 2008, p. 202)

It is important to establish that in some cases, Town Planning Officers are complicit in the unlawful rezoning practises described above. Some planning officers, neglecting their professional ethics, do act on the wishes of traditional authorities and landowners to make modifications to existing plans. Others also act on their own volition, often under the influence of a bribe in cash or kind to make illegal modifications to approved plans in order to qualify as authorized developments proposed by prospective developers who have offered the bribe. In both cases, such modifications do not go through the laid down processes of rezoning described earlier in this chapter and therefore do not reflect in the approved local plans. Instead, the resultant unauthorized development only becomes known to people who might be in the knowledge of the original proposed land uses after such development has occurred. Such decisions, despite being illegal in both customary and state laws, almost always go unchallenged. This is because within the prevailing culture of utmost respect for leaders, chiefs are revered to such as an extent that it considered disrespectful to question their decisions even when the intent is to hold them accountable, while existing processes to ensure accountability from public officials are either unknown to the ordinary citizens or cumbersome to follow.

Empirical evidence suggests that prospective developers are aware of their obligation to obtain development permit from local authorities and that while for some non-compliance is the result of indiscipline and disregard for regulations, for others, non-compliance results directly from institutional inefficiencies and unnecessary bureaucracy (see, e.g., Awuah and Hammond 2014; Yeboah and Obeng-Odoom 2010; Arku et al. 2016). The Local Government Act (Act 462) stipulates that MMDAs should communicate their decision on development applications to prospective developers within a maximum period of 3 months after submission. However, the approval process tends to be longer than stipulated. For example, in their study of non-compliance with building permit regulations in the Accra-Tema city-region, Arku et al. (2016) found that it takes an average of 4–5 years for prospective developers to receive decisions on the development applications. While unresolved land dispute could be responsible to delays, they attribute lengthy permitting process largely to unnecessary bureaucracy and institutional inefficiencies.

The grant of development permit often requires input from several public-sector agencies outlined under the section on stakeholders in the development process. Some of these agencies and departments only have offices in the ten regional capitals (e.g. Land Commission and EPA), while others with offices at the district level are not localized on the same premises to allow for easy access to them. Thus, depending on the nature of proposed development, prospective developers may be required to visit several of these institutions to obtain the required documentations. The series of administrative hurdles involved in the uncoordinated process certifications and the pervasiveness of corruption in the public sector in general are major sources of frustration to developers. Gross incompetence on the part of officials, coupled with poor, often archaic data management practices result in frequent loss of permit applications submitted to MMDAs; unnecessary bureaucracy and institutional inefficiencies impose cost (i.e. money, time, stress) on prospective developers. Under such circumstances, as Arku et al. (2016) capture it, there is a perception among prospective developers that 'it is "more convenient" and "less costly" to develop property without permits than to attempt to fulfil building permit requirements prior to land development', hence the pervasiveness of non-compliance with existing development management regulation in Ghana.

6.6 Towards Effective Local Spatial Planning and Development Management

The reforms leading to the renewed emphasis on spatial planning at all levels offer new opportunities to realize synergies between strategic development goals at higher levels (i.e. national and regional) and spatial planning at the local level. This alone constitutes a significant development in the contemporary history of spatial planning in Ghana. That notwithstanding, very important institutional bottlenecks remain unresolved. Firstly, it was argued that the existence of two different systems of planning—spatial planning established by the new spatial planning law (Act 925) and the tradition of development planning established by the National Development Planning (System) Act (Act 480)—has compartmentalized the task of planning at the level of MMDAs. While DPUs undertake the narrowly defined task of development planning realized through MTDPs, district TCPDs undertake the narrowly defined task of land use planning realized through DSDFs, SPs and local plans. Further restructuring would be required to bridge the spatial development gap as a means to strengthen local planning in all of its dimensions and to avoid the lack of policy co-ordination, duplication of efforts and waste of resources that currently characterizes planning at the local level. To this end, it would be necessary to harmonize the two laws establishing the competing planning systems (i.e. 1994 National Development Planning (System) Act (Act 480) and the 2016 Land Use and Spatial Planning Act (Act 925)). Ideally, this should merge the district TCPDs and DPUs into a single department mandated to plan all aspects of development at the local level.

Secondly, it is unclear how the current arrangement whereby district TCPDs are expected to function under one sector Ministry (i.e. Ministry of Local Government and Rural Development (MLGRD)) while its newly created LUSPA at the national level is situated under a separate sector Ministry (i.e. MESTI) would resolve the institutional challenges that were intended under the reforms that established the new spatial planning system. How district TCPDs are expected to continue to deliver effectively under the MLGRD despite not being recognized by the Local Government Act (Act 462) as one of the decentralized departments at the local level is quite confounding. Rather than mainstreaming the TCPD and spatial planning for that matter into the local governance apparatus, the rather confusing institutional configuration, if unresolved has the potential to continue to isolate and weaken the TCPD at the local government level. The current situation where district TCPDs are regarded vital only to the extent of contributing to the revenue mobilization priorities of MMDAs through the development-permitting process but precluded from receiving direct support (e.g. through budgetary allocations) from MMDAs is likely to continue. This will undoubtedly affect the department's ability to effectively deliver its spatial planning mandate at the local level.

Furthermore, one of the major challenges of effective spatial planning and development management is the prevailing land tenure system. The discussion highlighted how traditional authorities as trustees of land in the customary sector could frustrate land use planning powers and efforts legally vested in local governments. While the customary land sector poses enormous challenges to spatial planning and land management in Ghana, experts in the land management sector have argued that overturning this system for outright nationalization is impractical and unworkable, with the potential to create further problems (see, e.g., Kasanga and Kotey 2001). Since nationalization is not an option, the customary land sector must be embraced while the state seeks pragmatic solutions to achieving the legitimate interest of ensuring effective planning and management of land.

Previous land use planning laws (i.e. CAP 84) failed to explicitly recognize the enormous powers of traditional authorities and how their powers interface with the powers granted local governments and their TCPDs as public-sector institutions. Contemporary legislations including the decentralization law and new spatial planning Act have sought to address this by recognizing traditional authorities as one of the stakeholders in the spatial planning process. While these laws elaborate and entrench the powers of District Planning Authorities as deciders of the use to which land should be put, they fail to outline any specific roles for traditional authorities in whom landownership rights is vested. Instead, modern-day legislations continue to assume that planners through informal mechanisms would work out the details of any collaborations with traditional authorities needed to realize land use planning objectives at the local level. They further assume that all local governments would have the technical capacity to undertake land use planning tasks, when in fact, to date, the TCPD acknowledges that not all local governments have professional town planners to perform their spatial planning functions. As the discussion has demonstrated, spatial planning at the local level is essentially about land, implying that by virtue of their status as caretakers of some 80% of all lands that at some point in time might become the subject of land use planning, traditional authorities are no ordinary stakeholders. In view of this, it would be useful to rethink the role of traditional authorities in land use planning and development management beyond simply recognizing them as stakeholders.

The possibility of granting traditional authorities some basic land use planning powers at the local level could be explored. For example, spatial planning law could explicitly mandate traditional authorities to prepare planning schemes for areas under their jurisdiction before selling land. Such mandate will not by any means take away the authority and competences of public-sector institutions and their technocrats (i.e. professional town planners). Instead, the responsibilities granted by planning law would provide the legal basis to hold traditional authorities accountable in matters relating to spatial planning and development control. Beyond the potential accountability benefits, such an arrangement could complement efforts of local governments and even fill in the void, especially in cases where local governments lack professional town planners to undertake the task of land use planning, as traditional authorities could rely on the expertise of professional planners in the private sector to deliver land use planning imperatives.

In addition, it is important for MMDAs and traditional authorities to identify and prioritize land use planning in rural and peri-urban settlements that would become hotspots of physical development in future. The introduction of SDFs at the district level would offer the opportunity to achieve some level of planning for all areas including the urban, peri-urban and rural, and thus provide the basis to identify and prioritize areas where Local Plans would be needed. This, however, presupposes that local governments will have the capacity to formulate these Plans for all areas in Ghana. In reality, this might not be the case, implying that pragmatic approaches must be sought to achieve land use planning objectives in rural and peri-urban settlements that have the potential to experience rapid growth. Invariably, traditional authorities become the first to experience the signal that a given area would become a hotspot for development through increased demand for land. Thus, granting traditional authorities the basic planning powers described in the previous paragraph could be one of the effective ways of proactively planning for new areas.

Last but not least, it is hoped that the new reforms and the introduction of the new planning law will streamline the development management process and remove unnecessary administrative bureaucracies that impose cost on and frustrate prospective developers. To win public trust and confidence in the planning and development management process, planners must adhere to all ethical considerations of the profession and desist from corrupt practices which not only impugn the integrity of the planning profession but also poses serious risk to public health and safety.

References

Acheampong RA, Ibrahim A (2016) One nation, two planning systems? Spatial planning and multilevel policy integration in Ghana: mechanisms, challenges and the way forward. Urban Forum 27(1): 1–18

Arku G, Mensah KO, Allotey NK, Addo Frempong E (2016) Non-compliance with building permit regulations in Accra-Tema city-region, Ghana: exploring the reasons from the perspective of multiple stakeholders. Plann Theory Pract 17(3): 361–384

References

Awuah KGB, Hammond FN (2014) Determinants of low land use planning regulation compliance rate in Ghana. Habitat Int 41: 17–23

Kasanga RK, Kotey NA (2001) Land management in Ghana: building on tradition and modernity

Korah PI, Cobbinah PB, Nunbogu AM (2017) Spatial Planning in Ghana: exploring the Contradictions. Plann Pract Res 32(4): 361–384

Land use and spatial planning Act (2016) Act 925 The nine hundred and twenty fifth act of the parliament of the Republic of Ghana

Local Government Act (1933) Act 462 The four hundred and sixty two act of the parliament of the Republic of Ghana

National Development Planning (System) Act (1994) Act 480 The four hundred and eightieth act of the parliament of the Republic of Ghana. http://urbanlex.unhabitat.org/sites/default/files/urbanlex//gh_national_development_planning_act_1994.pdf. Accessed 24 Jul 2018

Town and Country Planning Act (1945) CAP 84. http://www.epa.gov.gh/ghanalex/acts/Acts/TOWN%20AND%20COUNTRY%20PLANNING%20ACT,1945.pdf

Ubink JM, Quan JF (2008) How to combine tradition and modernity? Regulating customary land management in Ghana. Land Use Policy 25(2): 198–213

Yeboah E, Obeng-Odoom F (2010) 'We are not the only ones to blame': district assemblies' perspectives on the state of planning in Ghana. Commonw J Local Gov (7). http://epress.lib.uts.edu.au/ojs/index.php/cjlg

Part III
Issues in Spatial Planning

Chapter 7
Policy Integration in Spatial Planning: Mechanisms, Practices and Challenges

Abstract Spatial planning embraces the task of bringing together the economic, social and environmental dimensions of visions of transformation at different spatial scales and across policy domains. In view of this, effective integration across policy domains and between spatial scales is indispensable in dealing with the inherently complex process of policy formulation and implementation at all levels. This chapter deals with policy integration as one of the key issues in spatial planning. It explores the concept of integration and its relevance in spatial planning globally. Next, the various mechanisms and instruments adopted to achieve coherence between various policies in spatial planning in Ghana are discussed, identifying the gap(s) between integration as embedded in the design of the planning system and integration in action. Mechanisms to deal with the challenges of policy integration in Ghana's spatial planning system are put forward.

Keywords Spatial planning · Multi-level governance · Policy integration
Vertical integration · Horizontal integration · Planning instruments · Ghana

7.1 Introduction

Invariably, the objective of spatial planning is intermediate to some wider policy goals relating to social welfare, environmental protection, economic growth and cultural conservation (Vigar et al. 2000). Given that divergent and often conflicting interests, visions and expectations characterize the processes leading to the formulation and implementation of policies, plans and strategies to achieve these goals, the need for integration across sectors and policy domains (i.e. horizontal integration) and between policy levels or scales (i.e. vertical integration) not only become crucial but also a prominent feature of spatial planning (Nadin 2007; Counsell et al. 2006). Spatial planning must embrace the task of integrating the economic, social and environmental dimensions of strategies at different spatial scales and across policy domains with the aim of ensuring that development outcomes are sustainable. Policy integration has become so crucial to the overall effectiveness of spatial planning in delivering its objectives that planning systems across the globe have had to be restructured, rescaled and modernized to respond to its imperatives.

Ghana is no exception in this reinvigoration and modernization agenda with respect to its spatial planning system. As has been established in the previous chapters, contemporary reforms in spatial planning have culminated in a new tradition of multi-scale spatial planning at the national, regional and local levels. Several spatial planning instruments at different spatial scales have been formulated to institutionalize the new spatial planning system. This includes including a National Spatial Development Framework, a Regional Spatial Development Framework for the oil region in the western part of the country; a Sub-Regional Spatial Development Framework for the second largest city, Kumasi and its surrounding districts as well as many Structure Plans and Local Plans at the city and neighbourhood scales, respectively. In addition, a tradition of development planning aimed at formulating medium-term plans to bring about socio-economic development with emphasis on poverty reduction had long existed. Also, sector Ministries, departments and agencies formulate policies and plans in areas such as education, health, transportation, environmental protection, housing, among others all of which have spatial implications.

With the proliferation of spatial planning instruments and the existence of many sectoral plans and policies arise the practical challenge of effectively synchronising visions, goals and strategies across policy domains and spatial scales over time. The purpose of this chapter is to examine the instruments and mechanisms of vertical and horizontal integration with respect to spatial planning in Ghana. To do this, the concept of policy integration will first be explained. Next, the various mechanisms and instruments adopted to achieve coherence between various policies in spatial planning will be discussed, identifying the gap(s) between integration as embedded in the design of the planning system and integration in action. On the basis of these, the challenges of policy integration will be identified and ways in which the system could potentially be improved to deliver its objectives in an effective and efficient manner will be put forward.

7.2 The Concept of Integration in Spatial Planning

The concept of integration is broad and used across a wide range of disciplines concerned with public policy. Policy integration is often used alongside other related terminologies such as policy coordination, co-operation, policy coherence and cross-cutting policy-making, joining-up among others, to imply a notion of holistic approach that avoids fragmented decision making by integrating different but interrelated policies (Meijers and Stead 2004). Stead and de Jong (2006, p. 4) offer a comprehensive definition in which they referred to integration as 'the management of cross-cutting issues in policy-making that transcends the boundaries of established policy fields, and which do not correspond to the institutional responsibilities of individual departments'.

7.2 The Concept of Integration in Spatial Planning

Within the context of spatial planning, integration is a deliberate and concerted process involving different actors. It involves the use of various formally established or informal mechanisms and instruments to synchronize cross-cutting and often conflicting goals towards a shared vision articulated in the form of policies, plans and projects, to influence the distribution of population, land use and economic activities in space. Thus, fundamentally, integration is an important task and strength of spatial governance at different levels (Albrechts 2006) and resonates with the modern system of political thinking and management theory which emphasize democracy, participation and pluralism (Osborne 2006). Integration in spatial planning is therefore essential to promote consensus building through participation, avoid policy conflicts, contradictions and redundancy, and to facilitate the realization of governments' overall policy goals (Stead and Meijers 2009; Peters 2006; Counsell et al. 2006).

7.2.1 Types and Dimensions of Policy Integration

There are two main mutually linked types of integration in the spatial planning literature: vertical integration and horizontal integration. These overlap with formal governance structures and the accompanying administrative institutions which have authority and competencies in planning derived from various legislative instruments. Vertical integration is linked with the rescaling or subsidiary principle of government (Davoudi and Evans 2008) and takes place between actors and policies at different tiers of government from national to local or vice versa (Allmendinger and Haughton 2010; Cowell and Martin 2003; Vigar 2009). Horizontal integration occurs between and across sectoral policies of the same level and the institutions (i.e. departments, agencies) that prepare them and see to their implementation (Vigar 2009; OECD 2001; Shaw and Lord 2007). Horizontal integration aims to eliminate overlapping and duplicity of policy goals, and effectively handling variety of issues that transcend the boundary of a sector or spatial unit in order to attain efficiency, effectiveness and responsiveness to community needs while saving public money (Cowell and Martin 2003; Peters 2006).

Both vertical and horizontal integration are linked in four dimensions, namely time, space, actors and issues (Underdal 1980). The time dimension is concerned with whether integration is pursued on a long-term or short-term basis; the space dimension concerns the geographical extent at which policy or integration is covered; the range or proportion of actors included in policy integration forms the actor dimension whille the range and aspect of issues and their interdependencies incorporated in the integration process form the issues dimensions.

Furthermore, Peters (2006) identified three main mechanisms of integration that are relevant to the discussion on spatial planning. These are markets, networks and hierarchical mechanisms. Market mechanism of integration is linked to the invisible hand aphorism of Adam Smith and follows that integration would occur automatically whether in the private sector or within public policy when there is exchange of goods and services as well as bargaining. Network mechanism of integration is concerned with the interaction that emerges between individuals and organizations within the same policy level. Indicated by Peters as a natural mechanism, it has the potential to avoid policy conflicts by creating a mutual ground of understanding and co-ordinating public policies. Hierarchical mechanisms are more of a vertical integration in which there is a system of authority, legal provisions and instruments to achieving policy goals at different spatial scales.

Integration mechanisms could be formally articulated rules and procedures, or informal practices adopted by various institutions and other stakeholders to build consensus, harmonize competing goals and co-ordinate wider policy-making and implementation activities. In some context, legislations dictate institutional mandates and responsibilities and spell out the imperatives of integration as well as the relevant rules, procedures and channels of policy coordination. A typical example is the so-called duty to cooperate arrangement in the UK which places legal duty on local planning authorities, county councils and public bodies to engage constructively, actively and on an on-going basis in plan preparation in the context of strategic cross-boundary matters (see Merritt and Stubbs 2012). Moreover, the various mechanisms could be organized by way of formalized intergovernmental and interdepartmental committees to take comprehensive, multi-sectoral, long-range view of spatial issues, define priorities and co-ordinate the plethora of sectoral policies (ESPON project 2.3.2, 2006). In the absence of formal rules and procedures, integration across policy domains may occur through a more voluntary or ad hoc and often less formal arrangements on 'as-and-when-is-needed' basis (Silva and Acheampong 2015). These arrangements usually make use of pre-existing informal networks or new ones that evolve among officials working in separate government departments to meet a perceived need for coordination (Silva and Acheampong 2015).

7.3 Integration in Ghana's Spatial Planning System: Mechanisms, Practices and Challenges

Vertical integration in practice, seeks to achieve coherence and consistency between layered spatial policies at the national regional and local levels. Horizontal integration derives from the need to dismantle the boundaries created by silo-based approaches where public agencies and departments plan and implement spatial policies in their respective sectors as though they were separate from other sectors. It emphasizes a cross-sector approach as a means to achieve coherence and consistency between spatially focused programmes, policies and plans of multiple government agencies and departments. The various mechanisms of integration are not deployed in isolation, but in practice, they work in tandem with each other to deliver the imperatives of both horizontal and vertical integration.

Broadly speaking, planning sits at the nexus of integration between spatial scales and across policy domains. As has been explained in previous chapters, Ghana has

7.3 Integration in Ghana's Spatial Planning System: Mechanisms …

evolved a unique system where a combination of path dependence and recent reforms has inevitably created two distinctly separate planning systems: an established development planning system and a newly instituted spatial planning system. It is therefore essential to identify the mechanisms of integration inbuilt into these planning systems and to examine how policy integration is achieved in practice.

The hierarchy of planning instruments formulated by institutions with authority and competences in planning constitutes the primary mechanisms of policy integration in both the development planning and spatial planning systems. The requirement for NPFs, Regional Development Plans and MTDPs under the established tradition of development is essentially for purposes of vertical policy integration. This is because MTDPs at the local level are required to be coherent and consistent with national policies. The NDPC the highest planning body established by the Development Planning Systems Act (Act 480) achieves vertical integration imperatives through process of issuing policy guidelines directly to MMDAs. MMDAs, in turn, are required to reflect national policies in their MTDPs. Thus, vertical policy integration between the national and local is achieved through formally established top-down and bottom-up policy harmonization channels between the NDPC and MMDAs.

Vertical integration mechanisms are also apparent in the newly established spatial planning system. Here, conformity and coherence between layers of spatial policies from the national to local levels are expected to be delivered through the three-tier system of SDFs. Thus, in logical sequence, an SDF should exist at the national level whose policies regional SDFs are required to conform to. District SDFs, in turn, are required to conform to the policy directives of their respective regional frameworks. Like the hierarchical mechanism established under the development planning tradition, within the spatial planning system, formally established top-down and bottom-up channels of integration inbuilt in the system ensure vertical policy integration between the national spatial policies formulated by LUSPA and sub-national spatial policies formulated by regions and MMDAs.

The above illustrates the ideal vision of creating a 'chain of conformity' between plans, through the hierarchy of spatial planning instruments formulated at the national, regional and local level. Ongoing efforts aimed at institutionalizing this new three-tier instruments of spatial planning have, however, not adhered to ideas of vertical policy integration inbuilt into the new spatial planning system. In fact, the WRSDF and GKSR-SDF discussed previously in Chap. 3 were formulated and adopted even before the NSDF presented in Chap. 2 was formulated. Indeed, the WRSDF was completed and adopted in 2012, three years before the NSDF would be completed and adopted. Also, the GKSR-SDF covering the eight administrative districts in the Ashanti Region was completed in 2013 without an SDF for the wider region. In addition, several district SDFs, Structure Plans and Local Plans within the geographic scope of these Regional SDFs were formulated in parallel with the preparation of the higher-level instruments to which they are supposed to conform.

A major challenge for achieving policy coherence at between spatial planning instruments at national and local levels at the initial stages of the experimentation process has been financing the plan formulation process itself. In the case of the WRSDF, it was clear that the recent discovery of oil in commercial quantities in the western part of Ghana had necessitated the preparation of the plan. The plan preparation was financed by a grant from the Norwegian government under the Oil for Development (OfD) programme—a five-year programme which began in 2010 with the aim of strengthening the environmental management of the oil and gas sector in Ghana. At the same time, as the WRSDF was being prepared, Tullow Oil and its partners, as part of their corporate social responsibilities, initiated the 'Town Planning: An Imperative for Sustainable Oil economy in Western Region' project under which funding was provided for the preparation of several lower-tier spatial development plans for selected urban centres in the oil and gas enclave of the Western region of Ghana. Similarly, the GKSR-SDF was funded by the Japan International Cooperation Agency (JICA) (see Acheampong and Alhassan 2016).

Consequently, in deciding the locations and scope of spatial development plans, some forms of compromise need to be reached whereby the prevailing interests of donor agencies and cooperate bodies, often disguised as financial and technical assistance, must be aligned with local needs and strategic imperatives. The relative strengths of these actors (i.e. local technocrats and international donor agencies and corporations) in influencing critical decisions with respect to where and how spatial plans are prepared is not entirely clear. That said, it is apparent that during the initial stages of experimentation, the funding sources have played a prominent role in determining the geography and scope of spatial plans. Consequently, the process of institutionalizing the new concept of hierarchical spatial planning has been quite chaotic, particularly in the oil and gas enclave in the Western Region, where different actors have sought to use spatial planning as a tool to assert their influence, and to pursue various cooperate interests in the emerging oil and gas economy. Although substantial funding for planning purposes has been released in the process, the concomitant proliferation of spatial development plans does not appear to have been well co-ordinated. Hence, the accompanying processes and products have also not necessarily been consistent with the coherence and conformity envisaged at the inception of the hierarchical spatial planning system.

The above assertion is no attempt to downplay the strategic and pragmatic considerations that could have equally underpinned decisions regarding which areas were covered by spatial plans. Rather, it highlights how international funding sources could determine the geographies of spatial planning and the need therefore to appropriate financing systems for plan formulation internally if such influences are to be minimized in the future. Possibly, the decision to start with relatively smaller spatial units below the national was not dictated solely by the sources of finance, but in tandem with a more pragmatic consideration to start experimenting at these scales before transferring the experience to other regions and the national level. That said, the way the experimentation process of the new spatial planning model has evolved highlights the gap between vertical policy integration in theory and how it could be achieved practically through the inbuilt mechanisms of the spatial planning system. It also provides useful lessons for future practices bordering on integration between layers of spatial policies. One of the key lessons is that vertical integration between spatial planning instruments could be realized through a more cyclical and incremental approach by which lower level instruments that have been prepared and adopted would be synchronized in a bottom-up style with their corresponding higher-level instruments.

In both the development and spatial planning systems, one of the main mechanisms for horizontal integration is the use of formally established bodies and technical committees that are cross-sectoral in composition (Acheampong and Alhassan 2016). At the national level of policy formulation and implementation, integration across policy domains and the accompanying institutions with competencies in planning is achieved through the co-ordinating role of the NDPC. The NDPC accomplishes this through what it calls a Cross-sectoral Planning Group that acts as the formal body to translate sectoral policies of the various ministries and agencies as well as the priorities of governments as articulated in their election manifestos into medium-term NPFs. The Cross-sectoral Planning Group is constituted by representatives from the various sector ministries and agencies of which the national TCPD is a key member.

The above constitutes horizontal policy integration at the national level as practiced within the tradition of development planning. Horizontal policy integration at the national level is also apparent in the newly established spatial planning system. At the national level, the newly established LUSPA, is expected to play similar co-ordinating role as the NDPC in spatial planning matters. The inbuilt formal mechanism to harmonize policies of sectors ministries, agencies and departments at the national level is apparent from the composition of the governing body of LUSPA. The reader is directed to Chap. 3 where the institutional representations of the governing body of LUSPA were outlined. Thus, in spatial planning terms, LUSPA, through its formally established governing body, LUSPA becomes the focal point for integrating the spatial dimensions of the programmes, policies and plans of the various public-sector ministries, agencies and departments.

At the regional and local levels, the new spatial planning law establishes the RSPC and DSPC, respectively. These committees are the equivalent of the governing body of LUSPA at the sub-national levels and the cross-sector institutional representations legally required of both is designed to provide a single platform to integrate their programmes, policies and plans into spatial plans. The regional and district TCPDs working through their RSPCs and DSPCs, respectively, have the mandate to formulate SDFs for their respective jurisdiction as required under the three-tier model of spatial planning instruments. They are also expected to co-ordinate all activities bordering on policy integration in the spatial plan making and implementation processes. Thus, at the sub-national level (i.e. regional and local) vertical integration mechanisms as embedded in three-tier model of planning instruments meet horizontal integration mechanisms to deliver the objectives spatial planning.

As has been pointed out elsewhere, regional spatial planning constitutes a relatively new endeavour in the contemporary history of spatial planning in Ghana. The new institutions established partly for the purposes of achieving effective policy integration are therefore yet to be tested in practice. At the local level where spatial planning has existed since its inception in the 1940s, district TCPDs and DPUs continue to undertake their separate planning mandates derived from the two planning laws that currently exist. These institutions like their parent bodies at the national level (i.e. NDPC and LUSPA) also adopt the formal mechanism of constituting technical committees and sub-committees with membership comprising the relevant sector agencies and departments to realize the imperatives of horizontal policy integration.

Given their scope and mandate, the public-sector institutions that form the core of cross-sector planning groups and technical committees at the national, regional and local levels are essentially the same. In fact, the core the institutional membership of NPDC's Cross-sectoral Planning Group and LUSPA's governing body are not different, although in both cases, other organizations both national and international are be involved in the policy formulation and implementation process on 'as-and-when-required' basis. That said, the prevailing arrangement as well as the scope of issues addressed in SDFs formulated by LUSPA and its regional and district TCPD on the one hand and plans formulated by NPDC and its DCPUs on the other hand, imply that the various institutions, especially the core ones must be represented on these two seemingly different planning bodies. Representation on and participation in this 'double-integration' platforms for policy and planning created at the national, regional and local levels could potentially result in institutional fatigue from the perspective of the participating agencies who are required to work with two seemingly different planning bodies who are essentially doing the same thing under different labels. Unnecessary duplication of efforts and resource wastages are expected would most likely characterize efforts to achieve policy integration under the prevailing institutional arrangements.

It is important to highlight that although the mechanisms for both vertical and horizontal policy integration described above are embedded in the spatial planning system and the accompanying formal institutional arrangements, in practice, integration is realized largely through informal means. In principle, existing legislations may enjoin other sector agencies to collaborate with the spatial planning bodies, and by so doing reduce silo-based tendencies for a more integrated approach to policy-making. In practice, however, it takes more than formally established rules to get all relevant institutions to participate in the spatial planning process. Informal policy integration mechanisms complement those choreographed by legislations and the various planning instruments by relying on unwritten rules and networks that exist through professional relationships, friendships and goodwill. Heads of the various departments and agencies may be aware of their duties, but their level of co-operation and commitment could depend on organization and people management skills of planning officials. Similarly, in some cases, it takes interpersonal communication outside of formal settings to resolve technical language differences among sectors and to break down artificial barriers and competition created by compartmentalization of policy-making into sector ministries, agencies and departments. Different institutional cultures as well as personal styles of individual planning professionals complement the formally established mechanisms to build consensus and achieve integration in the spatial planning system.

7.4 Towards Effective Policy Integration in Spatial Planning

The complex nature of institutional settings, the difficulty in sustaining collectivism and the lack of a clear definition of the nature and scope of spatial planning itself present major challenges to achieving effective policy integration in practice (Newman 2008; Allmendinger and Haughton 2010). Other sources of challenge to effective integration include barriers in technical language in different sectors, the lack of co-ordinating bodies and of financial allocation systems, increasing levels of competition among sectors and institutions grounded in the perception that some policy sectors are important than others, the absence of political will or commitment and a lack of awareness and expertise (Stead and Meijers 2009; Stead and de Jong 2006; United Nations 2008; Peters 2006).

The discussion of policy integration in the Ghanaian context shows that the planning system by design has several inbuilt features aimed at ensuring policy coherence at all levels. However, the gap between the ideal and the reality for policy integration remains wide with several inherently conflicting structures. One of the main causes of weak policy integration is the nature of the planning system itself. It was argued that the pervasive notion that the 'spatial' is distinctly separate from the 'socio-economic' and hence the need to address them under separate legal and intuitional arrangements under two separate planning systems not only compartmentalize the task of planning but create multiple platforms for integration between sectors and across policy domains in planning. The tendency for duplication of functions and the institutional fatigue thereof would ultimately stifle an integrated approach to planning. Planning systems throughout the world are not static and that although the current reforms particularly in the area of spatial planning are commendable, further reforms aimed at creating a unified planning system to accomplish the single task of integrated development planning should be pursued.

The discussion also highlighted the lack of funding as one of the prominent factors that has hampered effective policy integration with respect to the experimentation with the new three-tier model of hierarchical spatial planning. The main challenge here was how foreign donor agencies have provided the much-needed financial resources, but in doing so, have dictated and steered the planning process in directions that are not necessarily consistent with the coherence and conformity envisaged by the system. Systems to mobilize substantial funds domestically combined with programmes to building institutional capacity would be needed to establish control over the process of operationalizing the new spatial planning model. As mentioned elsewhere, the Land Use and Spatial Planning Act (Act 925) which came into force in 2016 requires that a Land Use and Spatial Planning Fund be established. It is hoped that over time, the fund would strengthen the capacity of LUSPA and its TCPD offices at the regional and local levels to be able to carry out their wider mandate, including ensuring coherence between spatial policies at all spatial scales and between sectors.

The inception of regional spatial planning and, in particular, the new concepts of city-regions and functional regions would require a great deal of co-ordination between neighbouring MMDAs in order to ensure horizontal policy integration imperatives. However, formally established mechanisms enjoining MMDAs to collaborate in matters involving several local governments are still lacking in the contemporary legislations of spatial planning. In the absence of formal rules and procedures, informal mechanisms may be used. However, the collaborative culture required for effective policy integration between local governments cannot be guaranteed through informal and ad hoc approaches initiated on as-and-when-needed basis. In view of this, the use of legally binding mechanisms to institutionalize and enforce a culture of strategic alliance among local governments in cross-cutting matters should be explored.

7.5 Conclusions

Both vertical and horizontal integration are important to achieve coherence and consistence between multiple layers of spatial policy and across sectors and policy domains. This chapter has identified the formal and informal mechanisms for policy integration in Ghana's spatial planning system. The multiple platforms of policy integration in planning, created by the two competing planning systems (i.e. development planning system and the new spatial planning system) would ultimately result in institutional fatigue, duplication of efforts and resources wastages, weakening policy integration in practice. Moreover, the absence of a tradition of strategic regional planning and a culture of strategic partnerships among local authorities, the lack of appropriate institutional arrangements and sustainable sources of finance and duplicitous institutional functions are key barriers to effective integration within the new concept of hierarchical spatial planning. A new paradigm of integrated planning under a unified planning system is considered crucial to creating the fundamental conditions for effective multi-level policy integration in Ghana. While the establishment of the Land Use and Spatial Planning Fund would help address some of the challenges of effective policy integration associated with funding, formally established collaborative mechanisms through legally binding enactments would be needed to institutionalize and enforce a culture of strategic partnerships among local governments in cross-cutting matters.

References

Acheampong RA, Ibrahim A (2016) One nation, two planning systems? Spatial planning and multilevel policy integration in Ghana: mechanisms, challenges and the way forward. Urban Forum 27(1): 1–18

References

Albrechts L (2006) Shifts in strategic spatial planning? Some evidence from Europe and Australia. Environ Plann A 38(6): 1149–1170

Allmendinger P, Haughton G (2010) Spatial planning, devolution, and new planning spaces. Environ Plann C: Gov Policy 28(5): 803–818

Counsell D, Allmendinger P, Haughton G (2006) Integrated spatial planning- is it living up to expectations? Town and Country Planning-London-Town and Country Planning Association, 75(9): 243–246. Available at: http://www2.hull.ac.uk/science/pdf/geogGH07d5.pdf. Accessed 20 Apr 2015

Cowell R, Martin S (2003) The joy of joining up: modes of integrating the local government modernisation agenda. Environ Plann C: Gov Policy 21(2): 159–179

Davoudi S, Evans N (2008) Territorial governance in the making. Approaches, methodologies, practices. Boletín de la A.G.E, 46: 33–52.

ESPON 2.3.2 (2006) Governance of territorial and urban policies from the EU to the local level, final report, ESPON, Available at: http://www.espon.eu. Accessed 20 Apr 2015

Meijers E, Stead D (2004) Policy integration: what does it mean and how can it be achieved? A multidisciplinary review. In: Berlin Conference on the Human Dimensions of Global Environmental Change: Greening of Policies-Interlinkages and Policy Integration, Berlin

Merritt A, Stubbs T (2012) Incentives to promote green citizenship in UK transition towns modernisation agenda. Environ Plann C: Gov Policy 21(2): 159–179

Nadin V (2007) The emergence of the spatial planning approach in England. Plann Pract Res 22(1): 43–62

Newman P (2008) Strategic spatial planning: collective action and moments of opportunity. Eur Plan Stud 16(10): 1371–1383

OECD (2001) Towards a new role for spatial planning. In towards a new role for spatial planning. OECD Publishing, Paris

Osborne SP (2006) The new public governance? 1. Pub Manage Rev 8(3): 377–387

Peters B (2006) Concepts and theories of horizontal policy management. Handbook of public policy, pp 18–21

Shaw D, Lord A (2007) The cultural turn? Culture change and what it means for spatial planning in England. Plann Pract Res 22(1): 63–78

Silva EA, Acheampong RA (2015). Developing an inventory and typology of land-use planning systems and policy instruments in OECD Countries. OECD Environment Working Papers, No. 94

Stead D, de Jong M (2006) Practical guidance on institutional arrangements for integrated policy and decision making, Geneva. Available at: http://www.unece.org/fileadmin/DAM/thepep/documents/2006/

Stead D, Meijers E (2009) Spatial planning and policy integration: concepts, facilitators and inhibitors. Plann Theory Pract 10(3): 317–332

Underdal A (1980) Integrated marine policy: What? Why? How? Marine Policy 4(3): 159–169

United Nations (2008). Spatial planning—key instrument for development and effective governance with special reference to countries in transition, New York and Geneva

Vigar G (2009) Towards an integrated spatial planning? Eur Plan Stud 17(11): 1571–1590

Vigar G, Healey P, Hull A, Davoudi S (2000) Planning, governance and spatial strategy in Britain: an institutionalist analysis. Macmillan, Basingstoke

Chapter 8
Public Engagement in Spatial Planning: Statutory Requirements, Practices and Challenges

Abstract Spatial planning is about people, and as such, the success of the activity depends partly on how well it deals with the varying interests, views, preferences and aspirations of individuals, groups, communities and businesses. This chapter focuses on public engagement in spatial planning in Ghana. It discusses public engagement and its value from the perspective of the normative theories of planning. Building on the theoretical foundations, public participation in practice is examined, identifying the statutory requirements for stakeholder engagement as well as the procedures and methods adopted to involve stakeholders in spatial planning. The theory-practice gap and the shortcomings of current approaches used for stakeholder engagement in Ghana are identified. Ways of improving on current practices, including the prospects of leveraging ICT to access citizens' vision of transformation are discussed.

Keywords Democracy · Public participation · Stakeholder engagement · Consensus · Equity · ICT · Spatial planning · Ghana

8.1 Introduction

The need for the democratic ideals including openness, collaboration, dialogue, consensus building and inclusiveness in spatial planning has long been emphasized. As Chadwick (1978, 25) succinctly indicated, 'planning is done by human beings for human beings'. There is a technical side to spatial planning which requires years of university education and professional training to obtain the necessary qualifications to function as a professional planner. The nature and scope of spatial planning, however, imply that in practice, the professional planner will encounter and must work with a wider range of stakeholders including communities, businesses, public-sector agencies who represent different and sometimes conflicting interests.

The formulation of spatial plans must be evidence-based, and engaging individuals, communities and businesses in the plan-making process constitutes one of the key methods of eliciting the critical evidence in a transparent and inclusive manner (Morphet 2011). The views, interests and preferences of stakeholders as well as local knowledge are best understood and incorporated into programmes, policies and plans through effective engagement processes (Innes and Booher 2004). Public engagement brings together various stakeholders and has the potential to provide a common platform to resolve conflicting interests and to build consensus. Through effective public involvement, social capital could be harnessed not only for purposes of plan formulation but also for implementation and development management process to ensure that development outcomes are consistent with the shared vision of stakeholders. Public engagement is therefore crucial if spatial development would deliver its objectives.

This chapter focuses on public engagement in spatial planning in Ghana. We will begin by examining public engagement and its value from the perspective of the normative theories of planning. Next, we will examine public engagement in practice, focusing on the statutory requirements and procedures and methods adopted in Ghana, and examining the theory-practice gap in stakeholder participation. On the basis of challenges, we will discuss ways in which public engagement could be improved in the spatial planning process.

8.2 Public Participation in Theory

Theoretically, stakeholder participation imperatives derive from normative planning doctrines that emphasize dialogue, political awareness, inclusiveness and empowerment as fundamental principles of planning as a decision-making process. According to communicative theory, planning is an interactive process in which the collective management of public concerns and co-existence in shared spaces is key (Healey 2003). Forester (1999) a leading exponent of the deliberative approach to planning also argues that for planning to manage common concerns and promote consensus building and co-existence, citizen participation should be made a pragmatic reality rather than an empty ideal. Practitioners through public engagement act as negotiators seeking desirable outcomes and mediators managing conflicting interests and aspirations that may arise in the planning process (Throgmorton 2000). Other theorists espouse the need for public engagement as essential to a tradition of planning founded on the principles and benefits of social learning and co-production (i.e. collectively generated knowledge and outcomes) among organizations, communities and groups (Friedmann 1987). Similarly, Davidoff's (1965) advocacy tradition aims to empower disadvantaged groups to actively influence the content and distributive outcomes of planning.

Models of stakeholder participation have been advanced with the aim to offer a framework to characterize the scope and nature of citizen participation in practice, identify the gap between ideal forms of participation and what is actually done in practice, and to challenges practitioners to aspire for greater levels of public participation. Among the dominant models which retain considerable contemporary relevance in spatial planning are the 'ladder of participation' proposed by Arnstein (1969) and Jules Pretty's (1995) typology of participation.

8.2 Public Participation in Theory

Fig. 8.1 Arnstein's (1969) ladder of participation

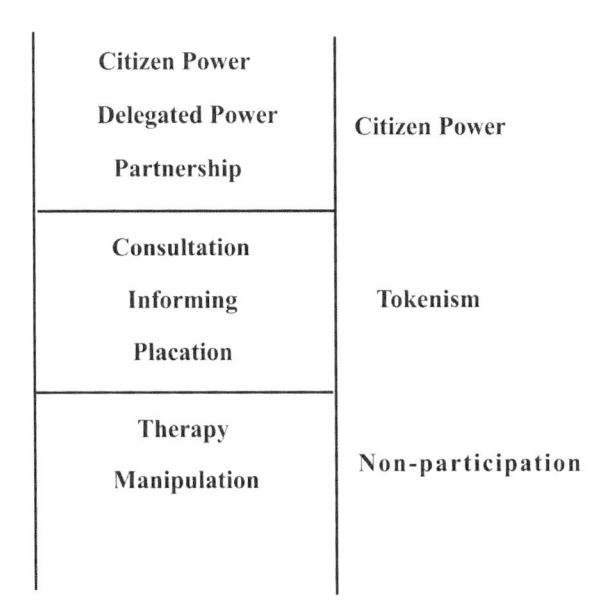

Citizen Power Delegated Power Partnership	Citizen Power
Consultation Informing Placation	Tokenism
Therapy Manipulation	Non-participation

Arnstein's normative model of participation employs the metaphor of a ladder to illustrate eight levels of participation that could be attained by stakeholders in decision-making. Each rung depicts the extent of citizens' power in and influence over determining the outcomes of decisions. The eight levels are further categorized into three main forms of citizen participation: at the bottom of the ladder is 'non-participation'—a category that comprises manipulation and therapy. In Arnstein's view, manipulation and therapy do not genuinely enable stakeholders to participate in decision-making. Instead, their real objectives, defined by powerholders and technocrats, are often to 'educate' or 'cure' participants (Fig. 8.1).

'Tokenism' appears in the middle of the ladder and has three gradations—placation, informing and consultation. According to Arnstein, these three forms of 'tokenism' when proffered by powerholders as the total extent of participation may allow citizens to hear and be heard. However, stakeholders under these conditions lack the power to ensure that their viewpoints will be heeded by the powerholders. Participation by information and consultation often take place through public hearings and stakeholder meetings. At such fora, stakeholders might get the opportunity to know about the content of plans and to share their views, but there is guarantee that the status quo would change. This is because, in most cases, planners determine when and how many times consultations should be carried out, and as such, it is possible that the number of consultations might not be adequate for stakeholders to fully participate. Also, it becomes difficult for stakeholders to follow-up on their views to ensure that they have actually been considered in the final plan. This is particularly so under placation where citizens, especially the disadvantaged, are allowed in principle to advice, but planners retain the right and power to decide.

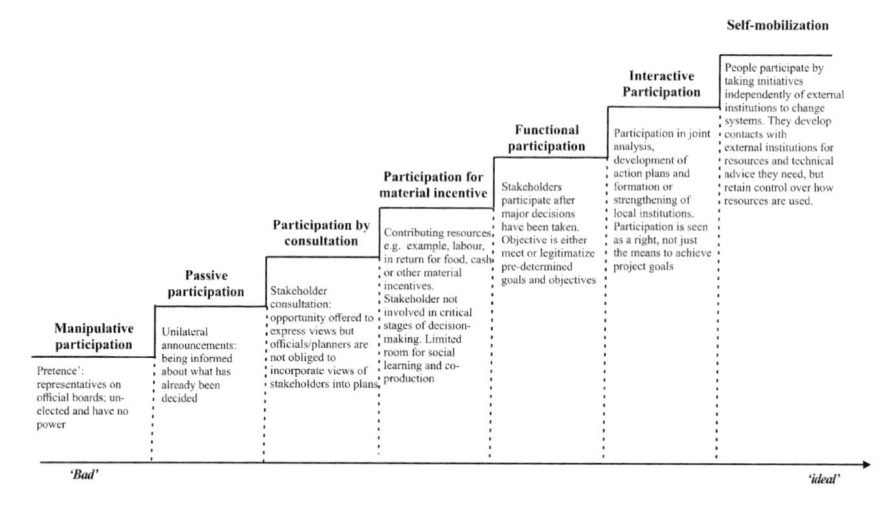

Fig. 8.2 Pretty's typology of participation. Adapted from Pretty (1995)

At the top of Arnstein's ladder of participation is 'citizen power' comprising three rungs, namely partnership, delegated power and citizen power. This final category could be considered the ideal forms of stakeholder participation as they allow stakeholders increasing degree of decision-making powers through dialogue, collaboration and consensus building. Through partnerships, for example, citizens can negotiate and engage in trade-offs with traditional powerholders. Delegation of decision-making powers not only imposes some responsibilities on stakeholders but also acts to empower communities to come together to address common concerns and interests.

Arnstein's model focuses on stakeholder participation from the point of view of the citizens at the receiving end (Cornwall 2008). It seeks to create awareness among stakeholders, primarily citizens about what to expect while equipping them with the lens through which they could unpack and interpret the underlying motives and objectives of powerholders and technocrats involved. On the contrary, Pretty (1995) proposes a model in which the focus is on the user of the participatory approaches (Cornwall 2008). Thus, while incorporating some of Arnstein's original ideas, Pretty approaches participation in its different forms from the perspective of duty bearers and powerholders such as government officials, consultants, civil society groups and other non-governmental organizations who are expected to create the platform for stakeholder engagements.

As illustrated in Fig. 8.2, the model Pretty puts forward is equally normative, identifying different forms of participation on a spectrum of 'bad' to 'ideal'. Manipulative participation and passive participation, which take the form of representations on committees and board without any real powers, and participation by information, are considered 'bad' forms of stakeholder engagement in Pretty's model. These forms of participation are similar to what Arnstein refers to as tokenism.

Perhaps, the most common forms of participation lie between the spectrum of 'participation by consultation' and 'functional participation'. In consultations, the platform is opened for stakeholders to express their views, but policy-makers are not obliged to take these views on board. This also suggests that participation by consultation is often a one-way process where feedback mechanisms between stakeholders and duty- bearers are almost non-existent or rarely used. There is therefore little room for public officials, including planners to be accountable to the citizenry. In 'functional participation', stakeholder meetings become only necessary as platforms to give legitimacy to predetermined courses of action. The argument for this form of participation is that it is efficient and less costly for policy-makers to come up with alternative courses of action and then use the consultative process as forums to build consensus around one of the options, which probably would be the best course of action (Cornwall 2008).

The last two forms of participation—'interactive participation' and 'self-mobilization'—according to Pretty are the ideal forms of participation that policy-makers including, planners should strive for. 'Interactive participation' emphasizes co-production, a bottom-up synergy between policy-makers and communities in jointly formulating a theory of change in respect of what issues ought to be addressed and how they should be addressed (Putnam 2000). This could involve joint identification and analyses of problems leading to the formulation of policies and implementation action plans, using the available resources. Participation approached from this perspective is not viewed as a means to an end but as a right, employing structured methodologies to facilitate collective generation of knowledge and learning (Hall 2005; Cornish 2006). Ultimately, the ideal form of participation according to Pretty's model is 'self-mobilization', where citizens can independently mobilize and initiate action towards bringing about desired changes in their communities.

Like many normative theories and models in planning, the models of Arnstein and Pretty have attracted several criticisms. Hurlbert and Gupta (2015), for example, argue that these models and the extant literature based on them often do not pay attention to the contextual challenges that make the so-called 'ideal' forms of participation challenging or even impossible in certain situations or where policy-making is more appropriately technocratic and created or implemented by expert bureaucrats. Instead, they present a utopian notion of what forms participation ought to take.

Indeed, as Curry (2012) recognizes, professionals often perceive five challenges in working with lay people: they are of the view that working with lay people is unnecessary within democracies; lay people lack expertise; they are not representative; there is commonly a lack of trust; and they complicate the decision-making process. Faced with such difficulty and given limited resources, participation is either avoided or may only be used to legitimize predetermined goals and objectives. In addition, some political science theorists such as Dahl (1989) have argued that broad-base, direct participation is unworkable in the modern bureaucratic state and that representative government by elites is the appropriate way. Also, as Morphet (2011) acknowledges, in some cases, consultation can be politically inconvenient for organizations, especially local governments, and that sometimes, it is much easier for local governments that certain decisions are imposed from a higher tier of government, such as the central government, instead of them having to make the tough and often unpopular decisions. Ultimately, the aforementioned normative models of participation are aimed at enabling policy-makers and stakeholders to reflect on prevailing practices in any given context, and to constantly search for pragmatic ways of widening engagement for all groups of people in the planning process.

8.3 Spatial Planning and Public Participation in Ghana: Practices and Challenges

The normative ideals of participation as expressed by the models of participation discussed in the previous section get absorbed, often partially into planning legislations, which in turn define the statutory requirements for public participation in practice. Thus, as the spatial planning system evolves, the statutory requirements for public participation might also change. In view of this, it is essential to take a historical perspective in analysing public participation practices in spatial planning in Ghana. In the sections that follow, we will discuss the statutory requirements for public participation as stipulated in the relevant planning legislations. In doing so, we will examine the gaps between normative theory of participation on the one hand and the statutory requirements and the accompanying practices on the other.

8.3.1 Public Participation Requirements and Practices Under the 1945 Town and Country Planning Ordinance (CAP 84)

The 1945 Town and Country Planning Ordinance (CAP 84) provided the first statutory requirements for public participation in spatial planning practice in Ghana. Under Section 12 and 14 of the Ordinance, local planning authorities were obliged to place notices of planning initiatives, in two daily newspapers and on public notice boards for a period of 2 months. For example, when a planning area was defined, qualifying the area to benefit from the preparation of a planning scheme, the Town and Country Planning Office, in whose jurisdiction the designated planning area falls, was required to make this known to the public by serving public notices through the avenues of newspapers and public notice boards.

This form of public participation seems to complement the centralized system of planning that was instituted in the early days of spatial planning in Ghana. Essentially, locally planning was not only centralized but also very technocratic. Government officials and their town planners would provide the technical reasons justifying the need for a plan for a particular area under the practice of declaration of planning areas (see Chap. 3). After such a unilateral decision had been made, local governments, through their TCPDs, were to inform the public that a plan was to be drawn for the area in question. Thus, public participation as statutorily defined was essentially passive, with stakeholders being informed about decisions that have already been taken by planning officials.

8.3 Spatial Planning and Public Participation in Ghana: Practices … 157

One could argue that by serving public notices in print media and on public notice boards, the government and technocrats had taken genuine steps to elicit the views and concerns of the public. This may be valid. However, it is also obvious that notice boards and newspapers would not necessarily have been effective in reaching the targeted audience. One would have to be able to read in the English language to get the message of the notices. Besides the limitation of the mediums in reaching their intended audience, citizens with concerns were expected to put them in writing to the TCPD. This implies that only a cross-section of the public, possibly the elites, had the opportunity to influence planning decisions at the local level. The larger mass of the population would therefore have been excluded from participating in planning decisions at the local level. Thus, following Arnstein's model, the form of participation stipulated in the planning law and practiced then sits at the very bottom of the ladder of participation, which could be delineated anywhere between 'non-participation' and 'tokenism'.

As has been mentioned elsewhere, CAP 84, despite losing its contemporary relevance, remained in force until it was replaced in 2016 by the Land Use and Spatial Planning Act (Act 925). This implies that in principle, while the old legislation remained in force, the statutory requirements it defined were to inform how planners engaged with stakeholders in contemporary planning practice. However, there is limited evidence that until 2016, the statutory requirements stipulated in CAP 84 were followed. Indeed, as has been explained earlier, the spatial planning system in Ghana was essentially dormant for the most part of the twentieth century until it was rejuvenated in the mid-twenty-first century. Given that spatial planning was virtually not working, and new planning schemes were rarely prepared or old schemes revised, the requirements for public participation also essentially became irrelevant. Instead, landowners, primarily traditional authorities, demarcated plots without the involvement of professional planners. Given that the existing law did not contain any enforceable clause that enjoined traditional authorities to consult with the wider public before land allocation decisions were made, there was no room for public participation even in its simplest form. Where planning schemes were prepared by local governments, under-resourcing with respect to staffing, equipment and finances, a problem which persists even to date, would have left the TCPDs impoverished to undertake their core functions, which includes providing the platform for stakeholders to participate in the planning process.

8.3.2 Democratization, Decentralization and Public Participation in Spatial Planning: Just More of the Same?

The notion of a bottom-up, consultative governance process underpinned the decentralization process which established local governments (i.e. MMDAs) as administrative units and development authorities at the lowest level of governance in Ghana.

Under the Local Government Act (Act 462), MMDAs are required to initiate and encourage joint participation with all relevant stakeholders to bring about development at the local levels of political administration. They are required to fulfil mandate through the political and administrative structures that constitute a local government in Ghana. MMDAs are at the apex of the local government structure. Under them, there are sub-district political and administrative structures including sub-metropolitan units, district, urban, town, zonal and area councils, and unit committees. These subordinate bodies of the assemblies have elected officials who perform functions delegated to them by law. Thus, the local government structure instituted a system of representative democracy where communities that fall under a local government are represented at the MMDA level by elected officials. Elected officials therefore provide the linkage between local governments and their constituent communities.

In previous chapters, we identified that in principle, the TCPD offices were not regarded as being part of the decentralized departments of MMDAs. Also, a new tradition of decentralized development planning accompanied the decentralization reforms and introduced legal and institutional structures for this type of planning that were different and separate from those of town and country planning. At the same time as decentralization was being pursued, the 1945 Town and Country Planning Ordinance (CAP 84) remained in force, implying that as far as spatial/land use planning was concerned, statutory requirements for public participation stipulated by this law were valid and binding. Thus, caught between the new wave of democratization and the old system of the centralized governance system, the TCPD at the local level struggled to assert its contemporary relevance. In the process, the land use planning system at the local level would attempt to imbibe the ideals and requirements for public participation in the new concept of decentralized governance while adhering to the statutory requirements for participation already stipulated in the 1945 Town and Country Planning Act (CAP 84).

For purposes of community participation, the land use planning system essentially kept the statutory requirements of CAP 84 and co-opted some of the administrative structures of the local government system. This gave birth to what became known as the Statutory Planning Committee (SPC). The core membership of the SPC included heads of technical departments within the MMDAs, elected officials representing various communities within the assemblies and other unelected opinion leaders who would be invited on an ad hoc basis depending on the nature of decisions being deliberated. SPCs therefore became the main platforms where the views and concerns of different groups were to be represented and fed into local plans.

When a new planning scheme is to be prepared or an old one is to be revised, the TCPD of the MMDAs, in conformity with the requirements of CAP 84, would have been required to inform the relevant stakeholders. The practice of using print media and service notices appears to have been abandoned over time. Instead, in rural settings, traditional authorities/chiefs could disseminate information about the plan through a town crier—a person employed to make public announcements in the streets or marketplace of a town. In urban areas, however, the common practice involves the MMDA under whose jurisdiction the planning scheme area falls identifying and formally inviting representatives of various interest groups they consider relevant to attend a stakeholder consultation meeting. The assumption ostensibly is that these individuals will convey the views and concerns of the various groups they represent at the stakeholder meetings. In reality, however, it is difficult, if not impossible, to ascertain the interests these individuals are actually representing at the SPC. This is because there are virtually not systems in place to ensure that the subject of letters inviting representatives to attend the stakeholder meeting is communicated to the groups being represented. Neither are there any systems in place to ensure that decisions taken at SPC meeetings are actually communicated back to the groups the committee members are meant to represent. When the plans are approved, and adopted, the core members of the SPC, excluding opinion leaders invited on ad hoc basis, continue to play key role in the development management process in the vetting of development applications. Thus, participation in the development management process is also realized through representation.

The consultations through representation described above are probably the weakest and most patronizing form of public participation that could exist. Indeed, in terms of practices and outcomes, this form of participation is no different from the statutory requirements and practices that were enshrined in the 1945 Town and Country Planning Ordinance (CAP 84). Spatial planning remains largely a technical activity with virtually no opportunities offered for the views, concerns and aspirations of individuals and groups from the wider communities that will be affected by the plans to be considered. Essentially, planning officials decide which group leaders and/or opinion leaders are invited to attend meetings of SPCs, and if for any reason local government decides not to send out formal invitations, there is no way ordinary people who would potentially be affected by decisions will get to know about the decisions in the first place. In fact, the overall notion of a bottom-up consultative governance and development process which occasioned the inception of decentralization in the late 1980s is still yet to be realized fully. The National Urban Policy acknowledges this:

> … [the] impact [of MMDAs] in terms of creating awareness and [ensuring] participation of people in the development process is limited. As a result, there is very little awareness of, and interest in, the laws and regulations on development controls, and (ii) no community participation in identifying and dealing with the unauthorized development found in many urban communities. (National Urban Policy 2012, 17, p.18)

8.3.3 Public Participation Under the New Spatial Planning System: Practices and Emerging Issues

Opening the planning process for greater stakeholder participation was one of the key goals of the reforms that instituted the new spatial planning system. As part of the reforms, the TCPD published its first manual—The New Spatial Planning Model Guideline—which documented features of the new planning system. In this document, the failures of the planning system to deliver its objectives were partly attributed to the lack of stakeholder involvement in the process. As the document indicates:

It is recognized that the plans so often fail in their realization, because key stakeholders are not involved in the formulation of the plans. In many cases, the stakeholders are not aware of the plan or of any planning proposals. The new planning model, therefore, proposes greater involvement of stakeholders, to include institutions and organizations, traditional rulers and large-scale landlords, real estate developers and individual plot holders. (MESTI, NSPS Module Guideline 2011, p. 8)

Acknowledging the failures of the planning system to engage with the wider public in the past, the NSPS, which was released in 2011 ahead of the promulgation of the new planning law, the Land Use and Spatial Planning Act (Act 925) in 2016, sets out the new statutory requirements and methods of public participation that ought to be adopted. As was explained in the previous chapters, the new spatial planning system introduced the concept of SDFs as spatial planning instruments to be formulated at the national, regional, sub-regional and district levels. At the bottom of the tier of planning instruments are Local Plans. Thus, the scope and form of participation required under the new spatial planning system will depend on the spatial scale of planning and the type of planning instrument that is being formulated. The stages and methods of participation required for the preparation of SDFs and Local Plans are outlined in Box 8.1.

Box 8.1: Stages and Methods of Consultation in SDF and Local Plan Preparation

Spatial Development Frameworks (SDF)

The SDF requires a minimum of three rounds of consultation. These are:
i. After data collection and during analysis stage when trends and issues identified;
ii. When determining the preferred scenario/option for development;
iii. To consider draft final plan.

At all three stages, adequate notice must be given to the general public to make meaningful inputs or, if published in the newspaper or exhibited in the Regional Coordinating Council or Assembly Public Data Rooms, adequate time provided for interested parties to make representation. Adequate notice for meetings of stakeholders will be 10 working days. Adequate time for comments following publication in the print or electronic media will be 20 working days. Copies of the plans at district level must be made available for a minimum of 20 working days in the District Public Data Room.

Reports on the public consultation process will be published and summarized in the local media. Full copies of the outcome should be made available at the offices of the RCC and affected MMDAs. The reports should identify how many responses were received and where from and the weight of opinion expressed and will form an annex to the SDF.

Local Plans

The Local Plan requires a minimum of three rounds of consultation. These are at the following stages:

8.3 Spatial Planning and Public Participation in Ghana: Practices …

i. Preliminary planning stage;
ii. Draft plan including proposals for phasing of plan;
iii. Final plan and report stage.

The stakeholders will be consulted and involved in the planning process through some or all of the following:

- One-on-one contact using interview guides and questionnaires;
- Public notices in newspapers;
- Mass media (radio, television) and Internet;
- Newsletter;
- Fliers, in particular, to encourage attendance at meetings;
- Public hearings and group discussions;
- Community meetings/consultations;
- Conferences, seminars, workshops;
- Placement of documents in the Public Data Room.

Reports on the public consultation process will be published and summarized in the local media. Full copies of the outcome should be made available at the offices of the MMDA, in their Public Data Rooms. The reports should identify how many responses were received and where from and the weight of opinion expressed and will form an annex to the Local Plan.

Source: The New Spatial Planning Model Guidelines, Published by MESTI and the Town and Country Planning Department in November 2011

As the NSPS Module Guideline state:

> The requirement is for a participatory planning system to be used in the preparation of all levels of plans. The plan preparation process will require periods for key stakeholders to air their views and opinions. The SDF preparation requires the participation of three groups: the key sector agencies; the MMDAs and the RCCs, and; the general public. (MESTI, NSPS Module Guideline 2011, p. 23)

With respect to local plans, the purpose, form and scope of participation envisaged under the reformed planning system are even more ambitious. As the NSPS Module Guideline indicates in relation to local plans:

> Participation as used here shall be a departure from what in the past has been mere consultation and information provision which often came as "fait accompli". In this context, it shall mean a process of active involvement that affords actors the opportunity to learn, and hence, own the process and break and transform past habits in order to achieve the desired objectives of the plan. Participation may involve information sharing, consultation and collaboration. Achieving this will require that simplified and interactive techniques such as Participatory Learning Actions be employed. Participation shall however not mean that the technical aspects of the plan preparation process that demands technical competence be sacrificed in the interest of involvement. (MESTI, NSPS Module Guideline 2011, p. 43)

It goes further to outline the category of stakeholders 'that shall mandatorily be involved in the planning process' at the local level as follows:

- The Assembly Executive Committee;
- All relevant sub-committees;
- Assemblymen for the area affected by the plan;
- Members of the Statutory Planning Committee;
- Heads of technical departments of the assembly including the:
- Budget office;
- Survey and Mapping Division of the Lands Commission;
- Land Title Registrar of the Lands Commission;
- Identifiable interest groups, including developers, landowners and users in the plan area and, in the case of small subdivision plans, the residents within the broader community who might be affected by the proposed development;
- Chiefs/elders/traditional rulers;
- Representatives of utility providers.

In practice, NSPS recommends that in the formulation of SDFs and Local Plans, a minimum of three rounds of stakeholder consultation should be held. During the experimentation phases of the new spatial planning system, the author had the unique opportunity of being involved in the formulation of some of the earliest generation of SDFs and Local Plans. In the sections that follow therefore, I will draw on my knowledge and experience in discussing public participation practices in the new spatial planning system. Some of the challenges involved in implementing stakeholder consultations as well as the strengths and weaknesses of the consultation practices will be highlighted.

Firstly, it is required of the planning agency to give adequate notices to all relevant stakeholders, often 10 days ahead of the consultation meeting to give adequate time for interested parties to make representation. This is the crucial point where the scope of the consultative process gets defined and attempts are made to resolve the essential tensions between the normative ideals of public participation on the one hand, and what is practically feasible given the spatial scale of planning and the available resources. The TCPD identifies who the relevant stakeholders are and write to officially invite them to the stakeholder meetings. The type of stakeholders invited could be classified into three groups, namely (i) government officials and head of technical departments; (ii) community leaders including traditional leaders and elected officials of communities and (iii) other stakeholders including opinion leaders, leaders of faith-based organizations, representatives of non-governmental organizations and civil society groups.

In the experience of the author, even under the new spatial planning system, participation in most cases is still by representation. Also, the scope of stakeholder consultations narrows as one moves up the hierarchy from, for example, when a local plan is being formulated to say when a regional SDF is being prepared. The first reason probably is because at the highest levels of planning such as at the national and regional scales, the issues addressed tend to be rather broad and are therefore quite technical. At this level, inviting representatives of various interest groups for the stakeholder consultation instead of consulting citizens may be justified. The other and perhaps the most important reason relates to the cost of the stakeholder consultations. Often, consultation meetings are held in the administrative capitals at the national, regional and district levels. This means that all stakeholders must converge at this central location. An example will help put things in perspective: the Western Region of Ghana, for example, had 22 local governments (i.e. MMDAs) as of the time of writing this book, with several communities, each having elected and unelected leaders, as well as other interest groups. Thus, during the formulation of the SDF for the Western Region, the number of stakeholders that would be invited to a typical stakeholder consultation meeting runs into hundreds of individuals from sector Ministers, heads of agencies and commissions to representatives of local governments and communities in the region. Consultations meetings were held in Sekondi-Takoradi, the regional capital, which implies that most of these individuals had to travel long distances to attend. It is an established culture for the transportation and feeding costs, and accommodation costs (in cases where consultation meetings are held over a couple of days) of all stakeholders invited, including government officials to be borne by the TCPD. Under such circumstances and given the constraints on resources, one would argue that it made practical sense to limit the number of stakeholders that were identified and invited to attend stakeholder consultations.

Secondly, the statutory requirements for stakeholder consultation under the new planning system are embedded in a procedural conception of spatial planning.

The plan formulation process begins with data collection to understand the current situation. The first stage of stakeholder consultations according to the NSPS Module Guidelines should be at the point where data has been collected and trends and key issues have been identified from the analysis of the data. This means that the planning authority, which could be MMDAs and their TCPD officials or consultants, is required to present their findings from the preliminary analyses to key stakeholders. This, in principle, is to offer stakeholders the opportunity at the earliest stage of the planning process to interrogate the planners' view and understanding of the local context and circumstances and for the planners to also benefit from local knowledge.

Outcomes of the first stage are expected to feed into the second stage of the process where the relevant stakeholders will be consulted again, this time to evaluate different scenarios of SDF proposals. At this stage of the process, all the technical issues are brought to bear. Scenarios of what the future could look like are formulated based on various technical analyses and projections in the areas of population growth, infrastructure demand, housing demand, sanitation requirements, opens spaces and the land use requirements of all proposals. The plan options are illustrated in maps and diagrams and accompanied by texts explaining the package of proposals contained in each option. A typical scenario evaluation workshop, which could take a day or two, follows in this order:

- The planning officials first present the proposals of the different options to all the stakeholders.
- After the main presentation, the stakeholders are divided into smaller groups to discuss in detail each of the scenarios. The discussion in each group is led by a facilitator, who would most likely be professional planner from one of the MMDAs represented. Another person records the proceedings of the consultation.
- In most cases, the group discussion starts with the facilitator going through the proposals of each of the options and starting with questions that are designed to let the participants think in terms of strengths and weaknesses or the costs and benefits of the various proposals. As the evaluation unfolds, different sets of questions and concerns could be raised and discussed among members of the group.
- At the end of the evaluation session, each group will decide the option/options of the plan proposals they would like to see implemented.
- The smaller groups converge for the final session where a representative from each group gives a brief discussion of their groups' assessment of the options, indicating what they identified as strengths and weaknesses and which option(s) they agreed on as to a group. The planning officials/consultants will make notes at this stage and collect the handwritten notes of each of the group's evaluation sessions.

From the author's experience, having been involved in the early plan-making and experimentation phases of the new spatial planning system, a major challenge at this stage of the consultative process is communicating the message behind the beautifully drawn diagrams and maps to the stakeholders. Technocrats outside of the planning domain even sometimes struggle to understand and appreciate the planning processes itself, the analytical tools employed and the technical jargons that get thrown at them at such meetings. Also, the planning team, in most cases, goes to the stakeholder meeting having a good idea which of the plan option(s) they wish to have approved and implemented at the end of the process. The tendency therefore exists for stakeholders to be led into agreeing with the planners' and legitimizing their preferred options without necessarily understanding the basis for their decisions. The scenario evaluation exercises are therefore held mainly to tick a box, and to demonstrate in various reports that the statutory requirements have been followed by the planning team. In such cases, the outcomes of stakeholder consultations are essentially the fulfilment of predetermined goals towards a single course of action.

The final stage of the consultative process is held to consider the draft final SDF. The planning team at this stage would have taken on-board all the views, suggestions and concerns raised at during the scenario evaluation workshops to generate the draft final plan. The draft final plan is presented to the stakeholders. Further questions, concerns and suggestions are welcomed at this stage, but often, they tend to be very minor and therefore do not require major alterations to the draft final plan. It is expected at this stage that the stakeholders will give their consent for the draft final plan to be accepted. From the author's experiences, all draft plans that made to the final stage were adopted by the stakeholders. While the stakeholders agree that the draft final plan is accepted, formal approval and adoption powers are vested in the technical committees of the respective levels of political administration.

In many aspects, the form and scope of stakeholder participation required under the new spatial planning system constitute an improvement over practices in the past. At the same time, translating normative principles of public participation into established political and professional practices has always been much more difficult to achieve (Wilson et al. 2017). Consequently, a number of challenges still remains with respect to enabling genuine forms of stakeholder participation in Ghana's spatial planning. Below, we identify some of the main challenges.

Firstly, the three stages of consultation instituted by the new spatial planning system appear to focus mainly on the plan formulation process. This suggests that participation ends at the stage where plans have been approved and adopted. Opportunities for stakeholders to participate in the development management process appear not to have been given much consideration per the statutory requirements of consultation outlined in the NSPS Module Guidelines and the Land Use and Spatial Planning Act (Act 925).

Secondly, both the Land Use and Spatial Planning Act (Act 925) and NSPS Module Guidelines make mention of the requirement for all MMDAs to set up a permanent Public Data Rooms, whether virtual or physical at an openly accessible place and opened to the public during normal working hours. MMDAs are also required to deposit detailed reports of all stakeholder consultation in their Public Data Rooms. Indeed, this facility could be one of the ways of enabling the wider public to get involved in the development process beyond the formulation of plans. To date, however, there is no evidence that a single Public Data Room has been established in any of the MMDAs across the country.

Thirdly, as we have established elsewhere, public participation under the new spatial planning system is still primarily by representation. Given that no formal channels exist to ensure that representatives bridge the gap between the communities and groups they are supposed to represent and decision-makers, it is difficult to establish whether wider community interests are being properly represented in the decision-making process. Thus, to a larger extent, the current practices of stakeholder engagement could be described as a formal, one-way process between community representatives and public-sector officials. This does not empower communities to initiate proposals and be co-producers of the places where they spend their lives, nor does it offer any real opportunities in formal and informal settings for individuals and stakeholder groups to properly engage with and influence the long-term planning, development and management of their communities.

Finally, as the discussion has shown, ensuring that stakeholders participate in the planning process is an expensive venture. The experimentation phases of the new planning system were largely funded by donor agencies including NORAD and the World Bank and executed by consultants with TCPD taking oversight responsibilities. It may be justified to provide allowances to some stakeholders, especially those from very remote areas to enable them attend. That said, the culture of paying to incentivize public and civil servants to attend stakeholder consultations ought to be reexamined given that this could be considered as aspects of their official duties for which they receive remuneration as government officials. The ability of the TCPD to sustainably fund the plan-making process including stakeholder consultations remains to be seen beyond the experimentation phases. The newly established Land Use and Spatial Planning Fund could provide some of the needed resources for this venture. However, it is early days yet for the fund to accumulate the much-needed financial resources to support spatail planning in Ghana.

8.4 Towards Effective Stakeholder Engagement: The Case for Technology-Mediated Participation Practices

The discussion in the forgoing sections has demonstrated that statutory requirements of public engagement and the accompanying professional practices have evolved over the years. The spatial planning system instituted by the 1945 Town and Country Planning Ordinance was largely undemocratic. The democratization process and the decentralization reforms that were introduced in the 1980s had virtually no impact on the scope and depth of stakeholder participation in contemporary spatial planning in Ghana. In the last decade, reforms in the spatial planning system have introduced new statutory requirements, which by intent is to offer stakeholders the opportunity to actively part take in the planning process. While recognition for active stakeholder participation is commendable in principles, in practice, the new era of active participation promised remains a policy rhetoric. The methods of participation being employed currently are still very much traditional, using the administrative structures of representative democracy instituted by the decentralization law. Also, the current focus of participation is essentially on the formulation of plans to the neglect of all the other important aspects of the place-making and development management process. The logistic, personal and financial resources for active, broad-based stakeholder participation raise concerns about the sustainability of this vital aspect of the spatial planning process going forward.

New and effective methods of ensuring that citizens continue to have a voice in decisions that affect them are needed. In recent years, technology is offering new and effective ways for citizens to become engaged in the planning system. Within the broader movement of smart cities, technology-mediated participatory urban planning platforms are being used to facilitate citizens understanding of planning issues and to raise awareness of the opportunities for them to influence planning policies at multiple scales (Wilson et al. 2017). Information distribution, transparency, co-creation of solutions to common problems and consensus building are some of the benefits that embracing technology-enabled participatory urban planning practices can offer (Bugs et al. 2010). For example, software applications that run on various electronic such as smartphones, tablets and computers can offer platforms for citizens to interact remotely with planning officials at any time, eliminating the space-time barriers associated with traditional methods of participation. Also, unnecessary bureaucracies and technocracies, created and entrenched by the ever-increasing complexity in organizational structures, could easily be avoided by citizens through technology-mediated stakeholder participation practices (Le Dantec et al. 2015; Wilson et al. 2017).

While the emerging practices of technology-mediated stakeholder participation are currently confined mainly to advanced democratic societies in Western Europe and North America, the potential exists in emerging economies, including Ghana that could be tapped to institutionalize a more effective and potentially cost-saving means of wider stakeholder participation. Indeed, globally, smartphones ownership and use are on the rise. Whereas an estimated 69% of adults in developed countries have adopted smartphones, in developing countries, the rate of adoption is 46% and is growing rapidly (Anthes 2016).

In Ghana, the adoption of smartphones and mobile phone subscriptions is increasing rapidly. The National Communications Authority (NCA) estimates that as of the second quarter of 2016, total mobile phone subscription nationwide was 36,613,987, representing a 13.1% increase in subscriptions in 2015 (NCA 2016). The estimated mobile phone penetration rate (i.e. the total number of mobile subscribers divided by the total population) over the same period was 131.9%, while Broadband Wireless Access penetration rate was 0.4% (NCA 2016).

What the above figures indicate is that increasingly, more people are adopting modern communication devices and Internet access, especially in urban areas is increasing. Indeed, the increased adoption of modern Information and Communication Technology (ICT) is directly responsible for the growing mobile banking sector of the country. The spatial planning system could benefit from the existing ICT infrastructure in not only the technical tasks of planning but also in facilitating unfettered citizen access to planning information at the metropolitan, district, city, town and neighbourhood scales. Investments in smartphone and computer-based applications would not only enable the larger mass of the public to interact directly with MMDAs and their planning officials but also, in the long term, reduce the huge financial resources required to implement the traditional methods of participation such as holding stakeholder's consultation meetings.

Moreover, technology-enabled participation methods could complement traditional methods to enable citizens to make their local authorities aware of place-based issues, especially those related to the development management process, thereby making social capital, a hitherto untapped potential, major asset to MMDAs in the plan-making and development management processes. These systems in their basic forms could evolve to become robust decision-support systems offering citizens user-friendly interfaces to interact with and provide feedback on, for example, various SDF or Local Plan scenarios that citizens wish to see implement. In addition, planning applications for rezoning, development and building permits could be made available on such online platforms, enhancing transparency in the ways decisions, especially those that would lead to changes in land use allocations, are reached.

8.5 Conclusions

In this chapter, the normative ideals and practical importance of stakeholder participation in the spatial planning process have been discussed drawing on theories of participation globally and the statutory requirements and practices in public participation in Ghana. Through the discussion, we identified that the statutory requirements and expected practices with respect to public participation have evolved since spatial planning was formally instituted in 1945. Despite major reforms being introduced, it has been argued that contemporary methods of public participation are very much traditional, relying mainly on elected representatives of communities and sometimes on opinion leaders and leaders of various groups. It was further highlighted that the system lacked the feedback mechanisms to ensure that views of the wider population are properly represented. Also, a major challenge identified with the traditional methods is the financial, logistic and human resources required to organize stakeholder meetings.

In view of the prevailing challenges, technology-mediated participation methods have been suggested to complement traditional methods in offering more citizens to actively engage in the spatial planning process. It is worth emphasizing that while modern technology has the potential to improve upon the current participatory processes, effective participation would depend on the commitment of planning practitioners to the democratic principles of inclusiveness, dialogue and consensus. Citizens could mobilize to demand for their voices to be heard, but it is important for practitioners to reflect on their duty as public servants to the populations they are supposed to serve, and on acknowledging the widening gap between the normative principles and ethics of their chosen professions on the one hand and prevailing practices on the other hand, seek new and innovative ways to expand the scope, depth and outcomes of public participation. For there cannot be successful planning stories to be told without a planning system that offers genuine, accessible and effective opportunities for citizen engagement. The future of spatial planning fundamentally depends on how informed the population are about planning issues.

References

Anthes E (2016) Mental health: there's an app for that. Nature 532: 20–23
Arnstein SR (1969) A ladder of citizen participation. J Am Inst Plan 35(4): 216–224
Bugs G, Granell C, Fonts O (2010) An assessment of public participation GIS and Web 2.0 technologies in urban planning practice in Canela, Brazil. Cities 27(3): 172–181
Chadwick G (1978) A system view of planning: towards a theory of the urban and regional planning process. Elsevier
Cornish F (2006) Empowerment to participate: a case study of participation by Indian sex workers in HIV prevention. J Community Appl Soc Psychol 16: 301e315
Cornwall A (2008) Unpacking 'participation': models, meanings and practices. Community Dev J 43(3): 269–283

References

Curry N (2012) Community participation in spatial planning: exploring relationships between professional and lay stakeholders. Local Gov Stud 38(3): 345–366

Dahl RA (1989) Democracy and its critics. Yale University Press, New Haven

Davidoff P (1965) Advocacy and pluralism in planning. J Am Inst plann 31(4): 331–338

Forester J (1999) The deliberative practitioner. The MIT Press, Cambridge

Friedmann J (1987) Planning in the public domain: from knowledge to action. Princeton University Press

Hall B (2005) In from the cold? Reflections on participatory research from 1970e2005. Convergence 38: 5e24

Healey P (2003) Collaborative planning in perspective. Plan Theory 2(2): 101–123

Hurlbert M, Gupta J (2015) The split ladder of participation: a diagnostic, strategic, and evaluation tool to assess when participation is necessary. Environ Sci Policy 50: 100–113

Innes J, Booher D (2004) Reframing public participation; strategies for the 21st century. Plan Theory Pract 5(4): 419–436

Land use and spatial planning Act (2016) Act 925 The nine hundred and twenty fifth act of the parliament of the Republic of Ghana

Le Dantec CA, Asad M, Misra A et al (2015) Planning with crowdsourced data: rhetoric and representation in transportation planning. In: Proceedings of the 18th ACM conference on computer supported cooperative work & social computing (CSCW '15). ACM, New York, NY, USA, pp 1717–1727

Local Government Act (1933) Act 462 the four hundred and sixty two act of the parliament of the Republic of Ghana

Ministry of Environment, Science, Technology and Innovation (MESTI) (2011) The new spatial planning model guidelines. Town and Country Planning Department. November 2011

Morphet J (2011) Effective practice in spatial planning. Oxon Routledge

National Communications Authority (NCA) (2016) Quarterly Statistical Bulletin on Communication in Ghana. Second Quarter Volume 1, Issue 2 https://www.nca.org.gh/assets/Uploads/Quaterly-statistics-03-11-16-fin.pdf

Putnam, Robert D (2000) Bowling alone: the collapse and revival of American Community. Simon and Schuster, New York

Pretty JN (1995) Participatory learning for sustainable agriculture. World Dev 23(8): 1247–1263

Throgmorton JA (2000) On the virtues of skillful meandering: acting as a skilled-voice-in-the-flow of persuasive argumentation. J Am Plann Assoc 66(4): 367–383

Town and Country Planning Act (1945) CAP 84. http://www.epa.gov.gh/ghanalex/acts/Acts/TOWN%20AND%20COUNTRY%20PLANNING%20ACT,1945.pdf

Wilson A, Tewdwr-Jones M, Comber R (2017) Urban planning, public participation and digital technology: app development as a method of generating citizen involvement in local planning processes. Environ Plan B Urb Anal City Sci 2399808317712515

Chapter 9
Urbanization and Settlement Growth Management

Abstract In the last seventy years, countries in the Global South, including Ghana, have experienced high levels of urbanization. Rapid urbanization manifests in two ways. Firstly, the phenomenon is evidenced by the large concentration of population in urban areas. Secondly, in spatial terms, urbanization results in the utilization of land for various uses including housing, infrastructure and economic activities, which can be monitored over time through land-cover transitions from non-built-up to built-up areas. Built-up area expansion trends, in turn, shape the physical size of cities and leave lasting impacts on the natural environment and livelihoods. The urbanization process therefore has serious implications for urban growth management and sustainable development. This chapter focuses on the interface between urbanization and spatial planning. It explores urbanization trends globally as well as trends in Ghana. Using remotely sensed data, this chapter presents historical analyses of spatio-temporal settlement expansions trends for major metropolitan regions in Ghana. The various tools and strategies that could be deployed through the spatial planning system to achieve sustainable growth management outcomes in urban areas in Ghana are also identified.

Keywords Urbanization · Peri-urbanization · Urban expansion
Growth management · Spatial planning · Ghana

9.1 Introduction

Urban areas emerge from several years of population concentrating in some defined location within a region or country. The processes by which relatively smaller settlements become urban include movement of people from surrounding rural areas to populate settlements that would later be defined as urban, as well as natural population increase resulting from births in urban areas. Over time, the increasing population and the attendant demand for land for various activities lead to the expansion of existing built-up areas into surrounding Greenfield areas. Several years of population growth and lateral expansion of existing urban areas, coupled with the growth of surrounding rural settlements, could lead to settlements of different sizes merging and thus widening the geographical extent of what may be defined as the limits of the urban. Thus, urbanization in very simple terms refers to the process, whereby settlements get bigger in terms of population size and physical extent relative to other settlements in a country or region.

Broadly speaking, people concentrate at specific locations for purposes of interaction. Interaction takes place in many different forms and between several actors. For example, by locating in areas with relatively bigger population, businesses get more people to patronize their goods and services. Over time, more businesses and population will be attracted to these locations, further generating additional benefits. New firms attracted to urban areas require labour with different skills. Similarly, the additional population will require jobs that match their skills. Thus, by locating in urban areas, job opportunities meet skills, creating additional benefits in economic growth. Given the large population size, it becomes economically viable to provide certain public amenities and services, which in turn could improve the well-being of the population. The amenities provided also serve as focal points of human interaction: people get to meet other people and build mutually beneficial networks. The effects are always not positive. There could be negative consequences with urbanization as evidenced by increased levels of crime, breakdown of social networks, housing shortages and the associated problems of homelessness and slum formations, unemployment and environmental deterioration in urban areas. Scholars have referred to the positive benefits of urbanization as *economies of scale* and the negative benefits as *diseconomies* (see, e.g., Duranton and Puga 2004; Glaeser 2010).

One of the main concerns of spatial planning is to formulate and implement interventions that can help optimize the benefits associated with urbanization while addressing the negative consequences associated with the growth of settlements. This goal is achieved through several urban growth management strategies. This chapter therefore focuses on the interface between urbanization and spatial planning. In doing so, we will explore urbanization trends at the global, continental, national and settlement scales. The chapter will rely on evidence in the form of natural population change trends from various sources. In order to appreciate the spatial manifestations of the urbanization process in Ghana, historical urban expansion trends, derived from satellite images, will be used to illustrate how the footprints of physical development in major metropolitan areas in Ghana have evolved in the last two to three decades. Various growth management strategies aimed at controlling physical expansion, attracting development at desirable locations and raising finance through the urban development process will also be identified. We will then identify examples of growth management strategies that are being adopted in Ghana and discuss the emerging challenges. Finally, we will explore the relevance and applicability of other of policy tools and strategies towards sustainable growth management outcomes in Ghana.

9.2 Global Urbanization Trends and the Processes of Urban Change

What constitutes the *urban* is a contested concept. Various countries have different ways of classifying urban areas from non-urban areas. In Ghana, for example, a locality with a minimum population of 5000 inhabitants is considered urban according to the official definition adopted by the Ghana Statistical Services. Despite the lack of a single, global definition of the urban in terms of population and physical size, the United Nations have used the official national census definitions of different countries to estimate how the proportion of population living in urban areas as defined by the individual countries have been changing over the years. The data shows considerable differences in the size of the world's urban agglomerations. It is estimated that nearly half of all urban dwellers reside in relatively small settlements of less than 500,000 inhabitants. While one in five of the urban population live in settlements of one million population or more, nearly one in eight reside in the world's 28 mega-cities of 10 million inhabitants or more (United Nations 2014).

In Fig. 9.1, the proportion of urban population globally and in Africa since 1950 has been shown based on the United Nations World Urbanization Prospects data compiled in 2014.

During the 1950s, nearly 30% of the global population lived in urban areas. In 2015, more than half (i.e. 54%) of the world's over 7.3 billion population lived in

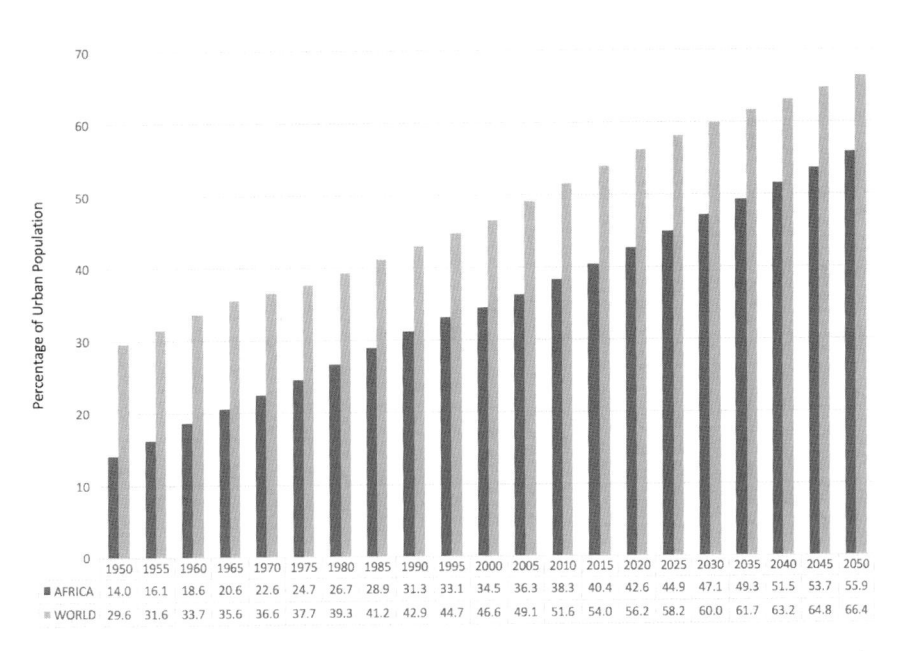

	1950	1955	1960	1965	1970	1975	1980	1985	1990	1995	2000	2005	2010	2015	2020	2025	2030	2035	2040	2045	2050
AFRICA	14.0	16.1	18.6	20.6	22.6	24.7	26.7	28.9	31.3	33.1	34.5	36.3	38.3	40.4	42.6	44.9	47.1	49.3	51.5	53.7	55.9
WORLD	29.6	31.6	33.7	35.6	36.6	37.7	39.3	41.2	42.9	44.7	46.6	49.1	51.6	54.0	56.2	58.2	60.0	61.7	63.2	64.8	66.4

Fig. 9.1 Percentage of the population living in urban areas (1950–2050). *Source* World urbanization prospects: based on world urbanization prospects data, United Nations (2014)

urban areas. By the 1950s, more than half of the populations in Europe (i.e. 51.5%), North America (i.e. 63.9%) and Oceania (i.e. 62.4%) were already living in urban areas (United Nations 2014). While the urban population in these areas have increased over the years, the highest levels of urbanization have occurred in the Global South. For example, in the 1950s Asia had only 17.5% of the total population living in urban areas. As of 2015, this had increased to 48.2%. In Latin America and the Caribbean, the urban population increased from 41.3 to nearly 80% between the 1950s and 2015. Across the African continent, similar patterns of rapid urbanization have also occurred. Whereas the African continent was the least urbanized in the 1950s, by 2015, the percentage of the urban population nearly tripled from 14 to 40.4% (see Fig. 9.1). It is estimated that by 2050, some 66% (i.e. 6.3 billion) of the world's over 9.5 billion population will be living in cities.

Although data on global population growth trends exists, data on built-up land cover is not available yet. Some attempts have been made to quantify built-up land changes in cities selected from different regions in the world. Angel and colleagues, in their 2016 edition of the Atlas of Urban Expansion, provide comprehensive data on spatio-temporal urban expansion trends between 1990 and 2014, using 30 m × 30 m resolution satellite data from a representative sample of 200 cities in the world. Their data covers all the mega-cities of the world as well as major urban areas in the selected countries. Also, the total urban population of the cities covered in their analysis represents approximately 20% of the global urban population. Thus, although the land-cover data does not cover all of over four thousand cities in the world, the cities represented in the sample provide an interesting and valuable trend analysis of the settlement expansion trends that could be observed globally.

Figure 9.2 shows a summary of the results of their work. The data shows that across the sample of 200 cities, total built-up land has been increasing over the 24-years period of analysis. In the 1990s, for example, total built-up land in the selected cities was over 4.7 million ha. Out of this, 78.63% were in areas defined as urban. By the year 2000, which also marked the peak of global urbanization levels, total built-up land in these cities had increased by 48.5% to 7 million ha, nearly 80% of which were in urban settlements. The quantum of built-up land further increased by 50.85% between 2000 and 2014, with 80.12% of the built-up land cover taken by urban settlements. On the average, an additional 3.42% of the existing built-up land in these settlements was developed every year. The average annual rate of built-up land expansions is also similar at 3.50%.

The growth of urban areas in terms of population and built-up land expansion has been fuelled by the demographic processes of natural population growth and rural–urban migration as well as the reclassification of previously rural settlements into urban centres (Potts 2012a, b; UN-Habitat 2010). From the point of view, urban agglomeration theory, the concentration of population and economic activities underpin the urbanization process and have historically been responsible for the emergence and growth of cities and large metropolitan regions (Jacobs 1969; Henderson 2002).

9.2 Global Urbanization Trends and the Processes of Urban Change

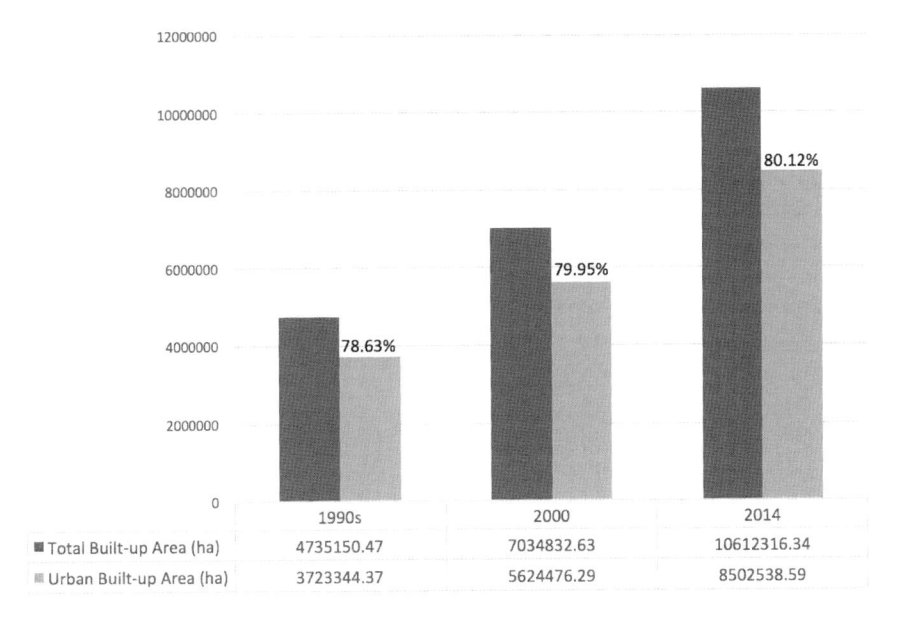

	1990s	2000	2014
■ Total Built-up Area (ha)	4735150.47	7034832.63	10612316.34
▨ Urban Built-up Area (ha)	3723344.37	5624476.29	8502538.59

Percentages shown on bars: 78.63% (1990s), 79.95% (2000), 80.12% (2014).

Fig. 9.2 Built-up area expansion for a globally representative sample of 200 cities (1990–2014). *Source* Based on data obtained from Atlas of urban expansion, 2016 edition (Atlas report can be downloaded at: http://www.lincolninst.edu/sites/default/files/pubfiles/atlas-of-urban-expansion-2016-volume-1-full.pdf. Raw data files were downloaded at website: http://www.atlasofurbanexpansion.org/data)

Urbanization generates external economies of scale, which not only enhance productivity and growth but reinforce the potential of existing cities as major attraction points for additional population and activities (Duranton and Puga 2004; Henderson 2002).

The urbanization process is also characterized by changing form and structure of settlements through the physical expansion of existing built-up areas into Greenfield land or surrounding rural settlements. The process of built-up land expansion in urban areas takes place in substantially different forms in different contexts. Consequently, different terminologies have emerged to characterize the nature of the expansion process. The term *sub-urbanization* has been coined to reflect one of the processes of urban change common in cities of developed countries. In broad terms, sub-urbanization is associated with low density, often fragmented and sprawling physical development on Greenfield land immediately surrounding existing built-up land of a city (Champion 2001; Pacione 2009).

In cities of the Global South, peri-urban development typifies the urbanization process. Like urbanization, the meaning of the term 'peri-urban', sometimes referred to as 'urban fridge', is contested, with no precise definition available in the literature. Two main definitional approaches have been adopted in conceptualizing peri-urban areas. The first approach conceptualizes peri-urban areas in terms of discrete spatial limits. Based on empirical observation, leading exponents of this approach suggest a distance of about 30–50 km, beyond the existing built-up land of major cities, as a reasonable generalization of the extent of the peri-urban zone (see, e.g., McGregor and Simon 2012; Webster 2002). Simon et al. (2004) estimate that the peri-urban zone of Kumasi, for example, stretches some 20–40 km radius around the city's main built-up area. The second definitional approach for peri-urban areas adopts an integrated and functional view by considering the urban–rural continuum. Based on this, the urban periphery is conceptualized as a transition zone between fully urbanized land in cities and areas in predominantly agricultural use (McGregor et al. 2011; Webster 2002). As transition zones between the urban and rural, peri-urban areas are characterised by mixed land uses some indeterminate inner and outer boundaries and are often split between many administrative areas (McGregor et al. 2011; Webster 2002).

Moreover, the physical expansion process observed through sub-urbanization and peri-urbanization may be characterized by different physical forms. In any given city or metropolitan region, urban expansion can be compact through the infill of existing open spaces in already built-up areas and/or redevelopment of built-up areas at higher densities (Angel et al. 2005; Wilson et al. 2003). Furthermore, urban expansion may occur either as contiguous extensions to existing built-up areas or spontaneous leapfrog development away from main built-up land, leaving swaths of undeveloped land that separate the new development from existing built-up areas (Torrens and Alberti 2000). This form of expansion normally occurs in linear direction along major road networks (i.e. ribbon development) and/or in radial direction around an already established built-up area (Sudhira et al. 2004).

While many positive benefits are derived from urbanization, it accumulated impacts also poses various challenges to urban growth management and sustainable development. The emerging challenges of climate change, environmental degradation and resource depletion resulting from decades of rapid urbanization (Watson 2009) pose serious threats to public health, the continuous supply of essential ecosystem services and food security (Eigenbrod et al. 2011; Baloye and Palamuleni 2015) at the city, national and global scales.

9.3 Urbanization Trends in Ghana

Like the rest of the world, especially countries in the Global South, Ghana's population has been increasing, although the rate of growth shows periods of rapid growth and periods of decline. With a population of nearly 5 million in the 1950s, the country has experienced a fivefold increase in population to nearly 27 million by 2015 (see Fig. 9.3). By 2050, it is estimated that Ghana's population will grow to about 45 million inhabitants.

Overall, the country witnessed the highest rates of population growth from 1955–1960 to 1960–1965 where the average annual increase was 3.16 and 2.95%, respectively. During the 1970s and early 1980s, however, the rate of population growth declined, peaking again in the mid-1980s, and falling steadily afterwards

9.3 Urbanization Trends in Ghana

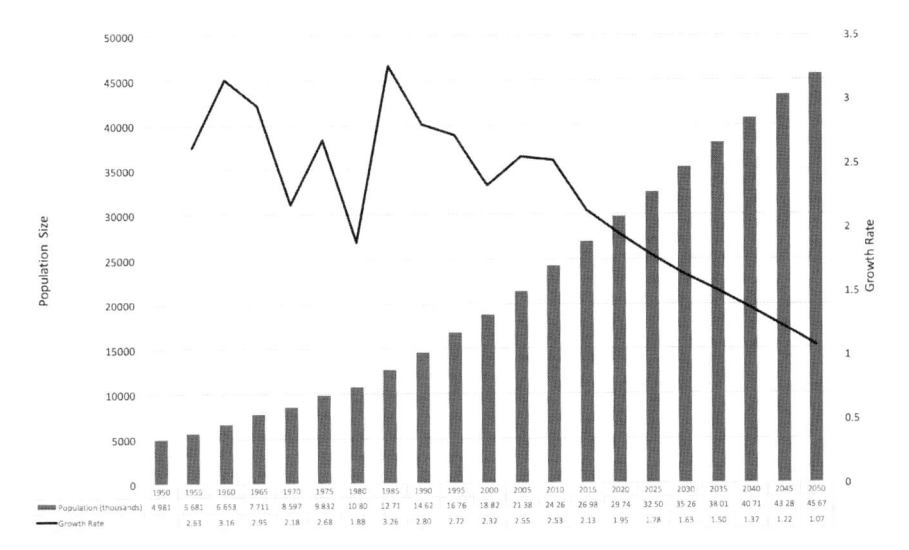

Fig. 9.3 National population growth trend (1950–2050). *Source* Based on world urbanization prospects data, United Nations (2014)

(see Fig. 9.3). The decline in rate of growth during the 1970s has been attributed mainly to fall in fertility levels as a result of increase in contraceptive prevalence in the country (Gaisie 2013; Blanc and Grey 2002). By the turn of the twentieth century, the rate of population growth had begun to slow down again to an annual rate of about 2.3%, and this is expected to continue as the data shows in Box 8.1. Thus, although in the coming years the population is expected to increase naturally, the rates of increase will be much slower than they have been in the past mainly because of factors affecting fertility rate, such as the effect on maternal age of increasing number of years spent in school, especially among women; increasing contraceptive prevalence; rising cost of living; and changing lifestyles among the increasingly urbanizing population.

9.3.1 Urban Population Growth

Across the African continent, Ghana is one of the countries that has experienced the fastest and highest levels of urbanization. Having only 15.4% of the population living in urban areas in the 1950s, by 2010, half of the country's population were living in areas classified as urban, compared to 41.6 and 38.3% of the population in West Africa and Africa (see Fig. 9.4). It is further estimated that by 2050, 70% of the country's over 45 million population will be living in cities. Although the proportion of urban dwellers on the African continent and in Ghana are expected to increase, there is emerging evidence that suggests that the pace of urbanization will

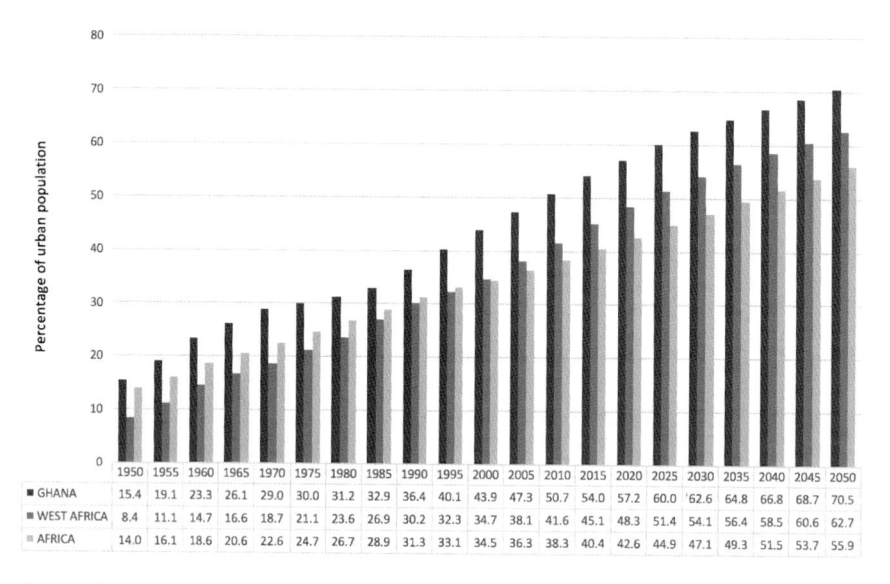

	1950	1955	1960	1965	1970	1975	1980	1985	1990	1995	2000	2005	2010	2015	2020	2025	2030	2035	2040	2045	2050
GHANA	15.4	19.1	23.3	26.1	29.0	30.0	31.2	32.9	36.4	40.1	43.9	47.3	50.7	54.0	57.2	60.0	62.6	64.8	66.8	68.7	70.5
WEST AFRICA	8.4	11.1	14.7	16.6	18.7	21.1	23.6	26.9	30.2	32.3	34.7	38.1	41.6	45.1	48.3	51.4	54.1	56.4	58.5	60.6	62.7
AFRICA	14.0	16.1	18.6	20.6	22.6	24.7	26.7	28.9	31.3	33.1	34.5	36.3	38.3	40.4	42.6	44.9	47.1	49.3	51.5	53.7	55.9

Fig. 9.4 Percentage of urban population in Ghana, West Africa and the African Continent (1950–2050). *Source* Based on world urbanization prospects data, United Nations (2014)

slow down (Potts 2012a, b). Indeed, the United Nations' prognosis shows the pace of urbanization—the annual rate at which the proportion of the population living in cities increases has been falling since the 2000s when levels of urbanization hit record high globally. Thus, in the next three to four decades, urban population growth in Africa, including Ghana, will occur below an average annual rate of one per cent (see Fig. 9.5).

In addition to the continental, sub-continental and national portraits of urban population growth trends discussed above, we look at population growth dynamics within Ghana, at the urban scale. Table 9.1 presents a summary of the population growth trend in the biggest cities in Ghana based on the available census data. The data shows that over the years, the population in the major urban agglomerations including Accra, Kumasi, Sekondi-Takoradi and Tamale have been increasing rapidly at rates higher than the national average. For example, between 2000 and 2010, the population of Accra grew at an average rate of 2.24% per annum. Kumasi, the second largest metropolis in the country, and the Sekondi-Takoradi, the administrative capital of the Western Region, experienced phenomenal annual rate of growth of 5.59 and 6.26%, respectively. While historically north-south migration flows have contributed significantly to urbanization levels Accra and Kumasi, in the case of Sekondi-Takoradi, the emerging oil and gas economy has been one of the major contributors of the rapid population growth.

9.3 Urbanization Trends in Ghana

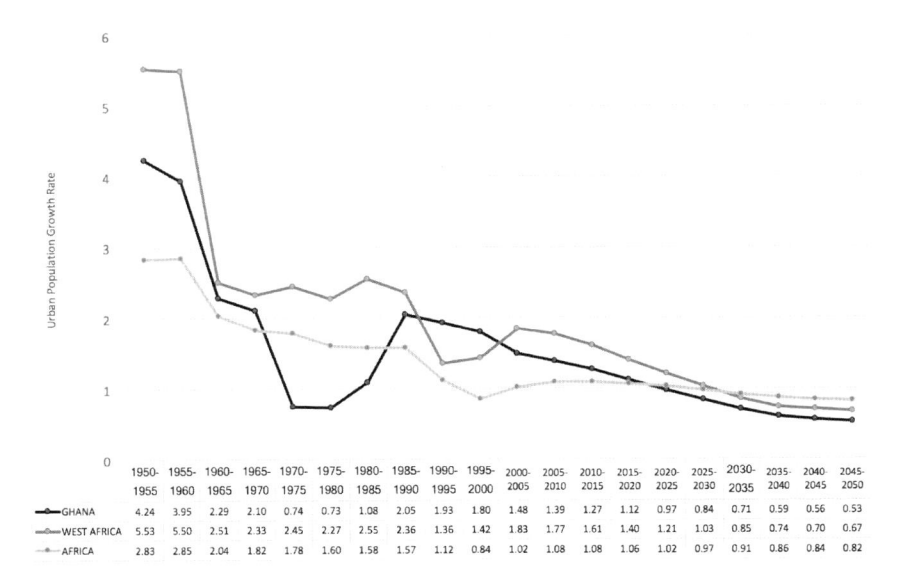

Fig. 9.5 Urban population growth rate in Ghana, West Africa and the African Continent (1950–2050). *Source* Based on world urbanization prospects data, United Nations (2014)

	1950– 1955	1955– 1960	1960– 1965	1965– 1970	1970– 1975	1975– 1980	1980– 1985	1985– 1990	1990– 1995	1995– 2000	2000– 2005	2005– 2010	2010– 2015	2015– 2020	2020– 2025	2025– 2030	2030– 2035	2035– 2040	2040– 2045	2045– 2050
GHANA	4.24	3.95	2.29	2.10	0.74	0.73	1.08	2.05	1.93	1.80	1.48	1.39	1.27	1.12	0.97	0.84	0.71	0.59	0.56	0.53
WEST AFRICA	5.53	5.50	2.51	2.33	2.45	2.27	2.55	2.36	1.36	1.42	1.83	1.77	1.61	1.40	1.21	1.03	0.85	0.74	0.70	0.67
AFRICA	2.83	2.85	2.04	1.82	1.78	1.60	1.58	1.57	1.12	0.84	1.02	1.08	1.08	1.06	1.02	0.97	0.91	0.86	0.84	0.82

Table 9.1 Historical population growth in major towns and cities in Ghana

	Population size				Growth rate		
	1970	1984	2000	2010	1970–1984	1984–2000	2000–2010
Accra metropolis	624,091	969,195	1,658,937	1795115	3.19	3.42	0.79
Kumasi metropolis	346,336	496,628	1,170,270	2,035,064	2.61	5.50	5.69
Tamale	83,653	135,952	202,317	371,351	3.53	2.52	6.26
Sekondi-Takoradi	143,982	188,203	289,593	539,548	1.93	2.73	6.42
Tema	60,767	100,052	141,479	161,612	3.63	2.19	1.34
Cape Coast	51,653	57,224	118,105	169,894	1.08	1.41	3.70
Koforidua	46,235	58,731	87,315	120,971	1.72	2.51	3.31
Obuasi	31,005	60,617	115,564	143,644	4.91	4.12	2.20
Winneba	30,778	27,105	40,017	57,015	−0.90	2.46	3.60

Source Based on historical census data provided by the Ghana Statistical Services

9.3.2 Spatio-temporal Settlement Expansion Trends

For several years, settlement-level population distribution derived from national census data has provided the single most important source for measuring urbanization levels in Ghana. However, population growth constitutes one of the dimensions of the urbanization process that could be monitored. In spatial terms, population growth drives land use distributions observed in settlements of different sizes and as such monitoring the physical expansion of towns and cities in addition to their population growth constitutes one of the ways of understanding and quantifying urbanization levels. However, data through conventional surveying and mapping techniques has been either unavailable, limited in scope, and expensive or time consuming to acquire to allow for accurate analysis of historical urban expansion at different spatial scales.

Advances in Geo-Information Science, especially in the areas of remote sensing and geo-computation, are offering new opportunities to obtain land-cover data at appropriate spatial and temporal resolution based on which urban expansion could be monitored and quantified. In this section, remotely sensed data derived from Landsat Satellite Thematic Mapper and processed will be used to demonstrate settlement expansion trends at the national level and for selected urban settlements. What the Landsat Satellite data helps us to do is to go back in time to trace how the physical size of settlements has been changing up to the current period, to quantify extent and pace of the expansion and on the basis of that information design and implement interventions through the spatial planning system to manage growth and to avert unsustainable urban expansion.

Table 9.2 provides information about changes in various land-cover classes at the national level observed between 1975 and 2013. The settlement class, at the national level, represents the total built-up land—the amount of land covered by human activities including roads, and buildings used for residential, commercial and industrial purposes and their immediate surroundings, as well as other built-up lands. The country's rapid population growth, especially in its major cities, is reflected in the pace of built-up land increase of the 38-year period covered. While 1460 km^2 of the total land area of Ghana (i.e. 0.61%) had been built in 1975, this increased to 3836 km^2 (i.e. 1.60%) in 2013 at a rate of 2.57% per annum. In the 25-year period between 1975 and 2000, built-up land increased at an average annual rate of 2.27%, while over the 13-year period 2000 and 2013 built-up land increased at an average annual rate of 3.15%. Thus, over nearly four decades, the total area of Ghana covered by settlements, including cities, towns and villages, has increased by approximately by 161%.

The data further shows that over the years, agricultural lands, which constitutes one of the direct impacts of population growth, have also been increasing. On the contrary, land classified as vegetation has seen substantial decreases of the years. While in 1975 nearly 75% of the total land area of Ghana was under natural vegetation, by 2000, it had decreased to by nearly 10% points to 65.17%, further reducing to 59% in 2013. Thus, the cumulative impacts of human activities through agriculture and physical development in towns and cities can be seen in the rate at which natural vegetation cover has been decreasing over the years.

In the sections that follow, we examine built-up land expansion trends for selected metropolises and their surrounding areas in Ghana. The data presented here was obtained from Landsat Satellite images, which had a spatial resolution of 30 m. The images were processed in GIS software to extract the built-up land for the areas of interest. The built-up land data is presented using the formal administrative boundaries of districts as they existed as of the time of writing this book. Interpreting

9.3 Urbanization Trends in Ghana

Table 9.2 Land-cover changes in Ghana between 1975 and 2013

	1975 Area (km^2)	1975 % of total area	2000 Area (km^2)	2000 % of total area	2013 Area (km^2)	2013 % of total area
Agriculture	31,680	13.25	67,648	28.30	78,780	32.96
Settlements (built-up)	1468	0.61	2564	1.07	3836	1.60
Water bodies	8372	3.50	8000	3.35	8508	3.56
Vegetation	177,800	74.38	155,780	65.17	141,040	59.00
Wetland– floodplain	4044	1.69	4516	1.89	6276	2.63
Others (open mine, rocky land, sandy areas and bare soil)	156	0.07	300	0.13	592	0.25
Cloud	15,512	6.49	224	0.09	–	0.00
Total	239,032	100	239,032	100	239,032	100

Source Based on Comité Permanent Inter-états de Lutte contre la Sécheresse dans le Sahel CILSS (2016). *Landscapes of West Africa—A Window on a Changing World*. U.S. Geological Survey EROS, 47914 252nd St, Garretson, SD 57030, United States

the data within the limits of administrative boundaries, however, obscures the nature and extent of the urban expansion underway in these areas. One of the major reasons for this is that, in recent years, reorganization of the local government system has led to the creation of new districts from existing ones, fragmenting hitherto large contiguous areas into small spatial units for purposes of political administration. This is particularly so in the cases of the Greater Accra Region where what used to be the boundary Accra Metropolitan Area has changed in recent years because of the creation of new administrative districts. It will therefore be useful in the context of analysing urban expansion trends to collapse some of the districts into single contagious spatial units in order to appreciate the true nature of the physical expansion process underway in these areas.

9.3.2.1 Urban Expansion Trends in Accra and Surrounding Districts (1984–2015)

The analysis of urban expansion trends covers an area of 886 km^2 comprising the Accra Metropolitan Area (AMA) and nine neighbouring districts in the Greater Accra Region. Built-up land-cover change in the area was observed for the 30-year period between 1984 and 2015. In total, 10 out of the 16 administrative districts in the Greater Accra Region as at 2012 represented this analysis, which together represents about 27.30% of the region's total land area. In terms of population size, there are estimated 3,345,622 people living in these 10 districts, representing 78% of the population of the Greater Accra Region according to the 2010 population and housing census.

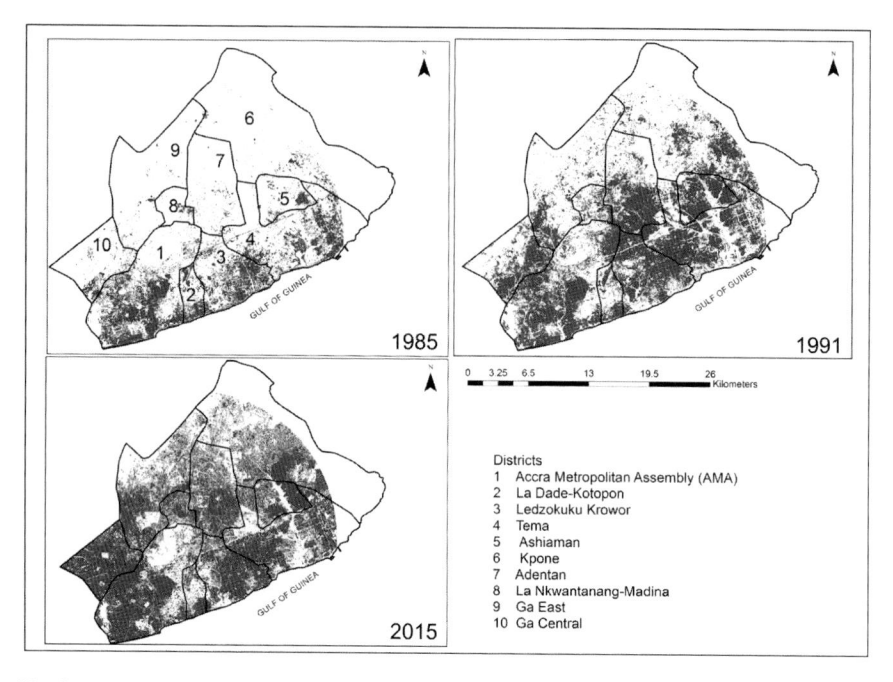

Fig. 9.6 Spatio-temporal built-up land-cover changes in the Accra metropolis and surrounding areas

Built-up land changes for 1985, 1991 and 2015 are shown in Fig. 9.6. A breakdown of urban expansion trends at the level of the individual administrative districts is presented in Table 9.3. The analysis shows that overall, the highest rate of urban expansion in the study area occurred over the seven-year period between 1984 and 1991, where the size of the built-up land more than doubled from 132.69 to 302.223 km^2 at an average annual rate of 14.70%. Although the quantum of built-up land increased by 33% between 1991 and 2015, the rate of urban expansion of 1.98% per annum over the 23-year period was significantly lower than what was experienced in the first seven-year period. Between 1985 and 2015, the size of built-up land in the area covered by the analysis increased by 45.38% at an average annual rate of nearly four per cent.

As has been explained previously, given the changes that have occurred over the years with respect to administrative boundaries of local government areas in the Greater Accra Region, presenting the analysis of spatio-temporal urban expansion trends in this region at the level of administrative boundaries obscures the nature and pace of the phenomenon. In view of this, the study area is further aggregated into three broad contiguous zones as follows:

- The coastal districts, comprising AMA, Tema, Ledzokuku-Krowo, Ashaiman and La Dade-Kotopon. These areas have been origins of historical urban expansion in the Greater Accra Region. Initially, AMA, Ledzokuku-Krowo and La Dade-

9.3 Urbanization Trends in Ghana

Table 9.3 Built-up land changes in Accra and surround districts between 1984 and 2015

Districts	Total area (km²)	Built-up land (km²)			Annual urban expansion rate (%)		
		1985	1991	2015	1985–1991	1991–2015	1985–2015
AMA	140.986	51.961	73.948	84.612	6.058	0.563	1.639
Tema	126.653	30.569	60.109	68.267	11.929	0.532	2.714
Ledzokuku-Krowo	64.214	25.243	41.783	44.828	8.762	0.294	1.933
Ashaiman	32.484	4.400	21.386	22.017	30.151	0.121	5.514
La Dade-Kotopon	18.524	7.003	7.605	8.795	1.384	0.608	0.762
Adenta	74.970	2.142	34.292	45.888	58.756	1.221	10.755
Madina	17.972	1.238	6.171	12.088	30.699	2.841	7.892
Kpone	237.099	4.329	33.259	57.574	40.471	2.313	9.009
Ga East	116.841	1.880	18.906	52.823	46.918	4.374	11.761
GA Central	56.308	3.925	4.764	5.279	3.281	0.429	0.993
Total	886.051	132.690	302.223	402.171	14.705	1.198	3.765

Kotopon districts formed one large contiguous area to the west, separated from Tema and Ashaiman to the east by large swathes of rural open space. Currently, because of outward expansion, the built-up land in these areas has essentially merged together to form one large contiguous zone. Thus, although they are individually administered as separate local government areas, in spatial and functional terms, the administrative divisions are only artificial and almost imperceptible on the ground. This zone covers approximately 382.861 km², representing 43% of the total extent of the area under consideration. An estimated 2,745,952 people lived in this zone according to official census figures in 2010, representing 82 and 64% of the population of the districts represented in this study and the Greater Accra Region, respectively.

- The suburban areas to the north comprising major settlements in the Adenta and Madina districts. As the summary results in Table 9.3 show, towns in these areas have been expanding very rapidly over the last 30 years. Together, the suburban districts cover an estimated area of 92.942 km², representing 11% of the total land area under analysis. An estimated 201,504 people lived in this zone according to official census figures in 2010, accounting for 6 and 5% of the population of the districts represented in this study and the Greater Accra Region, respectively.
- The peripheral districts comprising Kpone, Ga East and Ga Central. These districts cover an estimated area of 410.248 km², representing 46% of the total land area under consideration. An estimated 398,166 people lived in this zone according to official census figures in 2010, accounting for 12 and 9% of the population of the districts represented in this study and the Greater Accra Region, respectively.

Built-up land expansion trends aggregated at the level of the three contiguous zones delineated above are summarized in Table 9.4. The analysis shows that the coastal districts have historically attracted a significant proportion of physical devel-

Table 9.4 Aggregated built-up land changes in Accra and surrounded districts between 1984 and 2015

Zones	Total area (km^2)	Built-up land (km^2)			Annual urban expansion rate (%)		
		1985	1991	2015	1985–1991	1991–2015	1985–2015
Coastal districts	382.861	119.176	204.831	228.519	9.446	0.457	2.194
Suburban districts	92.942	3.38	40.463	57.976	51.248	1.510	9.937
Peripheral districts	410.248	10.134	56.929	115.676	33.329	2.998	8.455
Total	886.051	132.690	302.223	402.171	14.705	1.198	3.765

opment in the study area. Out of the total built-up land of 402.171 km^2 as of 2015, more than half (i.e. 56.82%) occurred in this zone. Within this zone, the size of the built-up area increased from 31% in 1985 to 53 and 57% in 1991 and 2015, respectively. While the rate of increase was higher at around 9.5% per annum between 1985 and 1991, between 1991 and 2015, urban expansion rate slowed down to nearly 0.5%.

The extent and pace of built-up land expansion in the suburban and peripheral districts demonstrate unfettered outward expansion of settlements in these zones. With only 3.38 km^2 of built-up land in the suburban districts in 1985, accounting for only 2.5% of total built-up land in the study area, by 2015, this had increased to 57.97 km^2, representing 14.4% of built-up land in the study area. Also, within suburban districts, only 3.64% of the land was built up in 1985 compared to 43.45 and 62.27% in 1991 and 2015, respectively. A similar trend can be observed in the peripheral districts where in 1985 their share of built-up land in the entire study area was 7.6% (i.e. 10.134 km^2) but increased to 28.7% (i.e. 115.676 km^2) in 2015. Within this peripheral zone, only 2.47% of the total land area was built up in 1985. This increased to 13.87 and 28.19% in 1991 and 2015, respectively. The analysis further shows that the fastest rates of urban expansion between 1985 and 1991 occured in the suburban districts at a rate of 51.2% per annum and in the peripheral districts at a rate of 33.32% per annum.

Summarizing, the analysis shows that the rapid increase in urban population has reflected spatially in the footprints of urban development. The spatio-temporal trend in urban expansion in Accra and the surrounding districts presented in this analysis shows that the entire area experienced the fastest rate of built-up land expansion between 1985 and 1991. While the largest proportion of physical developed that occurred in this period happened in the historical-core coastal districts, the fastest rates of outward urban expansion were actually underway in the five outlying districts categorized broadly in this analysis as the suburban districts and peripheral districts. In recent decades, the rate of urban expansion appears to have stalled in the coastal districts which marked the origins of historical urban growth while it continues rapidly in the outlying districts. The slow rate of urban expansion in this zone could be attributed to ongoing densification through redevelopment, especially in the central areas of Accra, the capital. This implies that while years of outward expansion have resulted in Greenfield land becoming limited, new developments attracted to these central areas are being accommodated on previously built-up land through redevelopment. Notwithstanding this, the overall trend shows years of unfettered outward expansion, and without any interventions this is expected to continue over the coming decades.

9.3.2.2 Urban Expansion Trends in the Greater Kumasi Sub-region (1986–2014)

In Chap. 4, where the focus was on regional spatial development planning, the Greater Kumasi sub-region (GKRS) was first introduced. We noted, among other things, that the sub-region was designated around 2010 as part of the experimentation with regional spatial planning, which hitherto was unexplored in the history of planning in Ghana. GKSR covers an estimated area of 2850 km^2 of urban, peri-urban and rural land in the Ashanti Region, one of the 10 administrative regions in Ghana. As of 2010, the GKSR had a total population of 2,764,091, representing about 58% of the total population of the Ashanti Region and 11% of Ghana's population.

In this section, an analysis of urban expansion trends in the GKSR, comprising the Kumasi metropolis, the second largest urban agglomeration in Ghana and seven surrounding districts, is presented. The analysis covers a period of 28 years from 1986 to 2014. As was done in the case of the Accra metropolis and the surrounding areas, the quantum and rate of built-up land change will be presented at the level of the eight administrative units that form the sub-region. The dynamics of urban expansion is examined further by dividing the sub-region into broad spatial units, namely;

- The sub-regional core, which comprises the Kumasi Metropolitan Area (KMA) and the Asokore Mampong Municipality. Until June 2012 when the Asokore Mampong Municipality became a local government authority under the Legislative Instrument L.I 2112, it formed part of the KMA. The sub-regional core covers approximately 234.342 km^2 (see Table 9.5). About two million people live in this zone of the sub-region, with an estimated 304,815 people, representing 15% of this total, living in the newly created Asokore Mampong Municipality. The sub-regional core is the most populous area within the sub-region, accounting for about 74% of GKSR's total population (see Table 9.6). Functionally, it is the main centre of economic activity and hosts the sub-region's Central Business District (CBD), which is located in the KMA.
- The peripheral districts, comprising the five administrative districts immediately surrounding the sub-regional core. This broad category does not in any way imply that urban expansion trends in these peripheral districts are uniform. Instead, as the analysis will show, taken individually, these districts exhibit different dynamics in terms of the quantum and pace of urban expansion that have occurred within them over the past 28 years. They are contrasted with the sub-regional core zone because

Table 9.5 Built-up land changes in Greater Kumasi sub-region between 1986 and 2014

District/ sub-region	Total area (km²)	Built-up land (km²) 1986	Built-up land (km²) 2001	Built-up land (km²) 2014	Annual urban expansion rate (%) 1986–2001	Annual urban expansion rate (%) 2001–2014	Annual urban expansion rate (%) 1986–2014
KMA	212.093	51.756	108.007	177.501	5.027	3.895	4.500
Asokore Mampong	22.249	5.903	9.033	15.807	2.877	4.398	3.580
Atwima Nwabiagya	596.979	6.746	9.768	41.822	2.499	11.837	6.733
Ejisu-Juaben	723.216	6.611	12.423	38.192	4.295	9.023	6.464
Kwabre East	134.822	4.289	10.154	32.752	5.913	9.426	7.530
Atwima Kwanwoma	290.721	3.862	9.213	30.305	5.968	9.591	7.635
Afigya-Kwabre	517.277	5.411	10.172	41.701	4.298	11.464	7.566
Bosomtwe	352.575	3.392	7.728	22.436	5.642	8.544	6.980
Greater Kumasi sub-region	2849.933	87.970	176.499	400.516	4.752	6.506	5.563

Table 9.6 Population distribution in the Greater Kumasi sub-region

Districts	Population size 1984	Population size 2000	Population size 2010	Annual population growth rate (%) 1984–2000	Annual population growth rate (%) 2000–2010
Atwima Nwabiagya	56,352	127,809	149,025	5.25	1.55
Ejisu-Juaben	78,783	124,176	143,762	2.88	1.48
Kwabre East	42,044	101,100	115,556	5.64	1.35
Atwima Kwanwoma	44,437	79,240	90,634	3.68	1.35
Afigya-Kwabre	39,971	89,358	136,140	5.16	4.30
KMA and Asokore Mampong	487,504	1,170,270	2,035064	5.63	5.69
Bosomtwe	41,283	66,788	93,910	3.05	3.47
Greater Kumasi sub-region	734,022	1,758,741	2,764091	5.13	4.62
Ashanti region	2,090,100	3,612,950	4,780380	3.48	2.84
Ghana	12,296,081	18,912,079	24,658,823	2.73	2.69

9.3 Urbanization Trends in Ghana

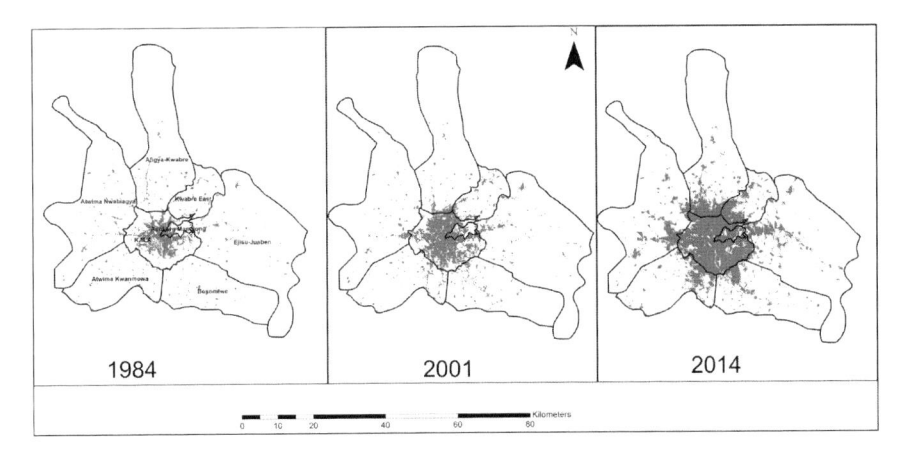

Fig. 9.7 Spatio-temporal built-up land-cover changes in the Greater Kumasi sub-region

while they were the least urbanized prior to the beginning of the twenty-first century, these peripheral zones, as the analysis will show, have become hotspots of urban physical development and expansion in the sub-region. The nuances in built-up land change for these districts can be observed by looking at the summary statistics presented in Table 9.6.

Urban expansion trends in the GKRS are mapped in Fig. 9.7. Table 9.6 presents a summary of built-up land changes, showing the quantum of built-up land change and the average annual rate of change for the 28-year period of analysis and for the first 15 years between 1986 and 2001 and the last 13 years between 2001 and 2014. In addition, Table 9.5 shows the estimated population sizes of the districts in the sub-region based on official census figures between 1984 and 2010 when the last census was conducted. As can be seen, the censal periods do not exactly match the years of the satellite images from which the built-up land-cover data was extracted for the analysis. The main reason for this difference is that often, the need for quality spatial data dictates selection of satellite images which are processed for this kind of analysis. This means that suitable land-cover data sets do not always match existing population figures, made available through decadal national population and housing census in Ghana. Notwithstanding this and given that population growth is one of the major drivers of urban physical development, population sizes for the districts, in the nearest time periods to the satellite data used for the land-cover analysis, have been presented with the aim to provide the necessary context for interpreting the urban expansion trends data for the sub-region.

It can be observed that built-up land in the GKSR has been increasing rapidly at an average annual rate of 5.6% over the 28-year period of analysis. Whereas in 1986 only three per cent (i.e. 87.970 km^2) of the sub-region's total land area was built up, by 2014, this had increased to 14% (i.e. 400.516 km^2). Over the 26-year period between 1984 and 2010—the closest years to the period of analysis for which reliable population census figures are available—the sub-region's population also increased rapidly at an average annual rate of 4.39%. Breaking the analysis down into the first 15 years (i.e. 1986–2001) and the last 13 years (2001–2014) shows that the annual rate of urban expansion in the sub-region was faster in the latter period (i.e. 6.51%) than the former (4.80%) and the entire 28-year period of analysis (see Table 9.5).

Moreover, at the scale of the two broad zones identified, it was observed that the sub-regional core, comprising the KMA and Asokore Mampong, experienced nearly a three-and-a-half-fold increase in built-up land between 1986 and 2014. Specifically, in 1986, only 57.659 km^2 (i.e. 25%) of total land in this zone was built up. By 2001 and 2014, total built-up land in the sub-regional core districts, relative to its total land area, had increased to 50 and 82%, respectively.

Much of the physical development that has happened in the sub-regional core has been accommodated in the area now defined as KMA, which excludes the newly created Asokore Mampong municipality. Indeed, out of the total built-up land of 57.659 km^2 which occurred in 1986 for the two districts combined, only 10% was accommodated in the now separate Asokore Mampong Municipality. Similarly, in both 2001 and 2014, only 8% of total built-up land in the sub-regional core occurred in Asokore Mampong, while the remaining 92% occurred in the KMA. What this suggests is that, although Asokore Mampong formed part of the KMA, its peripheral location meant that it urbanized at a relatively slower pace compared to the core areas of the metropolis. The larger share of its land would therefore have been undeveloped as one would expect in the early periods between 1986 and 2001.

The analysis further shows that the KMA, with annual urban expansion rate of 5.03%, recorded one of the highest rates of urban expansion between 1986 and 2001 (see Table 9.5). Between 2001 and 2014, however, the KMA recorded one of the lowest rates of urban expansion at an average annual rate of 3.89%. In addition, over the entire 28 years, KMA expanded its built-up land at a rate 4.50% per annum, the lowest rate recorded among all the districts except Asokore Mampong. Despite the diminishing rate of urban expansion relative to the other districts, the KMA in absolute terms was the most dominant in terms of the distribution of built-up land in the sub-region. What this suggests is that, although the quantum of built-up land increase in the metropolitan core of the sub-region was the biggest over the 28-year period, the rate and intensity of increase, observed between 2001 and 2014, were relatively slower compared to the remaining seven districts.

One possible explanation for the observed trend of urban expansion in the KMA relative to other districts in the sub-region is that, marking the historical origins of urban growth, the KMA initially attracted a significant share of all developments in the sub-region as evidenced by the relatively higher expansion rate between 1986 and 2001(see Acheampong et al. 2017; Oduro et al. 2014). Over time, as most of the land becomes built-up, and some of the new development occurs in previously built-up areas through redevelopment and infilling, lateral expansion would slow down, resulting in the observed falling trend in the rate of built-up land increase during the last 13 years. This is evidenced by trends in Asokore Mampong, for example, where between 2001 and 2014, however, the rate of expansion is increased to 4.39%, implying that while urban expansion in the KMA has stagnated over the last 13 years, Asokore Mampong as one of the peripheral districts has been attracting a significant

9.3 Urbanization Trends in Ghana

share of new physical development in the sub-region. Indeed, the size of the built-up land of Asokore Mampong increased from 5.903 km^2 in 1986—26.5% of its total land area—to 15.807 km^2 in 2014 representing 71.04% of its total land area.

Furthermore, it can be observed that all the remaining six peripheral districts recorded average annual expansion rates which were higher than GKSR average of 5.56%. Breaking the results down to the two broad years of analysis reveals a more nuanced trend of urban expansion among these peripheral districts. Over the first decade and half, three out of the six peripheral districts, namely Atwima Kwanwoma, Kwabre East and Bosomtwe, expanded rapidly than the sub-region at rates of 5.96, 5.9 and 5.64%, respectively. During the same period, Atwima Nwabiagya recorded the lowest rate of expansion at 2.49%. In the last 13 years, however, the dynamics changed considerably. Notably, Atwima Nwabiagya—which recorded the lowest expansion rate over the first period—emerged as the fastest urbanizing district in the sub-region with annual expansion rate of 11.84%, which was accompanied by a sixfold increase in its built-up land from 6.746 to 41.822 km^2. Similarly, Afigya-Kwabre—after expanding at a rate below that of the sub-region over the initial 15-year interval—cropped up as the second fastest growing district during the period between 2001 and 2014 with annual expansion rate of 11.46%. The accelerated rate of urban expansion, particularly over the last 13 years observed among the peripheral districts as compared to the generally slowed pace of expansion in the sub-regional core, shows that urban expansion in recent years has spilled over from the latter into the former.

Summarizing, in terms of the quantum of built-up land, the core districts of the sub-region, comprising KMA and Asokore Mampong, accounted for more than half of the total built-up land increase. As the metropolitan core where the CBD of the sub-region is also located, KMA has over the years attracted population and activities which explains its rapid expansion. The analysis, however, points to a general trend where urban expansion has since 2001 stalled in the sub-regional core but accelerated in the peripheral districts. This reinforces the fact that the GKSR is currently undergoing rapid peri-urbanization, with the highest intensity and fastest rate of urban expansion occurring in northern and north-eastern directions. The Atwima Nwabiagya and Afigya-Kwabre districts in particular have become the major hotspots of urban expansion in the sub-region.

9.3.2.3 Urban Expansion Trends in the Sekondi-Takoradi Metropolis (1990–2016)

Besides Accra and Kumasi, the Sekondi-Takoradi Metropolis (STM) is one of the fastest growing areas in Ghana. STM has an estimated population of 559,548 representing 23.5% of the population in the Western Region of Ghana. It covers a total area of approximately 191.663 km^2. By virtue of its role as the administrative capital of the Western Region, STM has over the years attracted population and economic activities, which in turn have transformed its spatial structure. In recent years, as a result of the discovery and exploitation of oil in commercial quantities in the West-

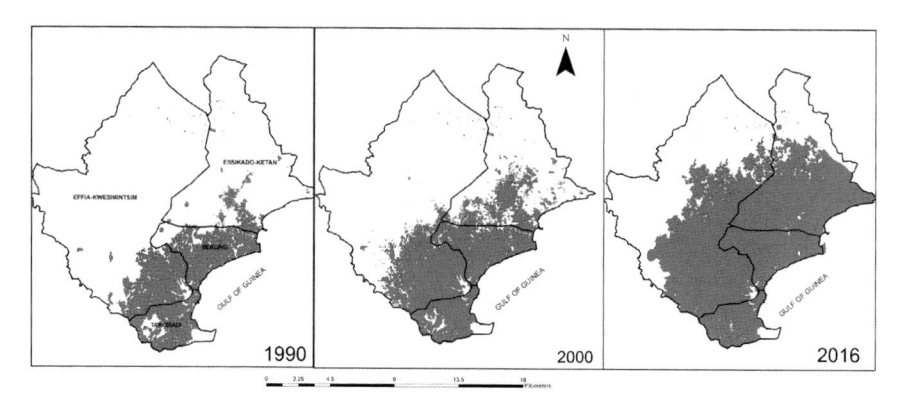

Fig. 9.8 Spatio-temporal built-up land-cover changes in the Sekondi-Takoradi Metropolis

Table 9.7 Built-up land changes in Sekondi-Takoradi between 1990 and 2016

Sub-metro	Built-up land (km^2)				Annual urban expansion rate (%)		
	Total area	1990	2000	2016	1990–2000	2000–2016	1990–2016
Takoradi	11.846	8.157	9.322	10.244	1.345	0.591	0.880
Sekondi	15.947	10.830	13.766	15.440	2.428	0.720	1.373
Essikado-Ketan	54.079	3.242	9.611	37.603	11.478	8.900	9.885
Effia-Kwesimintsim	109.791	12.001	21.338	48.678	5.924	5.290	5.533
STM	191.663	34.230	54.037	111.966	4.853	5.934	5.518

ern Region, STM being the most urbanized area in the Western part of Ghana has become one of the focal points of large-scale land acquisitions either for development of offshore infrastructure and economic activities related to the emerging oil and gas industry. Acquisition of land for speculative purposes is also evident in the metropolis and surrounding areas, with activities in the oil and gas industry being the main driver. Against this backdrop, nature and extent of urban expansion in the metropolis are examined over a period of 26 years between 1990 and 2016.

The analysis of urban expansion trends is presented for the STM administrative area, unlike the Accra and GKSR where the analysis covered several districts. Even so, the data has been disaggregated further into existing sub-metropolitan subdivision, making it possible to identify metropolitan-wide trends as well as differences in expansion trends at the scale of the relatively smaller sub-metropolitan spatial units. Urban expansion patterns in 1990, 2000 and 2016 are mapped in Fig. 9.8, while the summary statistics of the quantum and rate of expansion are presented in Table 9.7.

Urban expansion in the STM has been sporadic over the 26 years of analysis. As shown in Table 9.5, in 1990, only 18% (i.e. 34.230 km^2) of the total land area of the STM was classified as built-up. Over the next 10-year period, built-up land in the metropolis increased to 28%, which translated into total built-up area of 54.1 km^2. Similar to trends identified in Accra and the surrounding districts, as well as the

9.3 Urbanization Trends in Ghana

GKSR, urban expansion in STM intensified at the beginning of the twenty-first century where the total built-up land nearly doubled from 54.037 km^2 in 2000 to 111.966 km^2 in 2016, representing 28 and 58% of the total land area of the metropolis, respectively. The fastest rate of urban expansion also occurred over this period, where each year, as a result of physical development, built-up land increased by nearly 6%.

Moreover, it emerges from the analysis that, in STM, while the coastal towns marked the origins of historical development and urban expansion, over the years, owning to the physical barrier presented by the Atlantic Ocean, urban expansion has occurred in the eastern and south-western parts of the metropolis. Indeed, between 1990 and 2000, the historical coastal sub-metros of Sekondi and Takoradi accounted for 43% of all physical developments in the metropolis. The rates of urban expansion in these areas were, however, much lower at 1.35 and 2.43% per annum compared to other areas in the metropolis.

On the contrary, despite Essikado-Ketan to the east and Effia-Kwesimintsim to the north-west accommodating 18 and 22% of all physical developments, they did so at relatively faster rates of 11.5 and 5.92%, respectively. Similar trends could be observed between 2000 and 2016 where rate of urban expansion in the traditional coastal towns was significantly lower than the remaining two sub-metropolitan areas. This suggests that as far back as 26 years ago, urban expansion in STM had stalled in the historical coastal towns and started to spread in the eastern and north-western directions. In fact, as the summary statistics show, in 2016, 77% of total built-up land in the metropolis occurred outside the Sekondi-Takoradi areas. Should the current trends continue, urban expansion is expected to spread further outwards in the north-western direction.

By monitoring and quantifying urban expansion trends based on satellite data, the nature and extent and pace of urban expansion—the expression of the urbanization—process in three major urban agglomerations have analysed retrospectively. The analysis has demonstrated that urbanization rate in terms of population growth and the footprints of urban development has been rapid over the past three decades. Clearly, uncontrolled urban expansion at the current rate would have consequences on the natural environment, supply of ecosystem services and livelihoods while increasing the cost of urban development especially those related to infrastructure supply.

With knowledge of past urbanization trends, future trends could be anticipated and planned for with the aim of averting unsustainable urban expansion trends. To do so, however, requires the design and implementation of various instruments that can affect the location and timing of development while at the same time, providing opportunities to finance future developments through the urban development itself. This brings the discussion to the topic of urban growth management in spatial planning. In the section that follows, various growth management instruments deployed through the spatial planning system in different contexts will be outlined and discussed briefly. We will also identify how specific growth management instruments have entered the spatial planning system of Ghana and highlight the opportunities and challenges for designing and implementing effective instruments to manage the urbanization process in Ghana.

9.4 Urbanization and Urban Growth Management

Meeting the emerging needs of growing urban populations involves providing housing, social services, infrastructure and employment. This process, invariably, involves utilizing land. Indeed, as the discussion in the preceding sections has shown, rapid population growth is associated with consumption of more land. The associated impacts manifest physically in the amount of built-up land as well as in the overall emergent structure of cities.

While physical development and the utilization of land for that matter are essential to meeting the needs of growing urban populations, a failure to plan for and co-ordinate the development process can lead to unsustainable outcomes. For example, as cities expand their built-up areas into outlying Greenfield land, natural resources that provide essential ecosystem services such as water, food and clean air for humans get depleted, while habitats of other life forms are destroyed, threatening their survival. There are also financial implications to unfettered urban expansion. Compared to a more compact, consolidated development, urban sprawl requires huge investments in order to provide infrastructure and services. Where the financial resources are lacking, large areas develop without the basic amenities and services needed to support the resident population. This in fact is the case of many suburban and peri-urban communities in Ghana. In addition, as cities expand outward around a dominant functional core, as is the case in most Ghanaian cities, commuting distances tend to increase. In the absence of efficient public transit systems, automobile dependence increases leading to congestion, increased energy use and travel-induced environmental pollution.

Optimizing the benefits of urbanization while addressing the associated negative consequences is one of the major goals of spatial planning. To this end, various countries use their spatial planning systems to design and implement growth management strategies. Growth management strategies deploy various tools and instruments that affect the timing, the location and the nature and extent of physical development. The aim is to shape the decisions of actors in development process and to guide cities and towns towards desired patterns of growth. In essence, urban growth management instruments help to achieve some of the fundamental goals of spatial planning including ensuring efficient and optimal use of land resources; reducing the costs of providing urban infrastructure and services; protecting public health and safety by preventing and mitigating negative externalities; and raising revenues in the development process through value capture for future urban development investments.

Growth management instruments may be regulatory, incentive-based or revenue-based (see Table 9.8). It is not the objective of this chapter to provide a detailed discussion of the underlying principles as well as the effectiveness or otherwise of these instruments. Instead, in the sections that follow, a brief description of the three main types of instruments and some examples of policy tools under each of them are presented.

9.4 Urbanization and Urban Growth Management

Table 9.8 Types of growth management instruments

Regulatory instruments	Incentive-based instruments	Fiscal instruments (taxes, exactions and fees)
• Zoning	• Brownfield redevelopment incentives	• Dedications
• Green belts	• Capital gain tax	• Development impact fees
• Rate-of-growth controls (e.g. development moratoria)	• Conservation easements	• Land value tax
• Urban growth and service boundaries	• Historic Rehabilitation Tax Credits	• Linkage fees
	• Location-efficient mortgages	• Property tax
	• Special economic zones	• Real-estate transfer tax
	• Split property tax	• Special assessment tax
	• Transfer of rights of development	• Subdivision exactions
	• Use-value tax	• Tap fees

Source Silva and Acheampong (2015, p. 25)

9.4.1 Regulatory Instruments

Regulatory instruments are designed and implemented as containments measures to control urban sprawl, leapfrog development, preserve agricultural land and protect biodiversity. Examples of regulatory instruments in growth management included traditional land use zoning; designating Greenbelts, urban growth and service boundaries; and rate-of-growth controls. Containment measures achieve their intended objectives by regulating the timing, location, magnitude and extent of physical development.

Traditionally, land use *zoning* has been used to regulate land use and control physical development. Zoning involves categorization of land according to different uses. As mentioned elsewhere, land use plans contain zoning policies which identify permissible uses and development as well as prohibited development. Different zoning instruments may be deployed to achieve different objectives. Density zoning policies specify minimum and maximum allowable development density or floor area ratio in an area to achieved desired magnitudes of development. Up-zoning, which involves the rezoning of areas previously designated for low-density development, may be used to allow for higher density development. In some contexts, mixed-use zoning policies are introduced to relax traditional exclusionary zoning by establishing standards for blending various uses such as residential and commercial with the aim of achieving high-density, compact development. Rural cluster zoning is implemented in rural farming areas to ensure that houses are concentrated on small plots, leaving the remaining land as open space or making it available for farming (see, e.g., Bowler 1997; Brabec and Smith 2002).

Moreover, *greenbelts* have long been used to regulate physical expansion of urban areas. The use of greenbelts, especially in European cities, dates as far as the seventeenth century (Amati 2012). A greenbelt is a zone of open land dividing a city from its surrounding countryside. In principle, all development activities are prohibited in the areas within the greenbelt. Several cities, including Vienna, London, Barcelona, Budapest, Berlin, Tokyo, Toronto, Chicago, Sydney, Melbourne and Seoul, have established greenbelts. In Belgium, for example, some major towns retain continuous greenbelts which serve as buffers between the city core, industrial districts and outlying suburban areas or neighbouring agricultural areas (e.g. Sonian Forest and Boi de la Cambre). In order to ensure that development pressures are shifted away from the designated green areas, strong land use controls are enforced through comprehensive land use plans at the national and sub-national levels (see Silva and Acheampong 2015).

Greenbelts are seen as the most restrictive form of urban containment policy. Critics argue that greenbelts constrain land supply with attendant effects manifesting in increased land prices, uncompensated loss of development rights in the greenbelt area, housing shortage in areas within the greenbelt as well as leapfrog development into areas beyond the greenbelts (see, e.g., Bengston and Young 2006; Cheshire and Sheppard 2002; Dawkins and Nelson 2002). Notwithstanding these criticisms, in cities where greenbelt policies exist such as London, the restriction imposed has led to intensification of land through redevelopment of brownfield sites, minimizing the impacts that unfettered outward expansion would have had on farmland and the character of outlying rural areas.

Urban growth boundaries (UGBs) and urban service boundaries (USBs) are seen by some as more flexible alternatives to greenbelts in managing urban growth (Jun 2004). UGBs take the form of officially mapped dividing lines drawn around urban areas to limit encroachment into surrounding rural areas. Similar to greenbelts, UGBs promote densification within the designated boundary and restrict development on non-urban land outside the boundary (Nelson and Moore 1993). UGBs are also intended to discourage speculation at the urban or suburban fringe; protect open lands, including farms, watersheds and parks; and promote more compact, contiguous urban development. Unlike greenbelts, however, UGBs are more flexible growth management instruments set for given periods of time—typically 20 years—and are subject to revision (Jun 2004).

Like UGBs, USBs consist of a line drawn around a city or metropolitan area for the purposes of using infrastructure service provision to control the timing and pace of urban development. USBs achieve their objective by delineating areas beyond which certain urban services such as sewer and water will not be provided as a way of discouraging development in those areas. They are often linked with adequate public facilities ordinances that prohibit development in areas not served by specific public services and facilities. In some areas using USB, a tearing system is adopted to direct public infrastructure into new areas in a particular sequence. A typical example is the Priority Funding Areas initiative in Maryland in the USA which focuses infrastructure investment on the city centre (Bengston et al. 2004).

In Ghana, UGBs were designated for the first time in 2012 for the GKSR. The Comprehensive Development Plan for the Greater Kumasi Area identified, among other things, that excessive urban sprawl towards adjacent districts of the KMA, mainly along major radial roads, made it difficult to co-ordinate the timing of development and infrastructure of supply. Also, much of the development in these outlying areas was unplanned and therefore lacked the basic facilities and services needed to support the resident population. In view of this, the Plan established two types of UGBs in the GKSR. The objectives and physical extent of the GKSR growth boundary are presented in Box 9.1. Although this growth management policy has been established in the Plan, to date, clear strategies aimed at directing development towards desired locations as envisaged in the Plan remain to be seen.

The evidence suggests that effectiveness of UGBs depends on the size of the area where urban use is permissible (Lee and Linneman 1998; Yokohari et al. 2000). Thus, if the area where urban development is permitted is too large, which often happens due to political pressure, UGBs have only a limited effect on urban growth. If the area is too small to sustain development pressure, UGBs may be associated with adverse effects, such as increases of land prices, affordable housing problems within the boundaries and leapfrog type of development beyond the boundary.

Box 9.1: The Greater Kumasi Urban Growth Boundary

Objectives for Urban Growth Management for Suburban Areas
- Create better and attractive environments for residents and businesses in suburban areas of Greater Kumasi Conurbation by providing appropriate basic infrastructure and services.
- Make a compact conurbation area of Kumasi within Greater Kumasi sub-region.

Strategies for Urban Growth Management in Suburban Areas

Control and guide urban sprawl of residential area development in fringe areas adjacent to Kumasi City by seeking the following directions:

- Selectively providing basic infrastructure and services;
- Selectively controlling permits for land use and buildings.

In order to purse the above-mentioned strategic direction, two types of urban growth boundaries (UGBs) should be set for the objectives for urban growth management in suburban areas of Greater Kumasi sub-region.

Large urban growth boundaries (UGB-2);
Small urban growth boundaries (UGB-1);

These two UGBs should control urban development as follows:

- Within the UGB-1, provision of urban infrastructures/services should be implemented with high priority. Different infrastructures/services should be provided in an integrated manner.
- Outside the UGB-2, principally no urban developments should be allowed.
- Outside the UGB-2, preparation and approval of local plans (layout plans) should be strongly restricted.
- Outside the UGB-L, high priority should not be given to provision of urban infrastructures/services and it should be restricted.
- Outside the UGB-2, special large-scale planned urban development projects could be approved for implementation if proper infrastructure and services are provided by the project.
- Outside the UGB-2, rural–agricultural–natural environments should be conserved for rural life including agricultural production and natural environments in suburban contexts in vicinity to Kumasi City.
- Rural areas outside the large urban growth boundaries (Conurbation Boundaries) would contain the following land uses:

  - Rural towns;
  - Rural settlements;
  - Recreational areas;
  - Greenery open space;
  - Agricultural lands;
  - Woodland;
  - Water bodies.

9.4.2 Fiscal Instruments—Taxes, Exactions and Fees

As market-based mechanisms, fiscal growth management instruments may be deployed through the spatial planning system to capture some of the external benefits accruing from public investments, such as investments in infrastructure development while mitigating the potential negative external effects of new developments. Fiscal instruments include different types of *taxes* (e.g. property taxes, special assessment taxes, land value taxes) and *exactions*[1] (e.g. impact fees, dedication, fee-in-lieu, tap fees and linkage fees) levied on developments to raise revenues and/or to offset the impact of physical development through other means such as the dedication of facilities to an existing community that would be affected by the new development.

One of the most commonly used types of exactions is *impact fees*. New developments often have impacts on existing communities. For example, the development of a large shopping arcade is expected to attract more traffic and possibly put pressure on existing roads and parking facilities. Similarly, as population in cities increase, it becomes necessary to increase the stock of housing. This could be achieved through medium- to large-scale residential developments by private real-estate developers, as is the case in major cities in Ghana including Accra, Takoradi and Kumasi. New residential estates and the associated increase in population may increase pressure on existing community facilities such as schools and recreational parks. In such instances, impact fees may be applied by a local government on developers as a condition of development approval. The objective is to raise revenue which would then be channelled into the construction or expansion of existing facilities that benefit the contributing development (Evans-Cowley 2006).

Exactions may also take the form of *dedication,* by which developers may be required to donate land or public facilities for public use. For example, developers may be required to dedicate land or pay for at least a percentage of the costs of new public facilities that will be needed as a result of their proposed development. The *tap fee* is another form of exaction charged on utility connections in most countries to allow cost-recovery in tying new development into existing infrastructure network (Brueckner 1997; Ihlanfeldt and Shaughnessy 2004). Another way of paying for the secondary effects of new development is through *linkage fees* charged by local governments. This is a type of exaction levied to raise revenues from large-scale commercial, industrial and multi-family developments. The revenues are then used to finance new developments such as the provision of affordable housing, day care and facilities. In the USA, for example, linkage fees are primarily used by local governments in areas where the cost of housing is extremely high, such as California and Massachusetts (Evans-Cowley 2006).

---

[1] See Evans-Cowley's (2006) for a detailed discussion of the principles, rationales and types of exactions.

9.4.3 Incentive-based Instruments

While some growth management instruments are deployed to regulate physical development and/or raise revenues through the urban development process to finance new developments, others are delivered through the planning system to stimulate markets and to encourage and attract more desirable activities to locations of strategic interest. These are therefore incentive-based growth management instruments. Such instruments are used to encourage a wide range of actions aimed at improving conditions of the built environment and protecting the natural environment, which otherwise would not occur.

Incentive-based instruments may be used to attract investments for redevelopment of derelict buildings. Local government could also provide market-based incentives in order to ensure the preservation and/or rehabilitation of historic buildings. Incentive-based instruments in essence take the form of subsidies, tax credits, development rights and direct state action in the provision of land (through expropriation or compulsory acquisition) and infrastructure to attract investments.

A commonly used incentive-based urban growth management instrument is *Brownfield Redevelopment Incentive.* As the name suggests, this type of incentive is designed to encourage developers build on brownfield land located in inner-city areas. Contrary to development on Greenfield sites, brownfield (re)development poses a number of challenges to developers These include expensive land prices at inner-city locations, demolition cost of existing structures, clean-up/decontamination cost in previous industrial sites and limitations imposed by existing zoning regulations (McCarthy 2002). Brownfield (re)development, however, has several benefits including helping to regenerate areas experiencing decline; increased asset value of the site and the surrounding site; increased tax base for local government; and opportunities to create new employments. Also, by concentrating development in existing built-up areas, brownfield redevelopment helps to avert unsustainable urban expansion. Thus, the provision of incentives in these areas is considered essential in reducing development costs while helping to achieve sustainable development outcomes. In the UK, for example, tax incentives and other assistance such as dereliction aid and gap funding schemes are provided to eliminate barriers for brownfield development (Wong and Bäing 2010).

Another example of market-based growth management instrument designed to attract development in desired areas is *Transfer of Development Rights (TDR).* TDR programmes are used to achieve a wide variety of objectives including protecting agricultural lands, preserving wildlife habitats and controlling development densities in areas with limited infrastructure or public services (Johnston and Madison 1997). TDR programmes are designed to balance development across different areas in cities by reducing or eliminating development potential in places that require preservation while increasing development where growth is wanted (Pruetz and Standridge 2008). Such programmes are underpinned by the assumption that the development right for a parcel of land can be sold and used in another parcel located elsewhere in the city. In practice, TDRs require the designation of a preservation zone where additional development is not wanted and a receiving zone where development is wanted. The instrument then works by allowing the landowner of the designated preservation or sending zone to sell their development rights to a developer who will use these rights in an area designated as receiving zone. In the receiving areas, density bonuses which allow for higher density development than the base density established by existing zoning law are granted, creating incentives for developers to buy the development rights (Tavares 2003).

Other forms of incentive-based growth management instruments are use-value tax assessment, split-rate property tax and Historic Rehabilitation Tax Credits. To maintain agricultural uses in urban and peri-urban areas, farmland owners benefit from lower rate taxes than other uses in the form of use-*value tax* (Anderson and Griffing 2000). *Split-rate property tax* is used to encourage redevelopment of obsolete buildings and facilitate revitalization in older central cities by placing proportionally higher taxes on land than on built structures (Rybeck 2004; Banzhaf and Lavery 2010). This makes it costlier to hold on to vacant or underutilized centrally located sites. Finally, *Historic Rehabilitation Tax Credits* provide incentives for the public to preserve and rehabilitate historic places and cultural heritage (see, e.g., Mason 2005; Swaim 2003).

9.5 Conclusions

In this chapter, we set out to explore urbanization trends and to identify some of the policy instruments and strategies deployed through the spatial planning system to manage urban growth. Using global population growth data, we examined the size and rate of urban population change for the world, Africa, West Africa and Ghana from the 1950s all the way to the first half of the twenty-first century (i.e. 2050). The chapter also relied on national population and housing census information to examine population growth trends in major cities in Ghana. Besides urban population growth at these spatial scales, this chapter also explored historical urban expansion trends for selected cities across the globe as well as three major urban agglomerations in Ghana (i.e. Accra, the capital and surrounding districts, Greater Kumasi sub-region and the Sekondi-Takoradi Metropolis), using satellite data.

The discussion has shown that at the global scale, more than half of the world's population are now living in cities of different sizes and this number expected to increase to nearly 65% by 2050. A similar trend is forecasted across African where an estimated 56% of the continent's population are expected to live in cities by 2050. In Ghana, we expect some 70% of the total population to reside in urban areas by 2050.

Perhaps, one of the most important findings from the global urbanization trends estimation is that while urban population is expected to increase, the rate of change is projected to be much lower than we have experienced in the past. Indeed, in most developing countries, the rate of urbanization has slowed down since 2005, and this is expected to continue into first half of the twenty-first century.

Notwithstanding the above, historical analysis of urban expansion trends using Landsat Satellite data demonstrated that rapid urban population growth in the past decades has been accompanied by major impacts on the natural environment as evidenced by the rapid increase in the built-up area of cities. Focusing on three major urban agglomerations in Ghana, for example, we observed a general trend of several decades of unfettered outward expansion of these cities: sporadic sprawling of physical development into outlying Greenfield land was observed in all three urban agglomerations. Without any interventions, this is expected to continue into coming decades. It is also likely that other major towns and cities have experienced urban expansion characteristics similar to those observed in this chapter, raising major challenges for transitioning towards sustainable urban futures in Ghana.

Urbanization provides several opportunities, but there are also many sustainability challenges associated with uncontrolled urban population growth and the physical expansion of cities. In search of ways in which growth could be managed in order to optimize the benefits of the urbanization process, this chapter outlined some of the policy instruments at the disposal of spatial planning for sustainable urban growth management. We identified that various instruments exist to manage urban growth by affecting the location and timing of development, while at the same time providing opportunities to raise finance through the urban development process to finance future urban development. Three broad categories of instruments—regulatory, fiscal and incentive-based instruments—were outlined, identifying some specific examples that are being designed and implemented in different cities across the globe.

In the context of Ghana, we identified that except for traditional land use zoning used to regulate development, the application of modern growth management instruments has remained largely unexplored. In recent years, however, following the reinvigoration of the spatial planning system, new growth management instruments are being experimented with in large urban conurbations. The designation of urban growth and service boundaries in the GKSR, for example, constitutes a typical example of how Ghana's spatial planning system is starting to embrace new and modern instruments to manage years of rapid, uncontrolled urban expansion. Even so, the effectiveness or otherwise of these instruments remains to be seen given the wide gap that has historically existed between plan proposals on the one hand and their implementation on the other hand. Indeed, as was indicated in the discussion, despite urban growth and service boundary strategies being proposed around 2012 for the GKSR, there is no clear evidence on the ground to date demonstrating the implementation of the strategy.

Thus, overall, there appears to be a lack of clear and co-ordinated strategies to manage urban growth towards sustainable outcomes in Ghana. Consequently, one could expect the rapid and uncontrolled expansion of settlements to continue into the future. Under this laissez-fair, businesses-as-usual scenario, the associated problems of urbanization including depletion of environmental resources and ecosystem services, the lack of finance to provide for housing, infrastructure and services and the overall conditions of living in Ghanaian cities are expected to worsen.

An alternative future is plausible if our knowledge of past urbanization trends and the associated challenges will provide the impetus to think strategically about cities

9.5 Conclusions

and to act by designing and implementing workable growth management strategies. As the discussion in this chapter has shown, enormous opportunities exist to reverse current trends and reduce impacts of urbanization by embracing new and innovative growth management strategies. In doing so, adaptation through experimentation, rather than strait-jacket importation of some of the instruments outlined in this book, will be crucial. The adoption of any of the growth management strategies will require careful considerations for issues bordering on prevailing political-economy context, legality, landownership systems in different urban contexts as well as resolving essential tensions that often arise from efforts to contain urban growth while meeting the needs of the growing urban population especially when it comes to making adequate land available for the provision of affordable housing. In the next chapter, we will explore possible ways in which as place urbanizes and housing need increases, spatial planning could help secure decent housing for urban citizens of different socio-economic means in Ghana.

References

Acheampong RA, Agyemang FS, Abdul-Fatawu M (2017) Quantifying the spatio-temporal patterns of settlement growth in a metropolitan region of Ghana. GeoJournal 82(4): 823–840

Amati (2012) Greenbelts (encyclopaedia entry). Berkshire Publishing Group, Great Barrington, MA. Available http://hdl.handle.net/1959.14/213796

Anderson JE, Griffing MF (2000) Use-value assessment tax expenditures in urban areas. J Urban Econ 48(3): 443–452

Angel S, Sheppard S, Civco DL, Buckley R, Chabaeva A, Gitlin L, et al (2005) The dynamics of global urban expansion. DC: World Bank, Transport and Urban Development Department, Washington, p 205

Baloye D, Palamuleni L (2015) A comparative land use-based analysis of noise pollution levels in selected urban centers of Nigeria. Int J Environ Res Public Health 12(10): 12225

Banzhaf HS, Lavery N (2010) Can the land tax help curb urban sprawl? Evidence from growth patterns in Pennsylvania. J Urban Econ 67(2): 169–179

Bengston DN, Young YC (2006) Urban containment policies and the protection of natural areas: the case of Seoul's greenbelt. Ecol Soc 11(1): 1–13

Bengston DN, Fletcher JO, Nelson KC (2004) Public policies for managing urban growth and protecting open space: policy instruments and lessons learned in the United States. Landsc Urban Plan 69(2–3): 271–286

Blanc AK, Grey S (2002) Greater than expected fertility decline in Ghana: untangling a puzzle. J Biosoc Sci 34(4): 475–495

Bowler C (1997) Farmland preservation and the cluster zoning model. J Am Plan Assoc 63(1): 127–128

Brabec E, Smith C (2002) Agricultural land fragmentation: the spatial effects of three land protection strategies in the eastern United States. Landsc Urban Planning 58(2): 255–268

Brueckner JK (1997) Infrastructure financing and urban development: the economics of impact fees. J Public Econ 66(3): 383–407

Champion T (2001) Urbanization, suburbanization, counterurbanization and reurbanization. In: Paddison R (ed) Handbook of urban studies, SAGE Publications, London

CILSS (2016) Landscapes of West Africa—A Window on a Changing World. U.S. Geological Survey EROS, 47914 252nd St, Garretson, SD 57030, United States. https://edcintl.cr.usgs.

gov/downloads/sciweb1/shared/wafrica/downloads/documents/Landscapes_of_West_Africa_Republic_of_The_Gambia_en.pdf

Dawkins CJ, Nelson AC (2002) Urban containment policies and housing prices: an international comparison with implications for future research. Land Use Policy 19(1): 1–12

Duranton G, Puga D (2004) Micro-foundations of urban agglomeration economies. In: Henderson JV, Thisse JE (eds) Handbook of regional and urban economics vol 4, Elsevier B. V. pp 2063–2117 https://doi.org/10.1016/S0169-7218(04)07048-0

Eigenbrod F, Bell V, Davies H, Heinemeyer A, Armsworth P, Gaston K (2011) The impact of projected increases in urbanization on ecosystem services. Proc R Soc Lond B Biol Sci 278(1722): 3201–3208

Evans-Cowley J (2006) Development exactions: process and planning issues. Lincoln Institute of Land Policy, Cambridge, MA

Gaisie S (2013) Fertility trend in Ghana. Afr Popul Stud 20(2)

Glaeser EL (ed) (2010) Agglomeration economics. University of Chicago Press

Henderson V (2002) Urbanization in developing countries. World Bank Res Obs 17(1): 89–112. https://doi.org/10.1093/wbro/17.1.89

Ihlanfeldt KR, Shaughnessy TM (2004) An empirical investigation of the effects of impact fees on housing and land markets. Reg Sci Urban Econ 34(6): 639–661

Jacobs J (1969) The economy of cities. Random House, New York

Jun MJ (2004) The effects of Portland's urban growth boundary on urban development patterns and commuting. Urban Stud 41(7): 1333–1348

Johnston RA, Madison ME (1997) From land marks to landscapes: a review of current practices in the transfer of development rights. J Am Plan Assoc 63(3): 365–378

Lee CM, Linneman P (1998) Dynamics of the greenbelt amenity effect on the land market—The Case of Seoul's greenbelt. R Estate Econ 26(1): 107–129

Mason R (2005) Economics and historic preservation. The Brookings Institution, Washington, DC, pp 35–100

McCarthy L (2002) The brownfield dual land-use policy challenge: reducing barriers to private redevelopment while connecting reuse to broader community goals. Land Use Policy 19(4): 287–296

McGregor D, Simon D (eds) (2012) The peri-urban interface: approaches to sustainable natural and human resource use. Routledge, Abingdon

McGregor DF, Adam-Bradford A, Thompson DA, Simon D (2011) Resource management and agriculture in the periurban interface of Kumasi, Ghana: problems and prospects. Singap J Trop Geogr

Nelson AC, Moore T (1993) Assessing urban growth management: the case of Portland, Oregon, the USA's largest urban growth boundary. Land Use Policy 10(4): 293–302

Oduro CY, Ocloo K, Peprah C (2014) Analyzing growth patterns of Greater Kumasi metropolitan area using GIS and multiple regression techniques. J Sustain Dev 7(5): 13

Pacione M (2009) Urban geography: a global perspective. Taylor and Francis, Abingdon

Potts D (2012a) Challenging the myths of urban dynamics in sub-Saharan Africa: the evidence from Nigeria. World Dev 40(7): 1382–1393

Potts D (2012b) Viewpoint: what do we know about urbanisation in sub-Saharan Africa and does it matter? Int Dev Plan Rev 34(1): v–xxii

Pruetz R, Standridge N (2008) What makes transfer of development rights work?: success factors from research and practice. J Am Plan Assoc 75(1): 78–87

Rybeck R (2004) Using value capture to finance infrastructure and encourage compact development. Public Works Manag Policy 8(4): 249–260

Silva EA, Acheampong RA (2015) Developing an inventory and typology of land-use planning systems and policy instruments in OECD countries. OECD environment working papers, No. 94

Simon D, McGregor D, Nsiah-Gyabaah K (2004) The changing urban-rural interface of African cities: definitional issues and an application to Kumasi, Ghana. Environ Urban 16(2): 235–248

References

Sudhira HS, Ramachandra TV, Jagadish KS (2004) Urban sprawl: metrics, dynamics and modelling using GIS. Int J Appl Earth Obs Geoinformation 5(1): 29–39

Swaim R (2003) Politics and policymaking: tax credits and historic preservation. J Arts Manag Law Soc 33(1): 32–39

Tavares A (2003) Can the market be used to preserve land? The case for transfer of development rights. European Regional Science Association 2003 Congress

Torrens PM, Alberti M (2000) Measuring sprawl. Centre for Advanced Spatial Analysis—University, London

United Nations (2014) World urbanization prospects, the 2011 revision. Population Division, Department of Economic and Social Affairs, United Nations Secretariat

Un-habitat (2010) State of the world's cities 2010/2011: bridging the urban divide. Earthscan, Abingdon

Watson V (2009) Seeing from the south: refocusing urban planning on the globe's central urban issues. Urban Stud 46(11): 2259–2275

Webster D (2002) On the edge: shaping the future of periurban east Asia. Asia/Pacific Research Center, Stanford

Wilson EH, Hurd JD, Civco DL, Prisloe MP, Arnold C (2003) Development of a geospatial model to quantify, describe and map urban growth. Remote Sens Environ 86(3): 275–285

Wong C, Bäing AS (2010) Brownfield residential redevelopment in England. Joseph Rowntree Foundation, York

Yokohari M, Takeuchi K, Watanabe T, Yokota S (2000) Beyond greenbelts and zoning: a new planning concept for the environment of Asian mega-cities. Landsc Urban Plan 47(3–4): 159–171

Chapter 10
The Spatial Planning System and Housing Development: Prospects and Models

Abstract Providing adequate and decent housing constitutes one of the biggest challenges of urban development, especially in countries such as Ghana with high levels of historical urbanization. This chapter examines the urban housing problem in Ghana in relation to spatial planning. It identifies the demand and supply-side challenges of the housing market and examines policy responses in the past. A key focus of this chapter is to explore ways of providing housing directly through Ghana's spatial planning system. To this end, two models that are intended for application in securing land for new housing development and continuously generating housing supply through the urban development process are advanced. The theoretical principles behind the proposed housing development models required legislative backing and some practical implementation strategies are also discussed.

Keywords Housing · Housing finance · Affordable housing · Value capture Development exactions · Land-based finance · Spatial planning system · Ghana

10.1 Introduction

In the last seven decades, the world has witnessed a substantial increase in the size of human population. From an estimated population of 2.5 billion in the 1950s, the world's population reached almost 7 billion in 2015, and this is expected to increase to 9.5 billion by the first half of the twenty-first century (United Nations 2014). Moreover, as we discovered in Chap. 9, while population has been increasing globally, in countries in the Global South, the highest levels of urbanization have also occurred over the last seventy years. On the African continent, for example, the United Nations' population estimates show that about 40% of the population currently live in cities, and this is expected to increase to 55% by 2050. In Ghana, an estimated 54% of the population has been living in cities since 2015; it is projected that about 70% of the over 45 million people expected in the country by the first half of the twenty-first century will also live in cities.

One of the biggest challenges of population growth and the associated high levels of urbanization is providing decent and suitable housing that meets the varying affordability needs of the population. Access to decent shelter is a basic requirement for securing the overall health and well-being of society, and an important source of wealth which is linked to long-term poverty reduction at the household level (Moser 2007). Yet, one of the contemporary realities globally is the prevalence of homelessness and increasing number of people living in slums (Brush et al. 2016; Toro et al. 2007; Wilson 2011; UN-Habitat 2016). The UN-Habitat estimates that about a billion of the world's population live in slums, implying that nearly one in eight human beings in the world do not have access to decent housing and supporting facilities including water and sanitation (UN-Habitat 2016). Whereas in developing countries, some 30% of the population are estimated to be living in slums, across sub-Saharan Africa, slum dwellers make up 59% of the total population (UN-Habitat 2016).

The prevailing housing problems have both supply- and demand-side causes. On the supply side, the production of new housing and/or refurbishment of old stock requires huge capital investments to cover land acquisition, building materials and labour costs as well as the cost of providing infrastructure and services. On the demand side, there is rising housing unaffordability relative to incomes and limited access to affordable home finance options especially among low-income informal economy workers in developing countries, making it difficult, if not impossible, for the majority of the population to meet their housing needs.

Thus, both the challenges of providing adequate and decent housing for the ever-increasing population and the solutions needed to address them could be understood from the interplay between the workings of market forces on the one hand and public policy interventions designed to facilitate markets to unlock supply on the other hand. Like many goods, one of the channels of housing supply is through private markets. With private markets, as is the case in many advanced capitalist economies, an organized private property development industry responds to prevailing housing need and demand by acquiring land and developing dwelling units for individuals to buy or rent. In most developing countries, including Ghana, private property developers exist but their contribution to total housing supply is not significant. Instead, the most common practice in these countries is the self-build incremental housing development model, where individuals and households acquire land and spend several years to build their homes using local labour and their own financial resources (see, e.g. Ferguson and Smet 2010; Amoako and Boamah 2017).

State policy often complements the operation of private markets to narrow the gap between housing need and demand on the one hand and housing supply on the other hand. Private housing markets act on effective demand and thus, respond to the housing needs of those who have the ability to pay for housing either in the owner–occupier or rental markets. In the context of developing countries where the incremental housing model is common, access to land constitutes the most important step towards meeting ones housing need. With rising land values and chaotic urban land markets, individuals of different socio-economic means also struggle to meet their home-ownership aspirations. Moreover, given that the rental market in these contexts depends largely on total supply that could be secured through self-build

10.1 Introduction

incremental constructions, constraints in latter ultimately affect the availability and pricing of dwelling in the former. Thus, invariably, private markets, whether operating through organized private property development industry or through the individual-based, self-build incremental model, tend to fail to meet the housing needs of all, especially when it comes to housing poor and low-income households. In cases such as those outlined above, state intervention in the housing market is justified, for example, in providing 'bricks-and-mortar' subsidies to developers to build affordable housing for households or in providing mean-tested, demand-side subsidies directly to low-income households to enable them pay for housing.

Besides the conventional ways of enabling households to meet their housing needs outlined briefly above, the ways in which housing could be delivered directly through the spatial planning system have gained considerable attention in both academic and policy discourse (e.g. Burges and Monk 2008; Gurran and Whitehead 2011; Johnson 2007; Agyemang and Morrison 2018). Building on the previous research, we will explore in this chapter models for providing housing through Ghana's spatial planning system. Before delving into the spatial planning models for housing supply in Ghana, we will briefly examine the housing situation in Ghana, highlighting the demand- and supply-side challenges as well as the policy responses that have been deployed so far. On the basis of the challenges, two models of housing development through the spatial planning system in Ghana will be proposed. Practical strategy design approaches as well as legal frameworks and institutional competences that would be required to enable their implementation will be identified.

10.2 The Housing Situation in Ghana

The high cost of renting and home-ownership, overcrowding,[1] residence in makeshift dwellings[2] such as kiosks, tents and containers as well as the growing number of slum dwellers, all lend credence to a severe housing crisis in Ghana. Although the research and policy communities acknowledge the overwhelming housing crisis in the country, accurate estimates of the country's housing deficit are lacking. Based on a review of historical policy-related documents and academic research, Kwofie et al. (2011) estimated that the housing deficit—the gap between demand and supply—in Ghana as of 2010 was around 1.2 million housing units. On 18 May 2017, the Minister for Works and Housing was quoted to have indicated that his Ministry estimates that the housing deficit was as of the time, 1.7 million units (Ghana News Agency 2017). These estimates suggest that, over the next 10 years, between 120,000 and 170,000

---

[1] The 2010 population and housing census found that about 44.5% of all households reside in single rooms, mostly in traditional compound houses. In the Greater Accra Region for example, the average household size is 4 persons. Meanwhile an estimated 493,327, representing 47.5% of all households in the region occupy single rooms.

[2] About 1.8% of Ghana's population live in makeshift structures. In the Greater Accra region, the most urbanized region in the country, some 67,005, representing 6.5% of the total number of households in the region live in improvised homes such as kiosks, tents and containers.

housing units would be required annually to eliminate the deficit alone. Accounting for annual population increase would further increase this target.

Moreover, the *Ghana Housing Profile* report published by the UN-Habitat has attempted to forecast the country's housing need in terms of the number of rooms instead of dwelling units, although the focus is on urban areas. The report justifiably recommends quantifying the deficit in terms of rooms because in Ghana, most households tend to occupy rooms which are clearly defined (UN-Habitat 2011). The housing sector profile estimates that about 5.7 million new rooms would be required by 2020 to accommodate the country's population. Over a ten-year period, the report estimates that approximately four rooms must be completed in every minute of the working day to successfully meet this target.[3] Perhaps, another benefit of this approach to estimating the country's housing deficit is that in addition to exposing the real extent of the housing problem, it gives policy makers room to think about how this could be met within the wider context of sustainable urban development outcomes. For instance, the total land consumption for the estimated number of rooms will depend on the type dwellings that would be built, which when planned and built as high-density, multi-storey dwellings, would reduce land consumption and ultimately contribute to averting unsustainable urban expansion.

The huge housing deficit implies that over the years, the channels of housing supply have not been able to cope with the growing population and attendant growth of housing demand. In order to understand why this is so, it is important to understand the supply side of the country's housing market and to identify the relative contributions of each of the channels of supply to the total housing stock and the associated challenges. These are dealt with in the section that follows.

10.2.1 Channels of Housing Supply and Supply-Side Challenges

Housing production in Ghana is realized through three main channels. These are the self-build incremental construction by individual households, housing provided by the fledgling private real-estate industry and housing provided by the state.

The self-build incremental model is the primary path towards home ownership in Ghana and accounts for over 90% of the country's total housing supply. As such, this sector of the housing market is also responsible for the supply of both rental and rent-free housing in Ghana. In incremental housing, households rely mainly on their incomes as well as other informal/non-institutional finance sources including savings, loans and remittances from families and friends and the sale of assets for housing development (see, e.g. Gough and Yankson 2011; Konadu-Agyemang 2001;

---

[3]The baseline data for the estimates comes from the population and housing census data in 2000, household six distribution reported in the Ghana Living Standards Survey Round 5, the number of new urban households expected between 2010 and 2020 and maximum room occupancy of two persons per room.

Sheuya 2007; Obeng-Odoom 2010). The self-build incremental model is also used by Ghanaians in the diaspora. Indeed, quite a significant number of new housing development in peri-urban areas in Ghana have absentee owners, the majority of whom live abroad (Gough and Yankson 2011).

Typically, households and individuals with home-ownership aspirations spend several years to accumulate savings in order to be able to acquire land first and to subsequently spend years building their housing. The amount of time spent therefore becomes a function of several other considerations including the inflow of finances from the various sources and the competition from other essential household needs. Households building their own homes through this model initially tend to either rent or live rent-free in extended-family-owned housing, and from there channel some financial resources into completing their would-be new home. Often, however, households do not wait for the entire structure to be completed. Instead, they move into their new homes as soon as an initial number of rooms have been completed and roofed, adding to the existing structure over time in response to life cycle changes and as and when the household budget allows them to do so.

The self-build, incremental housing model, despite being the primary channel of housing supply is fraught with several challenges. One of the challenges stems from the chaotic nature of the urban land market where problems such the sale of a piece of land to multiple buyers (known locally as 'multiple sale of land') often results in high levels of uncertainty, widespread litigations and loss of financial resources (Gough and Yankson 2000). Even when households successfully acquire land, they face additional challenges in securing the needed financial resources to meet the formal land use planning and land title registration requirements,[4] as well as the actual construction of the house. While individual developers may qualify for bank loans to finance housing construction, the high interest rates on loans mean that for the majority of developers, especially those who are employed in the informal economy, traditional bank loans are simply inaccessible. Moreover, combining renting and self-building for at least the first few years can impose enormous stress and financial burdens on households, often requiring that households sacrifice non-housing expenditures in order to realize their home-ownership aspirations (Asante et al. 2017).

The economic liberalization programmes first introduced in Ghana in the early 1980s under the Structural Adjustment Programme (SAP), among other things, enabled private-sector involvement in the housing market. Indeed, one of the major objectives of SAP was to restructure the country's then ailing economy by reducing public-sector spending and promoting private-sector investments in different sectors of the economy including the housing sector (Arku 2009). In Ghana, private real-estate developers operate under the umbrella union of the Ghana Real Estate Developers Association (GREDA). While GREDA was established as far back as 1963, it was not until the liberalization era of the 1980s that the prevailing emphasis on private-sector-led economic development marked increased participation of its members in Ghana's housing market.

---

[4] A detailed discussion of the formal land use planning regulations to which developers are expected to comply has been presented previously in Chap. 5.

As of the time of writing, more than 150 private real-estate companies were registered members of GREDA.[5] Despite the growing number of private real-estate developers, especially in Ghana's largest cities including Accra, Kumasi and Sekondi-Takoradi, their contribution to total housing supply has been rather low. Up-to-date data on the stock of residential properties supplied through GREDA is also lacking. Notwithstanding this, in 2007, the Bank of Ghana estimated that GREDA had since 1988 produced some 10,954 new homes in Ghana (Bank of Ghana 2007). Other sources also estimate current total housing supply at around 50,000 units, comprising mainly detached and semi-detached dwellings distributed across the country's major cities (see, Owusu-Manu et al. 2015). Since entering Ghana's housing market, private real-estate developers have targeted the high end of the housing market, building housing which only upper-middle-income and high-income households living in Ghana and abroad can afford. One of the lasting impacts of private estate developers on the residential geography of major urban areas, especially Accra is the development of Gated Communities—privately governed, often walled residential enclaves with strictly controlled entrances and private security (see, e.g. Grant 2005).

A number of studies have examined the key challenges of the private real-estate industry in Ghana (Owusu-Manu et al. 2015; Bank of Ghana 2007). Some of the challenges are market-oriented and directly related to macroeconomic conditions. Market-related challenges of the industry include high cost of building inputs, inaccessible credits and collateral barriers macroeconomic instabilities and inflation which affect the cost of borrowing and lack of infrastructure (Bank of Ghana 2007; Owusu-Manu et al. 2015). Other studies have attributed the inability of the real-estate industry to contribute significantly to housing supply to institutional factors including the land tenure system, lengthy and cumbersome land acquisition and title registration and land use planning and title registration procedures (Owusu-Manu et al. 2015).

The third and final channel of housing supply is the state. Overall, the state's involvement in housing provision in Ghana has rather been disappointing. According to the 2010 population and housing census, housing provided directly by the state accounted for only 2.2% of the total housing stock in Ghana. Comparing the stock of state-provided housing in 2000 and 2010 shows that over the 10-year period, the share of government-owned housing increased but marginally from 2 to only 2.2% (Ghana, Statistical Service 2012). To put this in perspective, out of the total of 3,392,745 housing units in Ghana (as of the 2010 population and housing census) accumulated supply from the state over the years accounted for just 74,640 units: Only 6785 new dwelling units were added to the existing stock by the state between 2000 and 2010.

The lack of direct state involvement in housing development in contemporary times has its origins in neoliberal policies pursued in the 1980s under the Structural Adjustment Programme (SAP). Under SAP, many countries in sub-Saharan Africa, including Ghana sought to restructure their economies by balancing national budgets, liberalising trade and improving macroeconomic stability (Hilson and Potter 2005;

---

[5]Data on membership of GREDA is available at: http://www.gredaghana.org/members.htm.

Rothchild 1991; Gibbon 1993). The neoliberal ideals of limited government interventions in private markets, including the housing market underpinned SAP. Thus, while Ghana continued to experience rapid increase in population and high levels of urbanization throughout the 1980s, the prevailing political-economy climate under SAP, the state played a very limited role in the direct provision of housing in Ghana (Arku 2009; Yeboah 2005).

10.2.2 Demand-Side Challenges

On the demand side of the market, households and individuals meet their housing need from one of three tenancy arrangements, which are home-ownership, renting and rent-free tenancies. About 47.2% of all households own their accommodation; 31.1% rent while the remaining 20.8% live rent-free in family-owned houses (Ghana Statistical Service 2012). With the exception of the rent-free sector which is largely a non-market housing tenure in the conventional sense (see Acheampong 2016; Tipple et al. 1997; Korboe 1992), the two remaining sectors are market-based, and thus require households to spend some amount of their incomes on housing to either rent or own their homes.

For both renters and would-be owner-occupiers, one of the main challenges to meeting their housing need is the growing gap between disposal incomes and houses prices or rents. In Ghana, housing affordability is a challenge for households of different incomes, but as we will soon demonstrate, the problem is particularly acute among urban poor, low-income and lower-middle-income households in urban areas. One of the basic measures of affordability is the house price/rent to income ratio, which also reflects the proportion of households' incomes that goes into housing consumption. Large sample data on households housing expenditures is limited. The majority of existing studies rely on rather small sample sizes, which makes it difficult to measure housing affordability in both the buyers and rental markets.

The Ghana Living Standards Survey Round 6 (GLSS6)[6] is by far the largest study that has attempted to estimate household incomes and expenditures for housing and non-housing-related consumption. One of the major advantages of the GLSS6 data is the large sample size—some 16,772 households across the country are captured in this study. Before drawing on the information provided by the study, it is important to point out a number of limitations. While the survey results provide valuable information on households' earnings and expenditures for various consumption, its computation of the percentage of household income that goes into housing is rather crude. The data presented in the survey report aggregates housing expenditures for households in different income groups but does not distinguish between home-owners and renters. Also, housing expenditures across the different dwelling

---

[6]The Ghana Living Standards Survey Round 6 (GLSS6) provides the most recent data on income levels in Ghana. The survey report, published in 2014 is based on a national sample of 16,772 households.

Table 10.1 Annual household incomes housing expenditure ratio

Income-group (Quintile)	Mean annual gross household income GH¢ (US$)[a]	Housing expenditure as % of income
First (lowest)	6,571.8 (1449.15)	10.9
Second	10,698.0 (2359.02)	10.9
Third	14,823.5 (3268.73)	11.5
Fourth	16,909.7 (3728.76)	12.0
Fifth (highest)	25,200.9 (5557.05)	13.0
Ghana (average)	16,644.6 (3670.30)	12.4

[a]Based on exchange rate of GH¢1 = US$0.22
Source Based on Ghana Living Standards Survey Round 6 (GLSS6), Ghana Statistical Services (2014)

types (i.e. detached, semi-detached, compound) and different property sizes are not distinguished. That notwithstanding, the survey findings offer useful starting point to understanding household expenditures for households of different income groups in across the country.

Table 10.1 presents a summary of the mean annual earnings of households categorized into five income groups and the proportion of their incomes spent on housing. The data shows that on the average, households in Ghana spend about 12.4% of their incomes on housing. Also, housing expenditures as a percentage earnings increases as income increases. For example, while households in the lowest income quintile spend about 10.9% of their income on housing, among those in fourth and fifth (highest) quintile, this increase to 12 and 13%, respectively. According to the study, in urban areas,[7] housing expenditure as a percentage of earnings increases to 14.2%, further reaching about 16.7% in the Accra, the country's capital and largest city.

Given the limitations already highlighted in the GLSS6 data on the one hand and the lack of affordability thresholds which are based on empirical studies on the other hand, it is impossible to arrive at any meaningful conclusions with respect to housing affordability in Ghana using the data presented above. Notwithstanding this, we can draw on the evidence from empirical studies to appreciate housing demand challenges in Ghana. For example, we know that in the rental market, the standard practice is that landlords require would-be tenants to make upfront rent payment covering the entire period of their first contract, which is often between two and five years. This is despite the fact that the Rent Act, which has been in existence since 1963 requires landlords to take not more than 6 months of advance rent. Even so, previous studies have shown that renters struggle to accumulate the 6 months advance payment stipulated by the Rent Act, implying that the standard practice of landlords requiring between two and five years advance rent payment puts tremendous pressures on tenants (see, e.g. Arku et al. 2011; Asante et al. 2017). Besides the difficulty in finding suitable

---

[7]The average annual gross household income in urban areas is around GH¢16,580.8, which is slightly lower than the national average.

and affordable housing, the lack of tenure security and persistent threat of eviction is a major source of psycho-social distress for the majority of urban household and individuals who rent (Luginaah et al. 2010).

Furthermore, housing units supplied by the private real-estate industry is simply beyond the reach of the majority of Ghanaians. In Box 10.1, four examples of properties supplied in the private real-estate market in different locations in Accra are presented. The case properties presented are not meant to be representative of the market but to provide an idea of types of properties available and prices, which we then compare to the household earnings in Ghana. With Case Property 1 (i.e. two-bedroom semi-detached house located in Katamanso in Accra), the ask price is US$83,400.[8] Using the average national gross earnings of GH¢16,644.6 (US$3670.30) yields minimum house price to income ratio of 1:23. Among households in the lowest and highest income quintiles, the estimated house price to income ratios is 1:56 and 1:15, respectively. Case Property 4 (see Box 10.1) also helps us to appreciate the housing affordability problem for households who may consider renting a property from one of the private real-estate developers in Ghana. This particular property, which has a monthly ask rent of $425, if it is furnished, is actually higher than the average monthly earning in Ghana, which is around US$305.[9] For the households earning the highest quintile income (see Table 9.1), renting this property will require spending around 92% of their monthly income on housing alone.

Traditional mortgage finance and personal savings are the main sources of finance for home purchase from the private real-estate companies interviewed. Real-estate developers operate through lending institutions such as the HFC Bank and the Ghana Home Loans (GHL) to provide mortgage finance to buyers. Prospective mortgagors are required to provide a minimum down payment, which is often 20% of the total cost of the house they wish to buy. Thus, given the household earnings, traditional mortgage finance is far beyond the means of the majority of households especially those in the low-income and lower-middle-income brackets. While most buyers rely on bank loans for down payments, they do so at very high interest rates, which could be up to 24% of the principal (Acheampong and Anokye 2015).

Providing housing subsidies either in the form of 'brick-and-mortar' subsidies to private property developers to provide affordable housing or mean-tested subsidies directly to households are two of the possible options by which households could be assisted to meet their housing needs. However, initial assessment of the potential of conventional housing subsidies in helping households to have access to decent and affordable housing cast doubts on the feasibility of these instruments yielding any significant benefits (see, e.g. Acheampong and Anokye 2015). One of the reasons is

---

[8]House prices are quoted in US$ by property developers as a way of insulating against risks and uncertainties which result from the depreciation of the local currency. This also implies that anytime the local currency depreciates against the dollar, house prices increases for potential buyers who may not have dollar accounts to benefit from depreciation.

[9]This figure is arrived at by dividing the average gross national earning reported in the GLSS6 by 12 and converting it into dollars using the exchange rate of GH¢1 to US$ 0.22.

that given the highly informal and less organized nature of the real-estate industry in Ghana production side subsidies will be difficult to assess for effective targeting. In addition, the intended redistributive and equity gains of this approach would not be realized given that real-estate developers have consistently targeted households in the higher end of the market. Also, demand-side subsidies would be difficult to administer and might only work only with public-sector workers who are on the government payroll.

> **Box 10.1: Examples of Properties Supplied in the Private Real-Estate Market in Accra, Ghana**
> **Case Property 1:**
> **Location**: Katamanso, Accra.
> **Description**: Two Bedroom non-expandable semi-detached house. This includes a living area, kitchen, dining, front porch and full-sized bedrooms as well as provision for built-in wardrobes and space for landscape and garden.
> **Plot Size**: 35′ × 80′ (10.66 m × 24.38 m^2)
> **Price**: $83,400.00
> **Property Developer**: Regimanuel Grey Ltd.
> **Case Property 2:**
> **Location**: Katamanso, Accra.
> **Description**: Three-bedroom house located on full-sized plot with space for the addition of an optional garage and domestic quarters. Living/dining room opens onto the front terrace. A separate multi-purposed area caters for family living. A large kitchen includes a pantry and provides access to the backyard service area. Full-sized, well-ventilated bedrooms and provisions for built-in wardrobes. The master bedroom has its own private en suite bath.
> **Plot size**: 80′ × 100′ (24.38 m × 30.48 m^2)
> **Price**: $234,000
> **Property Developer**: Regimanuel Grey Ltd.[10]
> **Case Property 3:**
> **Location**: Accra, specific neighbourhood/area not specified.
> **Description**: Condominium with two/three-bedroom apartments. Comes with wardrobe, kitchen cabinets, oven microwave oven, gas hop, heat extractor, contemporary light fittings, sockets and switches, porcelain floor tiles.
> **Foot Print**: from 45 to 99 m^2
> **Price**: $68,833.00–$77,000.00
> **Property Developer**: Devtraco Limited.[11]
> **Case Property 4:**
> **Location**: Tema Community 25.
> **Description**: Two en suite bedrooms, a living and dining area, kitchen, built-in wardrobes are available in bedrooms while the kitchen features a fitted cabinet, gas and microwave ovens, as well as a heat extractor; there is 24-h security, gym, go-kart racetrack and parking.

Rent: From $375.00 per month (Unfurnished); from $425.00 per month (furnished).
Property Developer: Devtraco Limited.

10.3 Policy Responses in the Past

The policy goals and objectives for the housing sector at the national are reflected in the National Housing Policy (2015) and the National Urban Policy (Government of Ghana 2012a). Prior to the adoption of these policy documents, earlier attempts at providing a coherent housing policy had failed to materialize. For example, as acknowledged in the National Housing Policy (NHP), the National Housing Policy and Action Plan (1987–1990), which marked the first attempt at setting an agenda for addressing the housing problems in the country was never adopted by the state. Subsequent attempts, including draft Comprehensive National Shelter Strategy (1991–1992) and review or update of the shelter strategy in 1999, 2000, 2003 and 2005 all remained on the shelf, never to be adopted for implementation by the successive governments over the period (NHP 2015).

Thus, one of the main reasons for the severe housing crisis currently facing the country is that for a very long time, a comprehensive policy agenda at the national level has been lacking. In the absence of a national policy, successive governments adopted a piecemeal approach by which the state attempted to get involved in the housing market directly to provide affordable housing schemes and/or create the enabling environment to attract private investments into the sector. For instance, in 2005, the Government of Ghana in partnership with the private sector initiated its affordable housing programme to provide some 100,000 housing units across the country (Bank of Ghana 2007). This culminated in the commencement of the Asokore Mampong Affordable Housing project in 2006 in Kumasi, the country's second largest metropolis. The affordable housing scheme would be abandoned following the 2012 election which led to a change of government. Today, squatters have invaded the uncompleted dwellings at Asokore Mampong. Around 2005, financial incentives, including a 10% reduction in corporate tax from 55 to 45%, a five-year tax holiday for new private real-estate developers and tax exemptions on housing purchases were implemented to attract investment into the sector (Arku 2009; Owusu-Manu et al. 2015). An attempt in 2009 to initiate the construction of some 300,000 affordable housing units, costing an estimated US$10 billion, under what became known as the STX Housing Project, failed to materialize (Owusu-Manu et al. 2015).

Thus, the NHP (2015) and the National Urban Policy (2012) constitutes the most recent attempts to provide a coherent agenda, indicative of the state's intent to address

---

[10] Company website: http://www.regimanuelgray.com/project_katamanso.php.
[11] Company website https://devtraco.com/.

the prevailing housing crises. Box 10.2 outlines the policy goals and objectives of the NHP as well as the initiatives and strategies for addressing housing the country's housing crisis. The NHP contains four key policy goals which reflect state's intent to secure affordable housing, provide dwellings with reduced environmental footprints, ensure access to finance for housing development and promote a participatory approach to housing development at the local levels. The policy objective in the housing sector expressed in the National urban policy is to 'improve access to adequate and affordable low-income housing'.

Clearly, these national policy statements target the supply side of the market, outlining various policy objectives and strategies to unlock housing supply. This appears justifiable because the housing problem in Ghana currently is largely supply-side problem where the lack of housing units in the face of a growing population and housing demand is naturally pushing up prices in both the rental and buyers' markets. But as history has shown, the problem in Ghana has not been with the absence of policies alone but most importantly the lack of their implementation. For instance, while the National Urban Policy predates the NHP by at least three years, to date, evidence on the ground with respect to implementation of the various strategies proposed is hard to come by. Similarly, while the NHP has been in existence for at least nearly two years as of the time of writing, concrete strategies to translate the policy into actions were yet to be seen.

Box 10.2: Housing Policy Statements in Ghana

1. **National Housing Policy (2015)**
   Policy Goals

   - To provide adequate, decent and affordable housing that is accessible to satisfy the needs of all people living in Ghana;
   - To ensure that housing is designed and built to sustainable building principles leading to the creation of green communities;
   - To ensure that there is participation of all stakeholders in decision-making on housing development and allocation in their localities; and
   - To ensure adequate and sustainable funding for the supply of diverse mix of housing in all localities.

   Policy Objectives
   The main objectives of the policy are:

   - To promote greater private-sector participation in housing delivery;
   - To create an environment conducive to investment in housing for rental purposes;
   - To promote housing schemes that maximizes land utilization;
   - To accelerate home improvement (upgrading and transformation) of the existing housing stock;

- To promote orderly human settlement growth with physical and social infrastructure;
- To make housing programmes more accessible to the poor (Social Housing);
- To involve communities and other non-traditional interest groups in designing and implementing low-income housing initiatives; and
- To upgrade existing slums and prevent the occurrence of new ones.

2. **National Urban Policy (2012)**
   Policy objective

- Improve Access to Adequate and Affordable Low-Income Housing in Ghana.

Initiatives to achieve policy objective

- Provide a congenial environment for private-sector delivery of affordable housing.
- Implement recommendations on the promotion of indigenous building materials and appropriate construction technologies.
- Promote the provision of social or low-income rental housing through public and private partnership arrangements.
- Upgrade slums and dilapidated housing stock especially in urban areas selected as growth poles.
- Explore the introduction of non-conventional Housing finance and strategies that benefit low-income groups.
- Encourage the formulation of housing cooperatives.

Source Government of Ghana (2012b) National Urban Policy Action Plan, Ministry of Local Government and Rural Development (pp 32–33)

10.4 The Planning System and Housing Delivery: Models and Approaches

Realising the housing policy objectives outlined in the previous section would require land—one of the basic inputs in housing development. While the national policy statements identify a number of objectives and strategies to unlocking supply, both documents are lacking in specific strategies to ensuring that adequate land at suitable locations are secured and released for housing development in the long term. Moreover, as identified elsewhere in this chapter, housing finance constitutes one of the major challenges in providing decent and affordable housing for the ever growing population. While the emphasis of the national-level policies is largely on securing adequate supply of housing, the potential role of the planning system, which deals with the determination of the use of land at the strategic and local levels and the distribution of land use activities is conspicuously missing.

In this section, two models for delivering housing through the spatial planning system are put forward and discussed. The planning system implied here refers to the wider institutional and legal apparatus at the national and sub-national levels responsible for the formulation and implementation of human settlement development plans and policies. The models proposed here are intended to complement the existing strategies already identified in the national policy statements by addressing the critical issues of land supply for new housing development and avenues of revenue generation to finance housing development in Ghana. The key principles underpinning each of the models will be identified. An outline of the practical approaches for implementing the proposed models will also be provided, discussing the institutional and legal factors required to support them. It is important to establish from the onset that while the proposed models are intended to be implemented at the level of local governments in Ghana, equivalent national-level policy and legal structures, although not a requirement would further enable their successful translation into action. Also, while the proposed models are discussed separately and could be deployed individually to achieve specific objectives, they are not exclusive. Instead, the two models could complement each other in helping to address the country's housing problems. The models are outlined and discussed in the sections that follow.

10.4.1 Model 1: Supplying Land for New Housing Development

Rationale and Purpose Land is a basic requirement in housing construction. Any attempt at increasing housing supply so that there is adequate and decent shelter for households of different socio-economic profiles must therefore be accompanied by innovative ways of bringing forth land at suitable locations for housing development. The land supply model proposed here is intended to serve this purpose. The model explores and advances mechanisms of securing land at a relatively cheaper cost for mass housing development through the planning system.

Principles Two key principles underpin the proposed land supply model. The first principle is founded on the equity and redistributive imperatives of spatial planning. The second principle for the model derives from the concept of land value capture in urban development. The tenets of these principles are outlined as follows:

Traditionally, Ghana's spatial planning system, where it works, has been confined primarily to making land use allocation decisions in local plans and controlling physical development. An approved local plan, which emerges from the process, provides the basis for landowners to sell individual parcels of land designated for various uses including residential development. As established in Chap. 5, in most

10.4 The Planning System and Housing Delivery: Models and Approaches

cases, however, residential land is sold by landowners without any formal local plan. In both situations, decisions as to who has access to land for housing construction and the type of housing developed is left largely to the dictates of the formal and informal land markets (Acheampong and Anokye 2013). The emergent land prices are often beyond the means of the majority of individuals and households across who have home-ownership aspirations.

Contrary to the prevailing *laissez-faire* system whereby transactions in private markets determines access to and the distribution of land for housing development, it is argued that the potential exists to realize the equity and redistributive imperatives of spatial planning in the housing sector. This can be achieved through interventions aimed at securing land outside of the traditional functioning of private land markets and making it available for low-income housing development. Ways of translating this principle into action will be discussed in the next section. Before delving into this, we first outline the second principle underpinning the land supply model.

As mentioned at the beginning of this section, the second principle of the land supply model derives from the concept of land value capture. In simple terms, the concept refers to mechanisms used to reap a share of the increase in land value induced by public sector interventions for the benefit of the larger community (Walters 2013). A typical example is the increase in value of privately held land that results from government's investment in infrastructure such as transportation, water and sewage systems.

In the context of the land supply model advanced here, we focus on a slightly different source of land value appreciation for which the principles of land value capture could be applied. This is the windfall gains in land value that accrues to landowners as a result of planning decisions. Local government policies such as the preparation of local plans and the grant of development rights not only signal that an area is ready for physical development but such decisions also result in appreciation of land value. For example, by designating land which was previously undeveloped or used for farming for new residential development in a local plan, land values in the local plan area increases because of the new uses assigned to the land. Thus, the underlying rationale is that since the landowner did nothing to bring about this upliftment in value, the local government responsible for the local plan area could reap a share of the value (Alterman 2012).

In Ghana, MMDAs through the Local Government Act (462) and the Land Use and Spatial Planning Act (Act 925) enjoy unrivalled monopoly of being the only institutions that can determine the use to which individual land should be put and grant development rights for the designated uses. Thus, although land ownership rights largely rest with the traditional authorities and private landowners in the customary land sector, the power to determine the use of land rests with the MMDAs as development authorities at local level. Following the principles outlined above, the preparation of local (residential) development plans covering formerly agricultural or vacant land leads to an increase in the value of such land, which would otherwise not have occurred. Given that such 'windfall gains' in land values result primarily from local authorities' land use planning decisions, the appreciated value in the land could be captured by the MMDAs for the benefit of the entire community. The practical ways of achieving this is what the land supply model proposed in this chapter seeks to address.

Besides local planning actions and the grant of development rights by local governments, another justification for land value capture could be made for land value increase that results from wider socio-economic trends. For instance, socio-economic changes such as population growth and economic growth over time reflect in land values (Alterman 2012). In Ghana, for example, peri-urban land values appreciate rapidly as population increases and cities expand their existing built-up areas into surrounding settlements which hitherto, remained remote, inaccessible villages where except for speculative buying, new developments would not happen for several years. In such instances, it is justifiable to conclude that any increase in land value is a windfall gain accruing to private landowners from exogenous factors such as the increase in and the favourable macroeconomic climate. Local government could implement mechanisms of capturing some of the increment in land value.

Approaches, Institutional and Legal Requirements In order to secure land through the planning system, using the set of principles outlined above, there is need for effective partnerships between MMDAs as planning authorities on the one hand and landowners (i.e. traditional authorities, families and individuals) on the other hand. For this discussion, we will focus on traditional authorities, whom through the customary land tenure system holds over 80% of available land in Ghana (Kasanga and Kotey 2001). Indeed, in addition traditional customs of governance and land ownership at the community levels in Ghana, the national constitution recognizes traditional authorities as the custodians of all lands that fall under the customary land ownership system. Thus, recognising their indispensability as custodians of the land and engaging effectively with traditional authorities can provide a huge potential for the release of vacant land for affordable housing development.

As explained under the principles of value capture, the land supply model is expected to deliver purpose through the planning decisions of local governments. Thus, for this model to work, local authority would have to be performing their spatial planning mandate by which in consultation with traditional authorities, local plans would be prepared for privately held land ahead of the sale of land and physical development. As suggested elsewhere in this book, doing this would require a shift in paradigm from planning being a reactive intervention in regularizing existing development to being proactive in anticipating new growth areas and liaising with landowners to prepare planning schemes to guide development.

Following the underlying principles of the land supply model, the preparation of local plans and the grant of development rights for new identified growth areas will result in increment in the value of land which could be captured by MMDAs. The suggested mechanism for capturing the value is that MMDAs would acquire land banks in the designated development areas in the local plans, equivalent to the appreciated value resulting from the preparation of local plans. MMDAs would rely on the expertise of their land use planning departments and the Land Valuation Division of the Lands Commission in determining the value of increment and the amount

10.4 The Planning System and Housing Delivery: Models and Approaches

of undeveloped land that is commensurate with the land value increment resulting from the local governments' planning interventions. Where possible, MMDAs could acquire additional land within the local planning area relatively cheaply at their initial agricultural value. The land could then be earmarked for affordable housing development.

It is possible that landowners would acquire the services of a consultant to prepare local plans instead of using the land use planning departments of the MMDAs. In such cases, land covered by the local plan would still be subject to value capture in the form of land banks in accordance with the principles outlined in the land supply model. This is because such a local plan would require approval by MMDAs to be legally binding.

Once land has been acquired through value capture and/or at its agricultural value, the next critical step would be the strategies to make it ready for housing development. A number of approaches could be deployed to accomplish this. Local government would provide very basic services such as plot demarcations, drainage systems and the opening up of proposed streets without necessarily tarring them. The serviced land could then be made available to households to develop their own housing and the surrounding environments using the incremental construction model. MMDAs could also explore the possibility of partnering with private real-estate developers to construct and manage affordable housing schemes for either ownership or renting depending on which options are viable. The latter option is recommended as this would allow for the design and construction of high-density, multi-storey dwellings which could provide affordable housing to a relatively large number of households and individuals.

For the proposed land supply model to work, it ought to be backed by a legislative framework. The Land Use and Spatial Planning Act (Act 924) already contains some provisions for doing this. Under the subject of recovery of betterment, the Act stipulates the following:

> (1) Where the provision of a plan, the execution of public works, or a decision or an action of a district planning authority increases the value of a land within a district, the district planning authority shall, on the advice of the body charged with the valuation of public land, determine and publish in the *Gazette,* a percentage rate to be paid as a betterment charge by a person who sells or otherwise disposes of land in the district; (2) Financial gains from land transactions are liable to betterment charges; and (3) A sum of money recoverable under this section may be set off against a claim for compensation.

Thus, with the above legal provisions as the starting point, MMDAs in consultation with landowners (i.e. traditional authorities, families and large-scale private landholders) within their respective jurisdictions would work out detailed mechanisms for achieving the objectives of the proposed land supply model. Issues that could be addressed in the local government legislations include setting thresholds for the amount of land in a local planning that would be eligible for acquisition through the land supply model. The threshold could, for example, stipulate that half an acre of land is eligible for acquisition for say every 5 acres of undeveloped land in covered by the local plan.

It is worth establishing that the proposed land supply model is not the same as government exercising compulsory land acquisition powers. In fact, compulsory acquisition of land requires landowners receiving compensations from the state. On the contrary, by this model, local government are only leveraging their planning powers to acquire land for the wider community to benefit in the form of affordable housing development. Indeed, the practice whereby MMDAs pay for local plan preparations by taken land in lieu of the cost incurred is already being adopted across the country. The only challenge with the current approach is that no transparent systems exist for accounting for the land acquired in the process. Under the proposed land supply model, one possible way of ensuring transparency and accountability would be the documentation and publication of all records of land acquired through the model in mediums readily accessible to the public.

Moreover, local governments do not need large tracts of land on a given site to be able to provide housing through this model. In fact, per the existing land use zoning regulations and the minimum residential floor space standards permitted, up to eight single-family detached/semi-detached houses could be built on an acre of land. Increasing the density of development by opting for multi-storey apartment buildings multiplies the number of dwellings that could be provided on the same piece of land. Even so, there may be cases where a large contiguous site is needed, which cannot be acquired in a single local planning area. Where large contiguous land is needed, the local authorities could accumulate small parcels of land at different sites and sell them to acquire a relatively larger land elsewhere. Land acquired at a completely new area could be expensive since the principles of value capture would not apply in such a case.

Finally, it is worth mentioning that the success of this model in bringing about an increase in the supply of housing especially for low-income households would depend to a large extent on the recognition by MMDAs that such initiatives are underpinned by social considerations of equity and distributional gains rather than profit maximization in the economic sense. On the basis of this recognition, systems can be designed to ensure that low-income households are the primary beneficiaries. In the long term, the investment cost could be recouped through property taxation and revenues from rented properties, for example.

10.4.2 Model 2: Development Exactions and Affordable Housing Quotas

Rationale and Purpose Housing finance remains one of the major challenges to reducing the housing deficit in Ghana. Even when local governments have been able to secure land using the land supply model discussed in the previous section, they would require significant amount of investment to develop adequate number of dwellings for the growing population. Local governments have the options of relying on traditional debt financing instruments such as borrowing, equity and/or partnering with the private sector to raise capital for housing development. In addition to these conventional sources of financing housing, local governments could secure housing directly through the urban development process itself. Thus, the second model advanced in this chapter outlines spatial planning instruments and strategies to raising finance from large-scale land development initiated by the private sector for housing development and/or securing complete houses as part of such projects.

Principles Like the land supply model discussed previously, the housing supply model proposed here is grounded in the principle that local authorities should engage with urban land markets and leverage planning decisions and wider socio-economic trends, which result in the increase in the value of privately held land, to raise revenues for housing development. A range of instruments and strategies collectively classified under umbrella term of 'land-based financing' are available to local governments. In the Ghanaian context, two of these instruments namely revenue generation through development exactions for housing development and affordable housing quotas are proposed. Specific approaches will be discussed later in the subsequent sections. In the meantime, two other principles underlying the proposed housing supply model, which are linked to normative ideals of building inclusive communities as well as promoting mixed-land uses for sustainable urban development are outlined.

In recent decades, there has been growing support for socially diverse, mixed-income communities. As opposed to creating exclusionary residential enclaves such as gated communities for the rich and powerful, the idea of inclusive communities is premised on the recognition that having socially diverse communities offer a means to reconnect disadvantaged groups to mainstream society (Arthurson 2002). As observed in the pioneering work of Wilson (1987), social exclusion, lack of access to socio-economic opportunities and poverty are deeply rooted in the spatial concentration of different groups of people in urban areas, particularly at the neighbourhood scale. Urban poor and low-income households and individuals tend to live in neighbourhoods of concentrated socio-economic disadvantage such as public housing estates in developed countries (Musterd and Andersson 2005; Talen 2006) or in overcrowded family-owned compound houses, poorly serviced neighbourhoods and inner-city slums as is the case in cities in Ghana and most developing countries.

Dwellings are the basic building blocks of neighbourhoods. This implies that approaches to deliver housing that meets the tenure preferences and affordability needs of different groups of households and individuals could form an integral part wider strategy to bringing about socially diverse communities. This, in turn, could help facilitate access to social capital, essential facilities and services and economic opportunities especially among poor and low-income households, which otherwise would not be possible with segregated residential neighbourhoods.

Moreover, having mixed-use development is crucial to achieving sustainable urban development outcomes. Often, as a result of exclusionary land use zoning policies, large-scale commercial and industrial developments undertaken by private real-estate developers become separated from residential areas. However, inclusive zoning policies coupled with local planning policies that encourage mixed-use development such as mixing commercial and residential uses in inner-city locations could form part of wider strategies to providing housing for the urban population while reducing the physical separation between the place of work and place of employment, and attendant travel costs and unsustainable travel behaviours.

Approaches, Institutional and Legal Requirements Under the development exactions and affordable housing quotas model, three main relatively simpler approaches by which local governments can finance housing through the urban development process are put forward. The first two approaches focus on the private residential real-estate markets and see the interaction between local governments land use planning and development approval decisions on the one hand and private real-estate developers' investments as an avenue to meeting narrowing the gap between housing supply for and the prevailing demand among different income groups. The third approach focuses on how local governments could leverage medium to large-scale non-residential developments initiated by private developers to contribute to reducing the country's housing deficit. The three approaches are discussed below in connection with the underlying principles of the model outlined earlier.

As was previously indicated, the private real-estate industry has in recent years experienced substantial growth in Ghana. Over this period, membership of GREDA has increased to over 150 developers the majority of whom are based in Accra the capital but have sites in almost all the major towns and cities across the country. A previous study by Agyemang and Morrison (2018) which also explored the potential of unlocking housing supply using the spatial planning system mapped about 135 residential development sites across over 15 districts in the Greater Accra Region being developed by private real-estate developers. The study also found that residential sites varied from relatively small-scale development providing around 50 dwellings to very large-scale developments providing up to 1000 homes.

While these developers obtain their development rights from MMDAs and pay the statutory fees, to date, mechanisms of linking the grant of development rights to housing provision remain explored. However, following the principles of land value appreciation resulting from public sector actions, we know that the grant of planning permission to these private developers not only legitimize their development and provide certainty for the huge capital investments involved but also directly lead to increment in the value of land, especially in instances where the land was undeveloped or reserved for agricultural use.

On the above premise and using the powers granted by the decentralization law and the Land Use and Spatial Planning Act (Act 925), local governments could capture some of the value increase accruing to the private landholder by setting affordable housing quotas for new residential developments proposed by private real-estate developers. The objective of the quota, in principle, would be that in addition to targeting the higher end of the market, which private developers invariably do so as to maximize returns on their investments, a certain percentage of the total development proposed by the developer must be constructed to be affordable. Informed by the market power of less affluent groups, local governments would work with developers on cost-effective building designs and ways of minimizing overall costs of development in order to ensure that dwellings achieved through the quota system are genuinely affordable. For example, instead of single-family detached and semi-detached homes, the percentage of the development eligible for the affordability quota could be designed and built as high-density flats so as to realize more affordable homes on a given site. In return for the affordable housing quota, local governments could either reduce statutory fees that would have applied on the proposed development or completely eliminate such fees for the developer.

As an alternative to and/or in conjunction with the use of affordable housing quotas described above, local governments could also capture the value accruing from the grant of planning permission and other related exactions such as development impact fees in direct monetary payments from developers. The revenues generated could then be invested by local governments on their own sites acquired through the land supply model advanced previously and earmarked for affordable public housing development. In keeping with the goal of achieving inclusive communities as opposed to residential enclaves developed exclusively for the wealthy, however, local governments should aim, where feasible, to achieve their affordable housing supply quotas on sites proposed for large-scale residential developments by private developers. Moreover, the revenue generated through development exactions could be invested in maintaining dwellings and expanding infrastructure.

The final approach under this model identifies ways by which local governments, using their planning systems could realize housing supply through the grant of applications for medium to large-scale non-residential development. Commercial developments such as retail parks, shopping malls, and light industrial and office developments are compatible with residential uses. In fact, the planning standards and zoning regulations encourage mixed-use developments involving residential and non-residential uses and provide guidelines for such developments. The strategy of providing housing as part of these developments on the same site could be pursued by local governments. Using a procedure similar to the affordable housing quota system outlined previously, local governments would set out clear policies on the percentage of such large development that developers should design and build for residence. The policies could then be applied on a case-by-case basis to new developments as well as the redevelopment of existing sites and buildings.

In the country's, major cities (i.e. Accra, Kumasi and Sekondi-Takoradi), inner-city redevelopment schemes are gradually replacing existing residential uses with dominantly commercial uses. This is response to increasing land values and changing socio-economic conditions that have culminated in the increasing demand for land for commercial uses such as malls and offices. The ongoing redevelopment process not only has real impact in reducing the stock of housing in prime urban locations but also displaces populations, especially urban poor and low-income households from city centres. Reversing the ongoing replacement of inner-city housing with commercial development and the associated redevelopment-induced displacements, city authorities could leverage their spatial planning and planning permission powers to promote mixed-use developments that meet the growing demand for commercial uses while making sure that the housing needs of the population are being met.

10.5 Conclusion

This chapter has examined the housing situation in Ghana, identifying the demand- and supply-side challenges. The discussion has shown that there is a huge housing deficit, which has emerged from the combined failures of private markets and public policy to provide adequate and decent housing for households of different socio-economic means. The discussion has shown that individuals and households wishing to develop housing through the self-build incremental model are faced with problems including access to land and housing finance. Limited supply of housing in the face of increasing demand has also created severe affordability challenges for households in the rental sector.

Moreover, since the inception of the neoliberal era in the 1980s, the state's involvement in direct housing development has gradually decreased to a point where it is now considered virtually non-existent. While the resulting economic liberalization and various market-enabling interventions from the government have provided some favourable conditions for the private real-estate industry to thrive, the discussion has shown that homes delivered through this channel are far beyond the means of the majority of Ghanaians. In recent years, governments' attempts at partnering with the private sector to build housing have also not yielded any significant results in narrowing the wide gap between housing need and supply.

Against the backdrop of the prevailing challenges in the housing sector, especially on the supply side of the market, this chapter has proposed two new, non-conventional models through which additional housing could be developed directly through the spatial planning system. The two models reflect relatively simpler mechanisms of land-based housing finance strategies, which are founded primarily on the principles of land value capture. The key underlying principle is that local governments' planning decisions as well as wider socio-economic development trends induce increases in the value of privately held land, which could be captured to finance housing development. The proposed models are also linked to achieving broader normative goals of spatial planning including planning for equity and redistribution of resources, and promoting inclusive communities and high-density, mixed-used neighbourhoods.

In the first model, which is the land supply model, approaches of securing land at relatively cheaper cost for mass housing development are advanced. This would involve MMDAs acquiring land banks in designated development areas in local plans, which is equivalent to the appreciated value of privately held land induced by the preparation and approval of the local plans and the subsequent grant of development rights. The second model would involve local governments designing and implementing various development exactions and setting affordable housing quotas. These instruments would be applied the point of grant of planning permission to proposed residential development schemes by private real-estate developers as well as to medium to large non-residential new development and redevelopment schemes.

Provisions in existing legislative frameworks (i.e. the Local Government Act (Act 462)) and Land Use and Spatial Planning Act (Act 925) set out the spatial planning mandate and competencies of local governments and grant them the relevant powers

10.5 Conclusion

which could be leveraged to develop and implement the models proposed in this chapter. Ultimately, the proposed models are intended bring forth land and raise finance—the two most critical housing development inputs—and complement other strategies identified in the national housing policy to increase supply of decent and affordable homes for the growing population.

References

Acheampong RA (2016) The family housing sector in urban Ghana: exploring the dynamics of tenure arrangements and the nature of family support networks. Int Dev Plan Rev 38(3): 297–316

Acheampong RA, Anokye PA (2013) Understanding households' residential location choice in Kumasi's peri-urban settlements and the implications for sustainable urban growth. Res Humanit Soc Sci 3(9): 60–70

Acheampong RA, Anokye PA (2015) Housing for the urban poor: towards alternative financing strategies for low-income housing development in Ghana. Int Dev Plan Rev 37(4): 445–465

Agyemang FS, Morrison N (2018) Recognising the barriers to securing affordable housing through the land use planning system in sub-Saharan Africa: a perspective from Ghana. Urban Stud 55(12): 2640–2659

Alterman R (2012) Land use regulations and property values: the 'Windfalls Capture' idea revisited. In: Brooks N, Donaghy K, Knaap (eds) The Oxford handbook of urban economics and planning. Oxford University Press, pp 755–786. Available https://ssrn.com/abstract=2309571

Amoako C, Boamah EF (2017) Build as you earn and learn: informal urbanism and incremental housing financing in Kumasi, Ghana. J Hous Built Environ 32(3): 429–448

Arku G (2009) Housing policy changes in Ghana in the 1990s. Hous Stud 24: 261–272

Arku G, Luginaah I, Mkandawire P, Baiden P, Asiedu AB (2011) Housing and health in three contrasting neighbourhoods in Accra, Ghana. Soc Sci & Med 72(11): 1864–1872

Arthurson K (2002) Creating inclusive communities through balancing social mix: a critical relationship or tenuous link? Urban Policy Res 20(3): 245–261

Asante LA, Gavu EK, Quansah DPO, Tutu DO (2017) The difficult combination of renting and building a house in urban Ghana: analysing the perception of low and middle-income earners in Accra. GeoJournal 1–15. https://doi.org/10.1007/s10708-017-9827-2

Bank of Ghana (2007) The housing market in Ghana, sector study report prepared by the research department of the Bank of Ghana, http://www.bog.gov.gh/privatecontent/Research/Research%20Papers/bog%20housing.pdf (accessed 12 November 2013)

Brush BL, Gultekin LE, Grim EC (2016) The data dilemma in family homelessness. J Health Care Poor Underserved 27(3): 1046–1052

Burges G, Monk S (2008) Delivering affordable housing through the planning system: challenges and good practice. Hous Care Support 11(3): 4–8

Ferguson B, Smets P (2010) Finance for incremental housing; current status and prospects for expansion. Habitat Int 34: 288–298

Ghana News Agency (2017) http://www.ghananewsagency.org/social/ghana-has-1-7-million-housing-deficit-works-and-housing-minister-117041

Ghana Statistical Service (2012) 2010 population and housing census. Final Results. Accessed 7 July 2017. http://www.statsghana.gov.gh/docfiles/

Ghana Statistical Service (2014) Ghana living standards survey, round six. Main Report. http://www.statsghana.gov.gh/docfiles/glss6/GLSS6_Main%20Report.pdf

Gibbon P (1993) Social change and economic reform in Africa. Nordiska Afrikain-stitutet, Uppsala

Gough KV, Yankson PW (2000) Land markets in African cities: the case of peri-urban Accra, Ghana. Urban Stud 37(13): 2485–2500

Gough K, Yankson P (2011) A neglected aspect of the housing market: the role of caretakers in peri-urban Accra, Ghana, Urban Stud 48: 793–810

Government of Ghana (2012a) National urban policy. Ministry of local government and rural development. http://www.ghanaiandiaspora.com/wp/wp-content/uploads/2014/05/ghana-national-urban-policy-action-plan-2012.pdf. Accessed 24 July 2018

Government of Ghana (2012b) National urban policy framework. Action plan. Ministry of local government and rural development. http://www.washwatch.org/uploads/filer_public/38/d4/38d4a952-f123-479c-ae10-8623b91582d8/national_urban_policy_framework_ghana_2012.pdf. Accessed 24 July 2018

Government of Ghana (2015) National housing policy. Ministry of works and housing. http://www.gredaghana.org/policy/National%20Housing%20Policy.pdf

Grant R (2005) The emergence of gated communities in a West African context: evidence from Greater Accra, Ghana. Urban Geogr 26(8): 661–683

Gurran N, Whitehead C (2011) Planning and affordable housing in Australia and the UK: a comparative perspective. Hous Stud 26: 1193–1214

Habitat UN (2011) Ghana housing profile. United Nations Human Settlements Programme, Nairobi

Hilson G, Potter C (2005) Structural adjustment and subsistence industry: artisanal gold mining in Ghana. Dev Chang 36(1): 103–131.

Johnson MP (2007) Planning models for the provision of affordable housing. Environ Plan 34: 501–523

Kasanga RK, Kotey NA (2001) Land management in Ghana: building on tradition and modernity

Konadu-Agyemang K (2001) A survey of housing conditions and characteristics in Accra, an African city, Habitat Int 25: 15–34

Korboe D (1992) Family-houses in Ghanaian cities: to be or not to be? Urban Studies 29(7): 1159–1171

Kwofie TE, Adinyira E, Botchway E (2011) Historical overview of housing provision in pre and post-independence Ghana. In: Laryea S, Leiringer R, Hughes W (eds) Proceedings of West Africa built environment research (WABER) conference, Accra, Ghana, 19–21 July 2011, pp 541–557

Land use and spatial planning Act (2016) Act 925 The nine hundred and twenty fifth act of the parliament of the Republic of Ghana

Local Government Act (1933) Act 462 The four hundred and sixty two act of the parliament of the Republic of Ghana

Luginaah I, Arku G, Baiden P (2010) Housing and health in Ghana: the psychosocial impacts of renting a home. Int J Environ Res Public Health 7(2): 528–545

Moser C (ed) (2007) Reducing global poverty: the case for asset accumulation. Brookings Institutions Press, Washington D.C.

Musterd S, Andersson R (2005) Housing mix, social mix, and social opportunities. Urban Aff Rev 40(6): 761–790

Obeng-Odoom F (2010) Drive left, look right: the political economy of urban transport in Ghana. Int J Urban Sustain Dev 1(1–2): 33–48

Owusu-Manu D, Edwards DJ, Badu E, Donkor-Hyiaman KA, Love PED (2015) Real estate infrastructure financing in Ghana: sources and constraints. Habitat Int 50: 35–41

Rothchild D (1991) Ghana and structural adjustment: an overview. In Rothchild D (ed) Ghana: the political economy of reform. Lynne Rienner Publishers, Boulder

Sheuya SA (2007) Reconceptualising housing finance in informal settlements: the case of Dar Es Salaam, Tanzania, Environ Urban 19: 441–456

Talen E (2006) Design for diversity: evaluating the context of socially mixed neighbourhoods. J Urban Des 11(1): 1–32

Tipple AG, Korboe DT, Garrod GD (1997) A comparison of original owners and inheritors in housing supply and extension in Kumasi, Ghana. Env Plan B 24: 889–902

Toro PA, Tompsett CJ, Lombardo S, Philippot P, Nachtergael H, Galand B, MacKay L et al (2007) Homelessness in Europe and the United States: a comparison of prevalence and public opinion. J Soc Issues 63(3): 505–524

References

United Nations (2014) World urbanization prospects, the 2011 revision. Population Division, Department of Economic and Social Affairs, United Nations Secretariat

UN-Habitat (2016) Slum Almanac 2015–2016. UNON, Publishing Services Section UN-Habitat, Nairobi

Walters LC (2013) Land value capture in policy and practice. J Prop Tax Assess Adm 10(2): 5

Wilson WJ (1987) The truly disadvantaged, the inner city, the underclass and public policy. Univ. of Chicago Press, Chicago

Wilson S (2011) Planning for inclusion in South Africa: the state's duty to prevent homelessness and the potential of "Meaningful Engagement". Urban Forum 22(3): 265

Yeboah E (2005) Housing the urban poor in twenty-first century Sub-Saharan Africa: policy mismatch and a way forward for Ghana. GeoJournal 62: 147–161

Chapter 11
Integrated Spatial Development and Transportation Planning

Abstract The spatial distribution of population and activities on the one hand and patterns of flows and interaction on the other hand are strongly connected. This chapter deals with the interface between spatial development and transportation at the strategic, metropolitan/city and neighbourhood scales. It takes an integrated, cross-disciplinary perspective to discuss the relationships between the spatial structure of towns and cities and emergent mobility patterns and travel behaviours. The influence of urban spatial structure on mobility patterns in Ghana is explored. The need for an integrated approach for spatial development planning and transportation planning is highlighted, followed by discussion of specific strategies for achieving the imperatives of land use and transport integration.

Keywords Spatial structure · Transport · Land use–transport interaction Built-environment non-motorized transport · Spatial planning · Ghana

11.1 Introduction

The need for interaction would not arise without the spatial separation among activities, and without transportation infrastructures, it would be difficult, if not impossible to overcome distance in order to reach different activity locations. In essence, transportation systems and the spatial distribution of activities co-determine each other at different spatial scales. At a much larger scale (e.g. national and regional scales), transportation systems play the fundamental role of shaping the structure and organization of space. Strategic transportation infrastructures such as national highways, railways and airports connect settlements of different sizes, forming a network of activity nodes, which together reflect the distribution of activity centres as well as the physical and functional interconnections among them.

Moreover, from a complex systems' point of view, the physical structure of any settlement can be broadly classified into its land use and transportation systems. Thus, at the city and neighbourhood scales, there is a two-way dynamic relationship between the land use and transportation systems. The land use system at these meso- (i.e. city or metropolitan) and micro-(neighbourhood or community) scales reflect the location of various activities such as residential, employment, shopping, education and recreation. The transportation system, comprising the different network of roads, railways and waterways as well as the different modes of transport and ancillary transport infrastructure, provides the means of moving between the spatially separated activities—the land use system. Moreover, in a dynamic urban system, the sitting and development of new land use activities can affect the transportation system through the development of new infrastructure and/or expansion of existing ones. The new transportation infrastructure also tends to attract new land use activities and/or alter existing land uses.

While as constituent parts of cities the land use and transportation systems are strongly connected, historically, the theoretical traditions, knowledge, tools and skills in the academic disciplines of spatial planning and transportation planning have evolved separately. Indeed, in most universities, spatial planning and transportation studies are still considered separate disciplines and are often situated in different academic departments or in different government Ministries agencies and departments at the national and sub-national levels.

In recent years, however, emphasis on interdisciplinary research and collaboration is bringing the two fields together. In fact, over the last six decades, a whole field of land use and transport interaction (LUTI) studies has emerged as interdisciplinary domain bringing together the hitherto separate disciplines of land use planning and transportation planning. Considerable amount of research in the field has gone into understanding the extent to which built-environment factors such as the density and diversity of land uses influence travel behaviour (see, e.g., Cervero and Landis 1997; Naess 2013; Aditjandra et al. 2013). Other research efforts are also focused on the development of state-of-the-art dynamic simulation models as decision-support systems for integrated land use development and travel demand management policies (see, e.g., Batty 2013; Wegener 2004; Waddell et al. 2010). In addition, the need to coordinate spatial development and transportation underpins contemporary concepts and principles in planning and urban design, such as transit-oriented development, polycentricity, walkable neighbourhoods and bike-friendly cities, as well as the policy strategies employed to achieve them in practice.

Although the interdisciplinary field of LUTI studies endeavours to integrate land use and transport planning issues, in practice, the institutional competences for land use or spatial planning on the one hand and transportation planning on the other hand are situated in separate public-sector Ministries, departments and agencies. In Ghana, for example, LUSPA with NDPC at the national level and TCPDs at the regional and local government levels have competences in spatial planning and urban development issues. The mandate and competences for transportation issues on the other hand rest with separate institutions such as the Ghana Highway Authority at the national level and the Urban and Feeder Road Departments as well as the Urban Passenger Travel Units at the local government levels. Thus, in practice, as we learnt by exploring the landscape of multi-level policy integration in Chap. 6, these institutions formulate and implement programmes, plans and projects separately, only co-ordinating their

11.1 Introduction

activities on an ad hoc basis where there are overlapping functions and in matters of strategic interest.

This chapter is about the interface between spatial development and transportation. Drawing on the traditions of the field of LUTI studies, the chapter takes an integrated, cross-disciplinary view to discuss the relationship between the domains of spatial planning, which is responsible for the determination of the distribution of land use activities and transport planning, which is concerned with the planning and development of the physical infrastructure linking the land use activities as well as the management of emergent travel demand. The chapter is organized as follows; Firstly, we will explore the links between spatial development and transportation using a conceptual diagram to illustrate how the relationship manifests at the strategic (i.e. national and regional), city and neighbourhood scales. Relating the conceptual diagram to Ghana, the wider institutional arrangements and the different types of policy instruments through which policies affecting spatial development and transportation are articulated and implemented will be identified.

On the premise of the broad framework of understanding mentioned above, the remainder of the chapter will focus on spatial development and transportation issues in the context of urban areas in Ghana. A generalized framework will be put forward to depict the spatial structure of a typical metropolis in Ghana, which will then be applied to explore the structural conditions underlying observed patterns of spatial flows and interaction in metropolitan areas in Ghana. The main focus here will be on understanding how existing and emerging transportation problems such as traffic congestion, longer commuting times, increasing motorization and environmental pollution as well as low levels of active, non-motorized transportation such as bicycling are deeply rooted in the structure and organization of land use activities in cities, towns and neighbourhoods. Ways in which spatial development policies and transportation policies could be integrated to bring about sustainable development outcomes will be outlined and discussed.

11.2 Conceptualizing the Spatial Development and Transportation Nexus

The links between the spatial distribution of activities and transportation could be conceptualized at three levels, namely strategic level (i.e. national and regional), metropolitan/city scale and the neighbourhood scale (see Fig. 11.1). At each of these three levels, the interaction between the two systems often reflect in various programmes, policies and plans such as Spatial Development Frameworks, Transportation Plans, Infrastructure Plans, Structure Plans and Local Plans, which shape the distribution of various activities and the spatial linkages between them. The relationship further manifests physically in the spatial organization of settlements at the macro-scale and the various land use activities within them at the meso- and

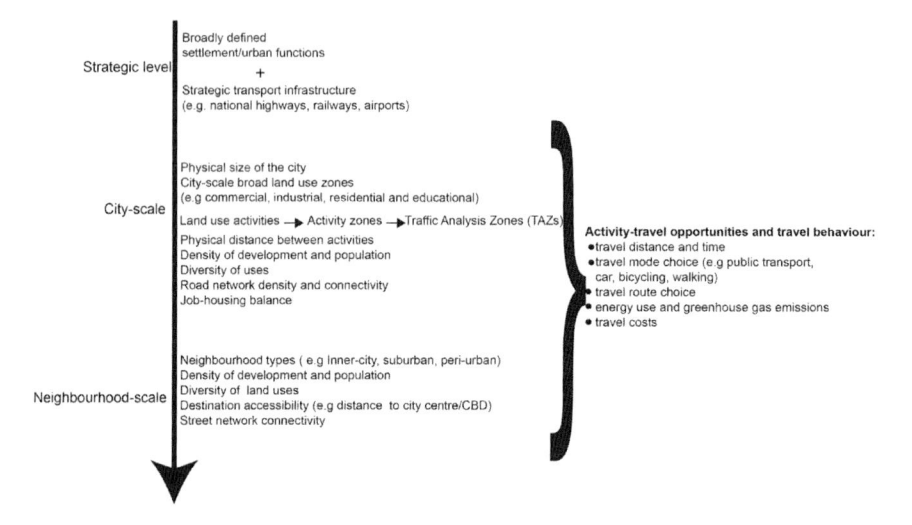

Fig. 11.1 A three-level conceptual illustration of the links between spatial development and transportation

microspatial scales where the transportation system provides the means by which people and goods move between various activity locations.

It is important to mention that in modern times, Information and Communication Technology (ICT) has become integral component of transportation systems, revolutionizing the ways people, goods and services are moved between different locations. With ICT, it is no longer possible to discern the physical traces of spatial interaction as the need for certain trips from one activity location to another may be replaced by over-the-Internet communications and transactions. Even in instances where conventional trips such as using the car, public transport or bicycling are still necessary in order for people to participate in spatially separated activities, ICT infrastructure and devices are enabling and changing traditional ways of travel (Mokhtarian et al. 2006; Ettema 2017).

ICT infrastructure and devices are enabling travel in many ways. For example, because of ICT, we are able to send and receive travel-related information instantly either on our smartphones or on information display screens at bus stops, airports and train stations. Moreover, continuous advancement in ICT has been one of the major driving forces of the emerging paradigm of shared-mobility services in urban areas. ICT systems and smartphone apps have now made it possible for cities to implement shared-mobility services such as car-sharing and bike-sharing schemes and Uber ride-hailing services schemes. With shared-mobility schemes, individuals instead of owning a car or a bike can register with a service provider, book and locate a car or bike for use it when needed emerging transportation technologies such as autonomous cars and connected autonomous vehicles are all leveraging ICT to revolutionize urban transportation.

11.2 Conceptualizing the Spatial Development and Transportation Nexus

Thus, in conceptualizing the complex relationship between the spatial distribution of activities and transportation at the strategic, city/metropolitan and neighbourhood scales, the emphasis is not only on interactions of the two systems that can be observed and anchored spatially. Instead, beyond the traditional view and understanding of spatial interaction and flows, it is important to account for the profound impacts that ICT has had and will continue to have on the movement of people, goods and services. In the sections that follow, the nature of the relationship between spatial development and transportation at the three levels is explained, illustrating these concepts with examples from the context of Ghana.

11.2.1 Spatial Development and Transportation at the Strategic Level

At the national and regional[1] levels, population and economic activities tend to concentrate in settlements of different sizes. The observed distribution of activity and population concentrations is often path-dependent, underpinned by a combination of historical, geographical, climatic, sociocultural and political factors. Overtime, the largest concentrations of economic activities and population that emerge from the interplay among these factors assume prominence in the national space-economy, becoming the key centres for economic, administrative and sociopolitical functions.

National- and regional-level spatial planning, working in tandem with other public policy domains such as transportation, telecommunication and economic development and finance, plan and fund the development of strategic transportation infrastructure to provide the spatial linkages between these regional activity centres. Strategic spatial planning, deployed through various instruments such as NSDFs, RSDFs and their accompanying Transportation Infrastructure Plans, plays the important role of formulating the overall spatial strategy that ultimately affect the location and development of different types of transportation infrastructure such as national highways, railways and airports. These transport infrastructures in turn become the critical networks that accentuate the economic importance of major activity nodes while providing the means to realize the benefits to economic growth that accrues from the improved spatial and economic linkages among them.

In Ghana, examples can be found on the manifestations of the strong connection between spatial development and transportation, especially in cases where the latter has played the strategic role of connecting resource-rich areas in the southern half of the country where relatively larger concentrations of population have also existed historically. A case in point is what has become known as the 'Golden Triangle'—a large economic zone covering the natural resource-rich regions in the southern part of the country (see Fig. 11.2). While the abundance of natural resources and favourable climatic conditions in areas within the 'Golden Triangle' offered the

---

[1]The term regional is used here in the generic sense to refer both administrative regions and functional regions, special regions and city-regions such as those discussed in Chap. 4.

necessary conditions for growth and prosperity, it was actually the establishment of strategic transportation infrastructures, mainly roads and railways that would facilitate the exploitation of resources and generate economies of scale (Songsore 2003). Moreover, through the expansion of transportation infrastructures connecting areas within the 'Golden Triangle' to other parts of the country, it would become possible to distribute resources from their initial points of exploitation to benefit other parts of the country. Today, more than half of Ghana's population (i.e. 56.2%)[2] live in the 'Golden Triangle'. This economic zone also accounts for an estimated 61.41% of the country's total gross domestic product (GDP).[3]

Outside of the 'Golden Triangle', a network of national truck roads connects major towns and cities across the country. Strategic transportation infrastructure not only articulates the economic importance of the individual towns and cities, but also facilitates mutual interdependencies among them, creating the network of flows and interactions, tangible and non-tangible which underpin the overall performance of national economy. In addition, the interdependencies between the major towns and cities and their surrounding rural areas are articulated and realized through existing transportation systems. For example, in Ghana, trunk roads and feeder roads connect regional and district capitals to their surrounding rural areas for raw materials such as agricultural produce while providing the means by which the surrounding rural areas can access the benefits of the economic, social and political functions provided by the major towns and cities.

Besides facilitating exchange between existing activity nodes at the national and regional levels, the development of strategic transportation infrastructures such as airports and inland ports could, in themselves, be seen as a strategy to develop new regional activity nodes and/or as a means of connecting newly designated, non-transportation functions such as the creation of new industrial or financial districts to existing regional activity centres. A typical example of transportation infrastructure generating new centres of economic activity while providing the needed linkages to existing activity nodes is the proposed development of the inland port at Boankra in the Ashanti Region of Ghana. Once completed, the inland port would become a major bulk-breaking point, facilitating the distribution of goods from the main port in Tema, Accra, to other parts of the country and neighbouring countries. The inland port, besides its primary bulk-breaking function, would also generate secondary land use activities including warehousing, real estate and commerce, thereby becoming a major economic activity node in its own right over time.

Historically, strategic transportation infrastructures such as trunk roads, railways and airports have been established to facilitate the physical movement of people, goods and services. In recent years, however, the development of ICT infrastructure, especially in the telecommunication sector, has seen rapid progress. Following the

---

[2]These are proportion of country's population living in the Ashanti (19.5%), Greater Accra (16.3%), Eastern (10.6) and Western (9.8%) regions according to the 2010 national population and housing census.

[3]Using data from the 2014 Ghana Statistical Services National Accounting Statistics, the NSDF estimates that the Greater Accra, Ashanti, Eastern and Western regions generated 22.39, 18.95, 10.48 and 9.59% of national GDP, respectively.

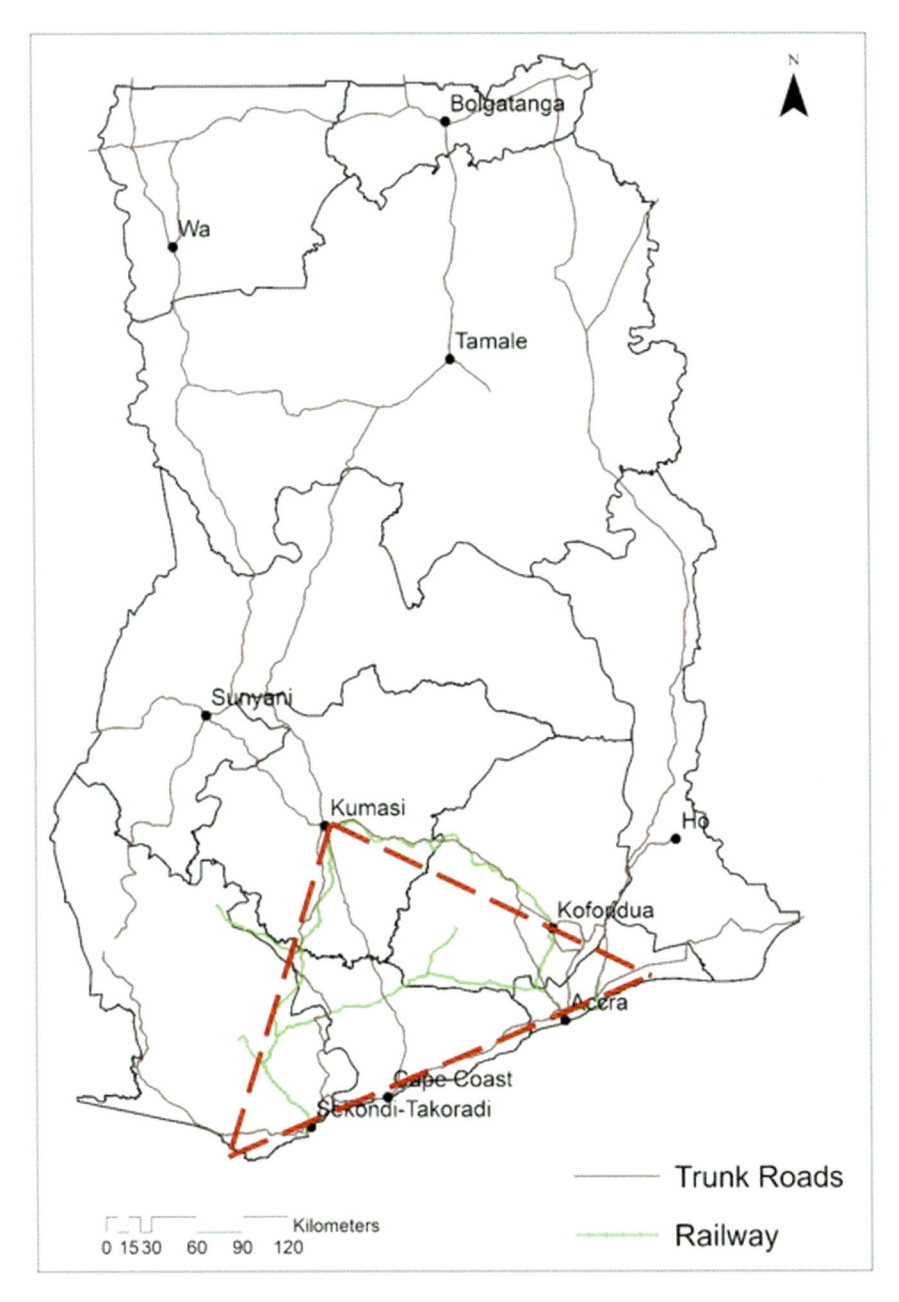

11.2 Conceptualizing the Spatial Development and Transportation Nexus   237

Fig. 11.2 Road and railway transport infrastructure networks in Ghana

formulation of the National Telecommunications Policy (NTP) in 2005, the country has attracted huge investments in the telecommunication sector from the private sector. Critical ICT backbone infrastructures such as fibre-optic networks are being expanded throughout the country (see Fig. 11.3). For example, in 2015, the 800-km Eastern Corridor fibre-optic network, which runs through 20 districts and over 120 town from the from the south eastern part of the country to the north-east, was completed. Several economic activities, especially those in the financial and service industries, have benefited from these developments. Beyond these sectors, the potential also exists to leverage the existing ICT infrastructure to improve the country's transportation systems, especially in urban areas.

11.2.2 Spatial Development and Transportation at the City and Neighbourhood Levels

At the city scale, spatial development manifests in the various land uses, including residential, commercial, industrial and educational activities (see Fig. 11.1). By designation, these land uses reflect the location of specific urban activities such as the place of residence, workplace, shopping, schools and recreation. Distinct neighbourhoods have their land use distributions, which from a bottom-up perspective could be considered sub-systems within the larger urban system. Neighbourhoods have different characteristics. They could be inner city, suburban or peri-urban, with variations in the density and variety of land uses as well as the socio-economic profiles of their residents. Thus, in terms of spatial structure, individual neighbourhoods come together to define not only the physical size of the city but also the attributes of built environment that is unique to different parts of the city.

Depending on the size of the city or neighbourhood, the land use structure may be divided into broad activity zones, with some with some zones having a dominant function such as commercial or residential and others having mixture of the various land use activities. These activity zones, in turn, provide the basis for the delineation of Traffic Analysis Zones (TAZs)—aggregate spatial units used in traditional transportation planning models to examine trip production and distribution patterns in cities (see, e.g., Martinez et al. 2007; McNally 2000).

The land use structure of observed at the city or neighbourhood scales is shaped by a number of processes. Two of the profound forces are the location decisions of urban actors and formal urban land use planning. The location decisions of several urban actors including individuals, households, private real-estate developers, firms and public-sector institutions, collectively determine the spatial distribution of land use activities. The emergent land use patterns, in turn, provide the structural conditions within which flows and interactions between locations occur daily and respond to each other over time (Acheampong 2017a; Pinjari and Bhat 2011).

The extent to which these urban land and property market decisions are shaped by spatial planning policies varies from one context to another. For example, in

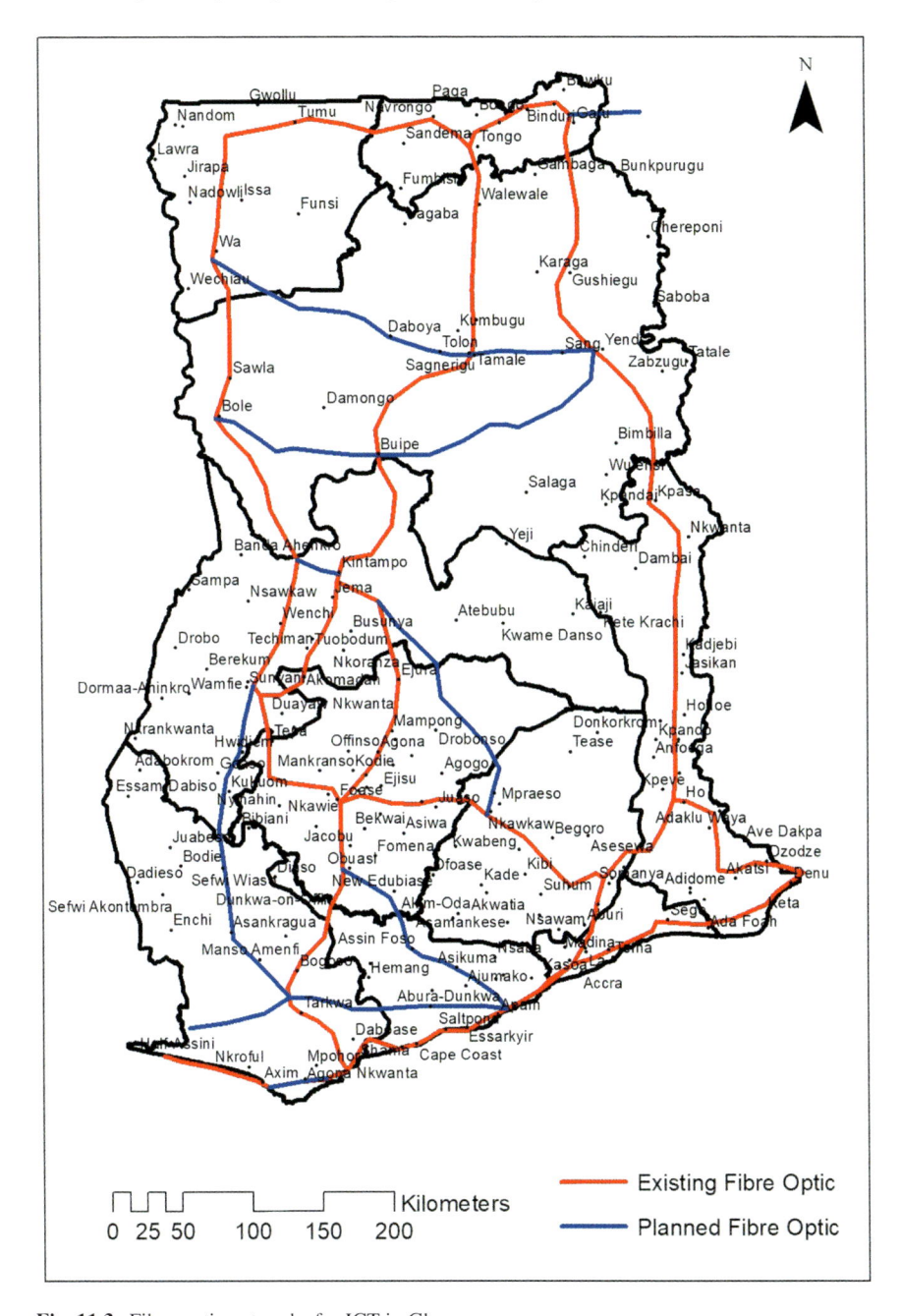

11.2 Conceptualizing the Spatial Development and Transportation Nexus 239

Fig. 11.3 Fibre-optic networks for ICT in Ghana

contexts where the spatial planning system efficient, planning tends to be proactive in identifying future growth areas, and using instruments such as zoning, designate various land uses to guide the type, extent and intensity of development in an area. Thus, urban actors' location decisions are constrained by predetermined land use planning regulations and standards which affect, for example, the density and variety of uses that could be realized within the planned area. In other contexts, such as Ghana where as a result of weak planning systems and complex land tenure systems, physical development precedes planning, the impact of formal spatial planning on the emergent location choices of actors and quality and quantity of physical development tends to be limited.

The urban transportation system comprises the various infrastructures and transport modes that enable interaction between activity locations. It includes the network of physical infrastructure such as roads, railways, walking and cycling paths and waterways. Within the system, the network of infrastructure facilitates movement of people and goods using the available transportation modes including buses, cars, trains, trams and bicycles. Moreover, facilities such as terminals, bus stops, car parks and bicycle racks complement the existing networks and available travel modes to ensure that the transportation system delivers its functions effectively.

It is important to note that transportation infrastructure is a major component of land use. Indeed, roads, railways, walking and cycling paths are built on land and therefore, by definition, could be considered as part of the land use system. At the same time, these networks of infrastructure could be considered as forming part of the transportation system when we consider the accessibility function they play by making it possible for people and goods to move between activity locations.

The nature of the interaction between the urban land use system and the transportation system has been conceptualized in various ways. One of the simple, yet insightful frameworks for understanding the land use–transport nexus is the 'land use transport feedback cycle' advanced by Wegener (2004). The framework sees the relationship as occurring through a complex two-way dynamic process. The distribution of land use determines the location of activities with the resulting spatial separation between the land use activities creating the need for spatial interaction or travel. The transport system creates opportunities for interaction or mobility, which can be measured as accessibility. For example, the transportation system facilitates access to employment, education, health care and social interactions. The distribution of accessibility in space, over time, co-determines location decisions and so results in changes in the land use system.

From the 'land use transport feedback cycle' framework, we learn that they key concept that connects the two systems is accessibility. In the land use and transportation systems, accessibility is a function of the quantum of activity locations such as the variety of shops and opportunities such as jobs that can be reached by individuals (Hanson and Giuliano 2004). Accessibility impacts land values. For example, when a new road is constructed or an existing one is expanded, the value of adjacent land tends to increase. Moreover, accessibility levels, whether perceived or objective, affect the location behaviour of households and firms, which in turn, shapes observed patterns of spatial interaction.

Another framework for conceptualizing the land use–transport nexus is the 'Brotchie triangle' (Brotchie 1984) The framework shows the relationship land use—transport links on the axes of spatial structure/dispersal and spatial interaction. The former being the degree of decentralization of land use activities such as employment and the latter taken as some measure of travel (e.g. trip distance and commuting time). The framework allows to plot various hypothetical combinations of spatial structure and their mobility implications, starting from a monocentric structure in which there is zero dispersion of jobs, to highly decentralized urban structures in which all jobs are as dispersed as population. Thus, for analytical purposes, the 'Brotchie triangle' can be used to depict the possible constellations of urban spatial structure and spatial interaction (Lundqvist 2003).

Evidence of the relationship between land use and transportation at the city and neighbourhood scales has been found through empirical research undertaken in various contexts. The evidence shows that built-environment attributes such as density, diversity of uses, destination accessibility and road network density and connectivity influence travel behaviour. For example, studies (see, e.g., Acheampong 2017a; Gim 2013; Naess 2013) have shown that people who live in inner-city locations with higher housing or employment destinies have lower levels of car ownership and relatively shorter commutes using public transport or non-motorized transport. Suburban residents on the other hand are more likely to use motorized transport, which is associated with higher energy consumption, pollution and urban traffic congestion. Moreover, the relatively shorter distances associated with denser residential neighbourhoods with mixed land uses have been found to be associated with increased use of the bicycling for transportation (Pucher and Buehler 2008; Beenackers et al. 2012; Cervero and Duncan 2003).

11.3 Urban Spatial Structure and Mobility Patterns in Ghana

Similar to other cities in around the world, urban spatial structure underpins the patterns of spatial interaction in towns and cities in Ghana. In the context of this discussion, the term spatial structure is used to invoke land use activity distribution at two spatial scales. Firstly, the term is used to refer to the configuration of land uses within the established boundaries of the city or metropolis, which creates the need for travel. In most cases, cities or metropolitan areas maintain functional relationships with areas outside their established administrative boundaries. In view of this, the term spatial structure is also used to connote the arrangement of activity centres at a much broader scale, comprising the main metropolitan or city and the immediately surrounding districts/and or towns which often fall under separate administrative arrangements. Thus, this second definition draws on what has previously been observed as the outcome of the process of peri-urbanization, whereby rural–urban transition zones become urbanized over time as result of both cities

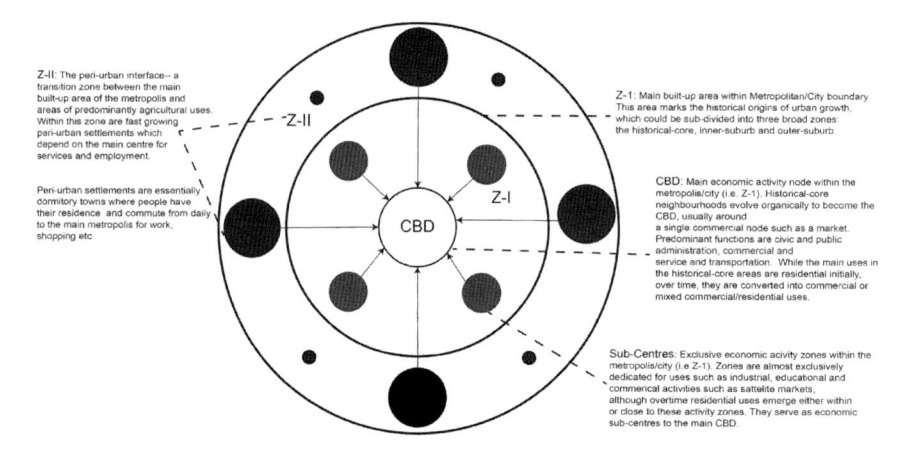

Fig. 11.4 A conceptual illustration of the spatial structure of metropolitan regions

and their surrounding villages expanding their built-up areas outwards, and in some cases, merging to form a contiguous landscape of mixed agricultural and urban land uses (Webster 2002; Simon et al. 2004; Cobbinah and Amoako 2012). The main city/metropolis therefore maintains functional linkages with the outlying peri-urban districts and/or towns, which can be observed from the commuting patterns between them.

11.3.1 A Conceptual Model of the Structure of Metropolitan Regions

In Fig. 11.4, a generalized diagram is presented to illustrate and unpack the spatial structure of cities and metropolises in Ghana in line with the dual meanings of the term outlined in the previous paragraph. The diagram is intended to provide a simplified model of a typical metropolis in Ghana, showing the configuration of broad land use activities internally, as well as the functional interdependencies that exist between the metropolis and activity nodes located in the surrounding peri-urban districts. Using one of the country's metropolitan regions as an example, we will further illustrate the extent to which the basic generalized model describes the structure of urban areas in Ghana. We will then focus on urban commuting patterns and travel behaviour, examining how these patterns and behaviours, and the associated travel-induced problems are embedded in the existing urban spatial structures.

The simplified model of urban spatial structure depicted in Fig. 11.4 shows the configuration of activity nodes within a metropolis, covering the area labelled as Z-1, and its immediately surrounding peri-urban districts and settlements, covering the area labelled as Z-2. Thus, Z-1 is intended to represent an imaginary boundary which we take as the formally established spatial extent of the metropolis while Z-2 represents the transition zone between the established boundary of the metropolis and the remote rural areas, where peri-urban settlements varying in physical size and population are located.

The components in the spatial configuration depicted above, the processes underlying its emergence and the interlinkages among the components of the structure are explained as follows:

i. **The main metropolitan area (Z-1)**: Z-1 encompasses areas, built and non-built within established administrative boundaries, which we take as the spatial extent of the metropolis. In Ghana, metropolitan areas are spatially anchored administrative entities. This means that their boundaries as defined by the legislative instruments establishing them are quite arbitrary, often justified by what is considered 'governable' as a single spatio-political entity rather than by technical considerations such as the spatial and functional linkages between places. Even so, we can adopt the administratively imposed boundaries in discussing spatial structure as it simplifies the task of delimiting the extent of the metropolis from the outlying areas.

Internally, areas within the metropolis can be categorized into three broad zones, namely historical-core, inner-suburban and outer-suburban zones. As implied in the name, areas within the historical core mark the historical origins of physical development. From this contiguous area of origins, the metropolis expands its built-up land outwards into surrounding areas over time, emerging in the process, the inner-suburban and outer-suburban zones within its boundaries. As will be explained shortly, the historical core evolves to assume the function of a Central Business District (CBD). Consequently, high-density, mixed-residential and commercial uses can be found in this zone within the metropolis. The inner- and outer-suburban zones on the other hand comprise neighbourhoods of predominantly residential uses developed around sub-activity nodes, which we will also describe shortly.

In general, the density of development and population tends to reduce as one moves outwards from the historical core to the suburban zones. In some cases, smooth transitions could be observed between the three contiguous landscapes of inner-core, inner- and outer-suburban zones. In other cases, physical constraints in the landscape may break physical contiguity, albeit the characteristics of spatial development allow to distinguish one zone from the other.

ii. **CBD**: Within the established boundary of the metropolis is an often centrally located CBD which serves as the main centre for socio-economic activities. As mentioned previously, historical-core neighbourhoods evolve organically, often around a single commercial node such as an open market to assume the role of a CBD.

The transition from dominantly residential historical-core neighbourhoods to a properly functioning CBD happens gradually over several years. CBDs become established and grow through a gradual process by which the interplay among market and public policy forces, including formal urban planning and design interventions, increasing demand for land for various activities and inner-city redevelopment necessitated by increasing land values and the changing socio-economic conditions, and infrastructure expansion transforms these historical-core neighbourhoods into functioning a CBD.

One of the major outcomes of this transformation is the displacement of population from central locations of the metropolis as residential uses become converted or redeveloped to serve the emerging commercial, service, civic and administrative uses. Redevelopment-induced displacement of inner-city populations therefore becomes one of the main driving forces of suburban and peri-urban expansion.

iii. **Sub-centres**: As explained previously, within the metropolis, residential areas develop around the main metropolitan centre (i.e. CBD) forming the inner-suburban and outer-suburban zones. While these zones are predominately residential, sub-activity centres, often exclusively dedicated for uses such as industrial, educational and commercial activities such as satellite markets can be found in the inner- and outer-suburban zones as well. While suburban populations depend on these sub-activity centres for employment, the metropolitan CBD continues to be the main focal point of administrative, commercial and civic functions, attracting a large share of the suburban population.

iv. **The peri-urban interface (Z-2)**: Z-2 constitutes the peri-urban interface—transition zone between the main built-up area of the metropolis and remote rural areas of predominantly agricultural uses. Within this zone are spatio-political entities administered by separate local governments called districts or municipalities within Ghana's decentralization set-up. Within these peri-urban districts are fast-growing towns located at varying distances from the main metropolitan area. In most instances, the closest peri-urban towns to the metropolis expand at the same time as the main metropolitan area expands, thereby merging to form a contiguous landscape of urban land. Over time, a new landscape emerges where spatial propinquity and the associated functional linkages transcend the artificial administrative boundaries separating areas into different local government jurisdictions. Thus, from a governance perspective, peri-urban districts exist as independent administrative entities. However, from the point of view of spatial and functional linkages, it becomes almost impossible to separate settlements at the boundaries of these districts and the main metropolitan area. Settlements in peri-urban districts depend on the main metropolitan centre for services, amenities and employment. In return, these peri-urban settlements essentially assume the role of dormitory towns where people have their residence and commute from daily to the main metropolis for work, shopping and other functions.

From the simplified spatial configuration described above derives the realities of spatial interaction patterns that typify metropolises and metropolitan regions in Ghana. Looking at the diagram representation, one may assume some form of a multi-nuclei configuration within the metropolis and in the overall spatial structure of the metropolitan region when the outlying area, comprising the peri-urban interface is taken into account.

In reality, however, the configuration depicted yields a monocentric structure. A monocentric structure becomes apparent when factors such as the distance separation between the sub-centres located within the metropolis as well as the overall patterns of flows created by the functional interdependencies among areas within the metropolis and between the metropolis and the outlying peri-urban districts are considered. Within the metropolitan boundary, activity sub-centres tend to be located a few kilometres relative to each other and the CBD. With the dominance of the CBD coupled with main arterial roads converging in the central area of the metropolis, the resulting configuration is a centralized one that is associated centripetal patterns of spatial interaction rather than a decentralized configuration that disperses spatial flows.

Moreover, as established previously, towns in the peripheral districts of the metropolis tend to be strongly dependent on the main metropolis, which results in the latter attracting a significant proportion of commuting trips that originate from the former. The monocentric structure at the scale of the metropolitan region is further accentuated by the convergence of main arterial roads in the central areas of the metropolis where major transportation infrastructure such as terminals is located.

11.3.2 Exemplifying the Generalized Model of Urban Spatial Structure the Case of the Greater Kumasi Metropolitan Region

In Chap. 5, we introduced the Greater Kumasi sub-region (GKSR) as a newly designated functional region, comprising the Kumasi metropolis, the second largest metropolis in Ghana, and its seven surrounding districts. Figure 11.5 shows the eight districts that forms the sub-region. In this section, we use the GKSR to illustrate how the generalized model of urban spatial structure advanced in the previous section can help explain the structure of metropolitan regions in Ghana. Using the labels previously assigned in Fig. 11.3, we will first examine the configuration of broad land use activities within the main metropolitan area of the sub-region (i.e. Z-1). We will also explore the spatial arrangement of major activity nodes in the sub-region, highlighting the functional linkages between the metropolitan core (i.e. KMA) and the surrounding peri-urban districts (i.e. areas in Z-2).

11.3.2.1 The Main Metropolitan Area Within GKSR (Z-1)

The main metropolitan area (Z-1) within the GKSR is the Kumasi metropolis. Administratively, this covers the area within the jurisdiction of the Kumasi Metropolitan Assembly (KMA). In 1995, under Ghana's decentralization programme, the KMA was established as one of the local government areas in Ghana by the Legislative

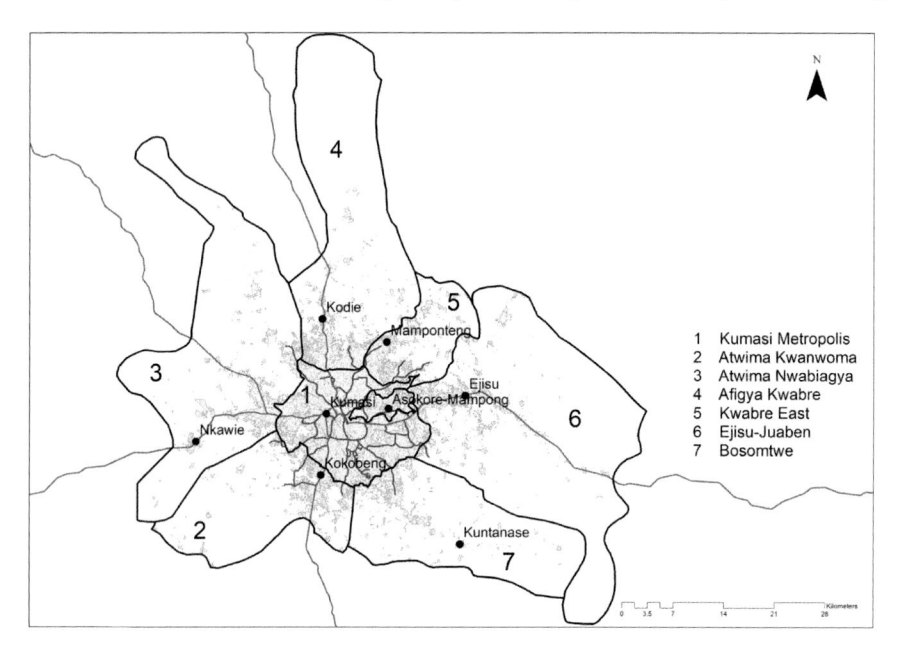

Fig. 11.5 Map of the GKSR showing local government administrative areas

Instrument (LI) 1614. The KMA covers an estimated land area of 212 km^2, representing about 9% of the total area of the GKSR.

KMA, the metropolitan core off the GKSR could be subdivided into three broad zones, namely, the historical-core, the inner-suburban and outer-suburban zones (see Fig. 11.6).[4] The broad urban zones are defined as successive circular zones extending outward from the central locations of the metropolis. The first zone, the historical-core, marks the historical origins of growth and built-up land expansion to the surrounding suburban and peri-urban areas. This zone covers traditional, inner-city neighbourhoods of the metropolis. These traditional neighbourhoods date back to the seventieth century when the Asante Kingdom was established by the *Asantehene* (King of the Asante State) at the crossroads of the Trans-Saharan trade routes. During this period, Kumasi, then a town of about 3000 inhabitants, became the political capital of the Kingdom (Amoako and Korboe 2011). Within the indigenous Kumasi Township were the historical settlements of Adum-Kejetia, Asafo, Bantama and Manhyia and Bantama. Thus, the historical-core zone comprises these centrally located, indigenous settlements that have existed over the past one hundred years. In terms of size, the historical-core zone covers approximately 22 km^2 of contiguous land (11% of the total land area of the KMA) within the inner ring-road system of the KMA.

---

[4]For a detailed discussion of the methodology used in delineating the three urban zones in the KMA, see Acheampong (2017a, b, pp. 55–57).

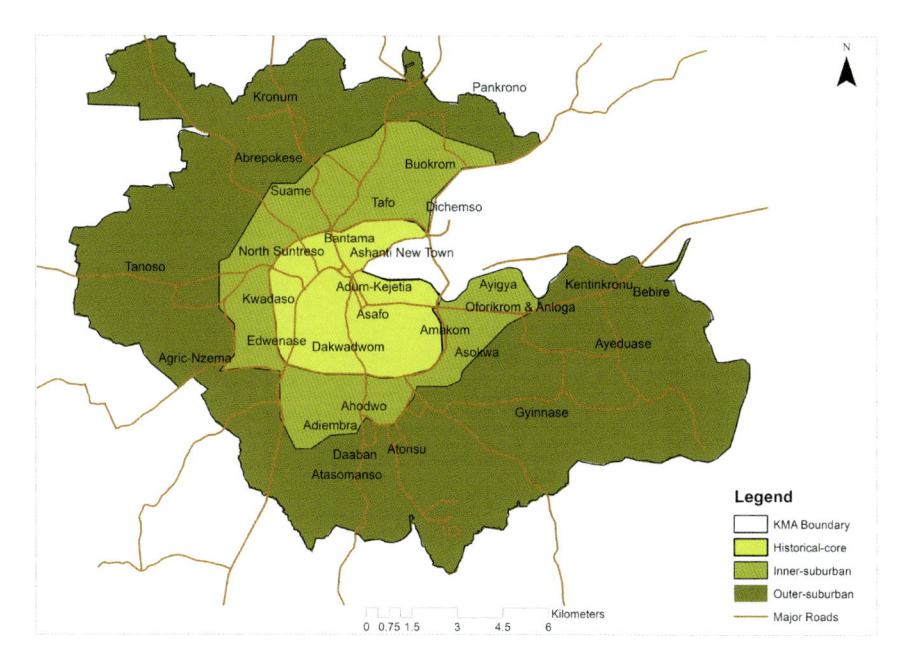

11.3 Urban Spatial Structure and Mobility Patterns in Ghana

Fig. 11.6 Three broad urban zones in the KMA

The inner-suburban zone covers a total land area of 38.7 km² of contiguous high-to-medium density development (i.e. 137–530 buildings per 0.25 km²) immediately surrounding the historical core. It comprises approximately 19% of the total land area of the metropolis. The outer-suburban zone covers a total land area of 145 km² of medium-to-low density contiguous urban land immediately surrounding the inner-suburban zone. The development density in this zone is generally low between 35 and 136 buildings per 0.25 km².

11.3.2.2 The Metropolitan CBD and Sub-centres, and Peri-urban Districts

The spatial distribution of broad land use activities within the KMA is depicted in Fig. 11.7. The existing land use configuration depicted here reflects the distribution of major employment zones and supporting residential areas, which have evolved largely from several years of incremental land development by several urban actors, including households and businesses.

Five major functional areas can be derived from the present physical structure of the KMA. The first is the main commercial and service node of the metropolis, which is centrally located at the Adum-Kejetia area—the CBD of the metropolis, and the GKSR as whole. The metropolitan CBD, which is located in the historical-core of the KMA, has evolved organically over several decades from the indigenous settlements

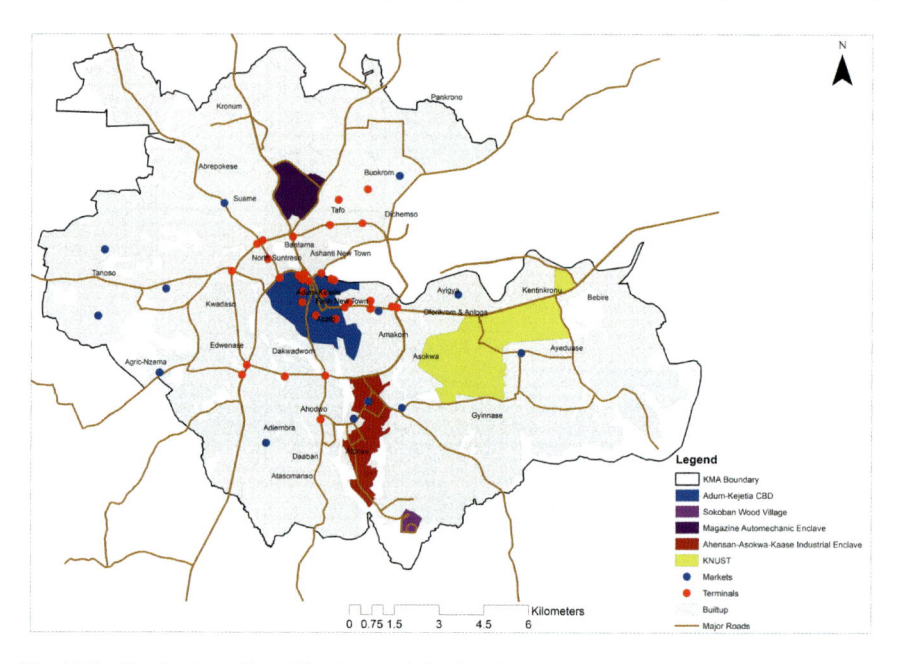

Fig. 11.7 Distribution of broad land use activity functions in the KMA

such as Adum, Kejetia, Asafo and Bantama. Within the CBD area are the Kumasi Central Market, one of the largest open markets in West Africa and the Kejetia transport terminal, the main hub of transportation in the metropolis. Indeed, the Kejetia transport terminal serves as the nodal point of interaction between the KMA and the rest of the country. Major administrative, service and commercial functions in the metropolis are concentrated in this area. Consequently, the dominant land use in this area is commercial and a mixed-residential-and-commercial land use pattern, which reflects a gradual transition from the hitherto dominantly residential functions that the historical settlements in this area provided.

In addition to the CBD are four major activity sub-centres, namely Ahensan-Asokwa-Kaase Industrial Enclave, Magazine Auto-mechanic Enclave, Sokaban Wood Village and the Kwame Nkrumah University of Science and Technology (KNUST), located at the southern, northern, south eastern and eastern parts of the metropolis respectively (see Fig. 11.7). Together, these nodes constitute the major employment areas in the metropolis.

Moreover, at the sub-regional scale, each of the seven peri-urban districts within the GKSR could be considered as activity sub-centres in their own right. The peri-urban districts have administrative capitals as well as other major towns which provide various functions to their resident populations, and to other smaller towns and villages in their respective jurisdictions. At the same time, through interaction between their major towns and the sub-regional core, these peri-urban districts maintain strong functional linkages with the KMA. In broad land use function terms, it

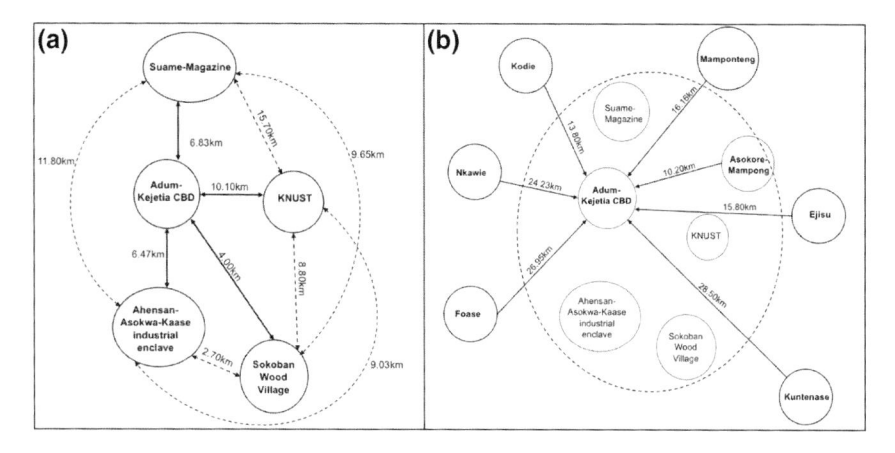

11.3 Urban Spatial Structure and Mobility Patterns in Ghana

Fig. 11.8 A conceptual illustration major land use activity functions in the KMA and road distances between them

is typical for the major towns in these peri-urban districts to provide housing for their resident population who commute daily to the sub-regional core (i.e. KMA) for work and to access social services. Thus, major towns in these peri-urban districts are essentially *dormitory towns* that provide residential functions to a significant proportion of the daytime population observed in the KMA, the sub-regional core.

11.3.3 The Influence of Prevailing Spatial Structures on Commuting Patterns and Travel Behaviours

The prevailing configuration of land use activities serves as the spatial anchors of trip production and distribution in the sub-region. In other words, spatial structure shapes commuting patterns and travel behaviour at the neighbourhood, metropolitan and sub-regional scales. Figure 11.8 illustrates the distribution of major activity zones in the KMA (Fig. 11.8a) and the GKSR (Fig. 11.8b). In Fig. 11.8a, we show the five main employment zones in the KMA and the road distances among them. Likewise in Fig. 11.8b, where the focus is on the sub-region, we show the road distances from each of the peri-urban districts with the GKSR to the CBD of the KMA, which is at the core of the sub-region. In both diagrams, the road distances are calculated by averaging distances given all possible linkages provided by arterial roads from the centroid of each point to the centroid of the other.[5]

---

[5]The methodology involved using Google Maps to identify alternative routes via arterial road(s) linking one point (e.g. Suame Magazine) with another (e.g. KNUST). Where more than one route was found, the distances were averaged as a way of taking into account differences in travel distances which result depending on route choice. Differences in distances were found to be negligible in all cases where road distances were averaged.

Within the KMA, the simplified illustration of the land use activity distribution and the network distances between them allows us to begin to explore the underlying causes of patterns of spatial interaction and how this in turn manifest in urban transportation problems such as longer commuting times and congestion. About 90% of work trips that originate from the KMA are distributed to various destinations within the metropolis, with the activity nodes depicted in Fig. 11.8a being the most important in terms of trip production and attraction.

The spatial arrangement with the five major activity sub-centres depicted above resembles some form of a polycentric configuration. In reality, however, the existing structure is far from a polycentric configuration. Instead, the prevailing metropolitan structure could be characterized in two main ways: firstly, the metropolitan CBD exerts the strongest influence relative to the other activity nodes. While contributing to only 2% of work trip origins in the metropolis, in terms of trip attractions, an estimated 26% of all work trips terminate in the CBD (i.e. Adum-Kejetia area). Secondly, as shown Fig. 11.8a, all the major activity sub-centres in the metropolis are located in relatively close proximity to each other, with the closet being 2.7 km apart and the farthest being 11.8 km apart. A recent study found that nearly 70% of all work trips in the metropolis are undertaken using motorized forms of transport, while the remaining 30% is done using non-motorized transport of which walking constitutes the significant share and bicycling mode share is less than 1% (Acheampong 2017a, b). Given the high levels of motorization and a limited number of arterial roads, motorized trips to the different destinations in the metropolis must use essentially the same road networks. Thus, the existing land use structure together with the lack of alternative roads from one destination to another creates a centripetal effect, especially during peak-hour periods where the capacity of existing roads is exceeded, resulting in congested road networks. Consequently, a 5-km car trip for example, which should ordinarily take less 10 min, assuming a travel speed of 50 km/h, could take up to an hour or more during the peak hours of travel.

At the sub-regional level, a pattern of spatial interaction, similar to that of the KMA, emerges. As we have already established, major settlements in the peri-urban districts tend to act as dormitory towns some of the day time population in the KMA. It is estimated that nearly one-third of all trips that arrive in the KMA are generated from locations outside the metropolis, with the major towns in the surrounding peri-urban districts (e.g. Ejisu, Abuakwa, Kokobeng and Esereso) contributing to more than half of these external flows. The relatively longer commuting distances means that commuters must use motorized forms of transport, either public transport or private cars. Indeed, the evidence suggests that with a generally poor quality of public transport services on the one hand and the increasing distances resulting from more people looking for cheaper housing in peri-urban areas, vehicle ownership has been increasing in recent years. Across the country, it is estimated that private vehicle to population ratio increased from 50 vehicles per 1000 population in 2010 to about 70 vehicles per 1000 population in 2015 (Ministry of Transport 2016). Out of the total fleet of about 1,952,564 registered cars as at 2015, 60 and 14% were in Accra, the country's capital and Kumasi, the second largest city, respectively.

Within the KMA and the GKSR as a whole, the layout of transport networks follows the traditional spoke-and-wheel structure, where the main avenues converge at the metropolitan core, with the existing ring-road system serving as the wheel that connects the spokes. The existing network of arterial roads not only distributes population among the activity centres within the metropolis, but also connects the metropolis to these outlying districts in the sub-region. Thus, the centripetal tendencies that exist at the metropolitan scale also manifest at the sub-regional scale, further contributing to network congestion and longer travel times.

The spatial structure of the GKSR described above could also be observed in other metropolitan areas (i.e. Accra, Sekondi-Takoradi and Tamale) and their surrounding districts. The newly designated city-regions and urban networks including the Accra city-region, Sekondi-Takoradi urban network and Tamale urban network (see Chap. 4) all exhibit spatial structures similar to the one described in the forgoing sections of this chapter. In Chap. 9, for example, we demarcated three contiguous zones in the Greater Accra Region. Although this demarcation was done purposely to help us to characterize and quantifying urban expansion trends, the delineation of the area into coastal districts (e.g. AMA, Tema, and Ashaiman), suburban districts (Adenta and Madina) and peripheral districts, comprising Kpone, Ga East and Ga Central, reflects strongly a spatial configuration and networks of flows and spatial interaction that are similar to the spatial structure and interlinkages advanced and exemplified in this section, using the GKSR.

11.4 Towards Integrated Spatial Development and Transport Policy and Planning

In the preceding sections, we have explored the nexus between urban spatial structure and patterns of flow and spatial interaction at a conceptual level. We have also demonstrated, broadly, how the land use and transportation systems interact within a typical metropolitan setting and its surrounding peri-urban districts to determine travel patterns and mobility choices. Building on these insights, we will in this section begin to identify specific strategies for achieving an integrated approach to spatial development planning and transportation planning in Ghana. We will begin the discussion by looking at integration at the policy and institutional levels. Next, we will explore practical ways in which spatial development planning at metropolitan/city and neighbourhood scales could be used to promote sustainable mobility. This section will also identify possible ways of achieving intermodal integration as well as examine the prospects of emerging mobility concepts such as shared-mobility services in meeting the travel needs of the growing urban population while managing travel demand towards sustainable outcomes.

11.4.1 Land Use and Transport Integration at Policy and Institutional Levels

Bringing together the domains of land use and transportation has both institutional and policy dimensions. At the beginning of this chapter, we identified that while the land use and transportation systems are strongly connected at the strategic, metropolitan and neighbourhood scales, in practice, the portfolio of programmes, polices and plans affecting these domains tend to be handled by separate institutions.

As we have already identified elsewhere in this book, in Ghana, authority and competencies for spatial planning rest with LUSPA at the national level, RSPCs at the regional level and TCPDs at the level of MMDAs. The institutions of spatial planning have historically oscillated between various sector Ministries mainly as a result of the eclectic and a plethora of cross-cutting issues embraced by spatial planning. Transportation on the other hand appears to be a clearly defined sector as far as the institutional arrangements for policy-making and implementation at the national and sub-national levels are concerned. At the strategic level, transportation is recognized as one of the key sectors of national development for which institutional arrangements separate from what exists within the spatial planning system exist to address sector issues. There are two main sector Ministries responsible for transportation issues nationally. These are the Ministry of Transport and the Ministry of Roads and Highways.

The Ministry of Transport is responsible for policy formulation and co-ordination in a variety of strategic areas including maritime, railway, road safety and aviation. The mandate of the Ministry is achieved through two broadly defined implementing agencies, namely road transport services agencies (e.g. Metro Mass Transit Limited, Driver and Vehicle Licensing Authority (DVLA) and National Road Safety Commission (NRSC)) and maritime agencies (e.g. Ghana Maritime Authority, Ghana Ports and Harbour Authority and Ghana Shippers Authority). The overall responsibility of the Ministry of Roads and Highways is rather narrow in scope. Specifically concerned with the road sub-sector of transportation, the Ministry of Roads and Highways initiates and formulates policies and programmes for road transport infrastructure. The Ministry discharges its functions via two implementing agencies, namely Department of Urban Roads, which is responsible for the administration, planning and development and maintenance of urban road networks and the Department of Feeder Roads, responsible for rural road networks. Box 11.1 provides a summary of the responsibilities of the two transport-sector Ministries.

Box 11.1: Ministries and Agencies in the Transportation Sector
1. **Ministry of Transport**
    The Ministry of Transport (MoT) is responsible for the overall governance of the transportation sector. It is mandated to formulate policies and co-ordinate plans and strategies that have a bearing on transportation from other Ministries

and sector agencies. It also monitors, evaluates and conducts research, skills training and development in the transportation sector at the national level.

MoT has oversight responsibility for two broadly defined sector agencies, namely

(i) **Road Transport Services Agencies (RTSAs)**: RTSAs include the Government Technical Training Centre; Metro Mass Transit Limited (MMTL); Driver and Vehicle Licensing Authority (DVLA); National Road Safety Commission (NRSC); and Intercity STC Limited;
(ii) **Maritime Agencies**: This includes the Ghana Maritime Authority, Ghana Ports and Harbour Authority and Ghana Shippers Authority.

Programmes, policies and plans formulated by the MoT are implemented by these sector agencies. Below an outline of the main functions of the Ministry, as performed through its implementing agencies are outlined:

Maritime sub-sector: MoT, through the Ghana Ports and Harbours Authority (GPHA) to build, operate and manage ports and harbours in Ghana. Through the Ghana Maritime Authority, the Ministry also regulates monitor and co-ordinate activating relating to safety and security of the marine and inland waterways in Ghana. MoT also exercises oversight responsibility over inland water transportation services such as the movement of passengers and cargo on the Volta Lake.

Road transport services and road safety: In this domain, MoT, acting through the NRSC, undertakes planning development and implementation of road safety programmes and activities. Through MMTL and STCL, MoT provides intercity and intra-city road transport services as well as urban-rural services.

Railway: In the railway sub-sector, MoT is mandated to promote the development and maintenance of rail infrastructure and services, regulate and grant railway service operation licences and set and enforce safety and security standards for the construction and operation of railways in Ghana. The Ministry undertakes these functions through the Ghana Railway Development Authority.

Air transport: GACL under the MoT is mandated to plan, develop, manage and maintain airports and aerodromes in Ghana.

Driver licensing: DVLA, one of the implementing agencies of the MoT, is mandated to establish standards and methods for the training and testing of drivers of motor vehicles and motor cycles, and to issue driving licence, inspect, test and register motor vehicles.

2. **Ministry of Roads and Highway**

The Ministry of Roads and Highway is responsible for the road sub-sector of transportation in Ghana. The Ministry is mandated to initiate and formulate road infrastructure policies and programmes; undertake development planning in consultation with National Development Planning Commission (NDPC); and co-ordinate, monitor and evaluate the efficiency and effectiveness of the performance of the sector.

Objectives of the Ministry are to:

- Create and sustain an accessible, effective and efficient transport network that meets user needs
- Integrate land use, transport planning, development planning and service provision
- Create a vibrant investment and performance-based management environment that maximize benefits for public and private-sector investors
- Develop and implement comprehensive and integrated Policy, Governance and Institutional Frameworks
- Ensure sustainable development in the roads' sub-sector
- Develop a multi-disciplinary human resource base to facilitate the implementation of our programmes.

The Ministry of Road and Highway fulfils its mandate at the local level through two implementing agencies, namely Department of Urban Roads and Department of Feeder Roads. Whereas the former is responsible for the administration, planning, development and maintenance of road networks in urban areas, the latter's mandate covers rural road networks. The Department of Urban Roads has Road Units in metropolitan areas and selected municipalities in Ghana.

From the aforementioned institutional settings, institutional-level integration of land use and transportation issues must involve co-ordination among the institutions responsible for spatial development planning on the one hand and those responsible for transportation on the other hand. Figure 11.9 shows the institutional setting for achieving integration between spatial development and transportation at the national and sub-national levels.

The policy direction for achieving the goals of land use and transport integration is articulated in the National Transport Policy (NTP). Among the seven transport-sector development goals, the NTP recognizes the need to *'Integrate land use, transport planning, development planning, and service provision'* (see Box 11.2). The NTP further identifies specific strategies for achieving integration such as establishing consultation mechanisms between the relevant public-sector agencies and departments, producing guidelines to practitioners to facilitate integration and ensuring that adequate land is reserved in Master Plans and local plans for transport infrastructure development.

Box 11.2: The National Transport Policy: Sector Goals and Strategies for Land Use and Transport Integration

11.4 Towards Integrated Spatial Development and Transport …

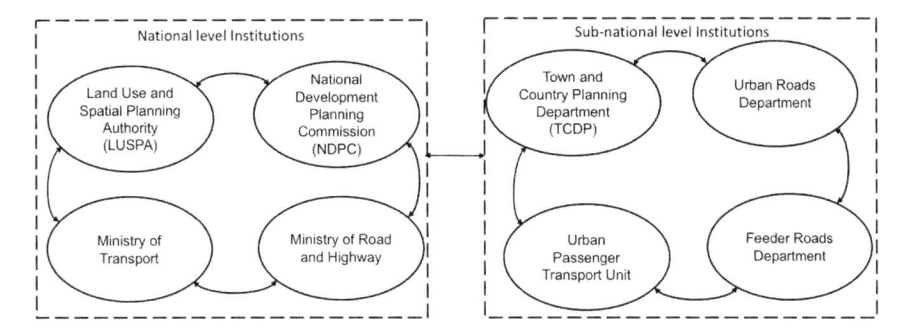

Fig. 11.9 Institutional setting for integration between spatial development and transportation policy and planning

Vision:

An integrated, efficient, cost-effective and sustainable transportation system responsive to the needs of society, supporting growth and poverty reduction and capable of establishing and maintaining Ghana as a transportation hub of West Africa.

Transport-Sector Goals:
- Establish Ghana as a Transportation Hub for the West African Sub-Region;
- Create an accessible, affordable, reliable, effective and efficient transport system that meets user needs;
- Integrate land use, transport planning, development planning, and service provision;
- Create a vibrant investment and performance-based management environment that maximizes benefits for public- and private-sector investors;
- Develop and implement comprehensive and integrated Policy, Governance and Institutional Frameworks;
- Ensure sustainable development in the transport sector;
- Develop adequate human resources and apply new technology.

Strategies for land use and transport integration

- Establish consultation mechanisms between transport-sector MDAs, with MLGRDE and MMDAs to implement:
  – Integrated land use and spatial planning;
  – Decentralized management, financing and maintenance of local transport;
  – infrastructure and services;
  – Urban Transport Policy
- Establish consultation mechanisms between transport-sector MDAs and other sectoral Ministries

- Produce practical guidelines for development and transport planners to facilitate effective integration
- Ensure proper acquisition and protection of land for transport infrastructure development
- Incorporate into Master Plans of cities, provision of inter- and intra-modal and 'break-bulk' facilities to improve the transfer of goods and passengers from one mode to another.
- In collaboration with the MMDAs, ensure the provision of independently managed lorry parks and other transport interchange facilities to encourage competition and improved customer service
- Ensure consistent application of the 'Road Utility Co-ordination Manual' by passing appropriate legislation.

By identifying the need for an integrated approach to planning and implementation of spatial development and transportation strategies and projects, the NTP provides the foundations to begin to bring together the network of institutional actors, programmes, plans and policies needed to achieve this policy goal. Yet, the policy only demonstrates a partial and narrow understanding of the principles and pathways towards integrating the domains of spatial development planning and transportation planning in practice. This can be seen from the scope of strategies that have been outlined in the NTP to achieve the goal of integration (see Box 11.2). Whereas some transport-related infrastructure development strategies are identified potentially with the intent to address gaps in infrastructure supply, specific strategies in domain of spatial planning and urban design, and how they relate to specific expected outcomes are missing in the policy. In other words, the NTP does not provide any strategic direction on how national- and sub-national-level spatial development planning could, for example, provide visions of and experiment with spatial development models that could actually promote sustainable mobility choices in towns and cities.

Moreover, as we have identified elsewhere, at all levels of decision-making in Ghana, there is always a wide gap between policy intent on the one hand and policy implementation on the other hand. The area of land use and transport integration is no exception. For example, while the NTP identifies the need for institutional co-ordination, in practice, there are no clearly defined mechanisms for achieving this among the relevant institutions. At best, co-ordination happens on ad hoc basis. At worst, the form of co-ordination intended in the NTP does not happen at all. In most cases, the standard practice involves TCPDs preparing local plans in which reservations for transport infrastructure are made. Often, such plans are never implemented or are only partially implemented. In subsequent time periods, often several years following the formulation of the local plan, by which period physical development would have already occurred, the Urban Roads Department move into develop road networks. It is therefore not uncommon to find land reserved for transport infrastructure either encroached upon by unauthorized uses or without any following consultation procedures, rezoned and developed into new uses completely different from the intended use in the original local plan existing plan.

Furthermore, while a National Urban Transport Policy is lacking, MMDAs do not also have comprehensive transport plans that would, among other things, enable them to co-ordinate spatial development with transport infrastructure investments and vice versa. At the same time, existing land use plans for major towns and cities are outdated while many suburban and peri-urban neighbourhoods are expanding rapidly without any local plans to guide their development. In these unplanned residential neighbourhoods, the chaotic development that emerges either result in limited provision being made for access road networks or some developments completely lacking accessibility. This clearly is a major challenge for any attempt to co-ordinate spatial development and planning and investment in transportation infrastructure.

The NTP itself is lacking in any clear strategies for embedding the imperatives of integration in the decision-making and professional practices of actors in the spatial development planning and transportation planning domains. Perhaps, the plan to issue guidelines for achieving an integrated approach to land use and transportation planning would serve this purpose. However, to date, no such guidelines have been issued.[6] In the absence of guidelines, it is expected that spatial planners and transportation planners through their training would not only appreciate the interconnections among issues in their respective disciplinary and professional domains but would also be equipped with basic technical competencies that would enable them to evaluate the potential impacts of their plans. This level of expertise is, however, not always available. Indeed, in the last six decades, the area of land use and transport integration has emerged as a technical sub-field in its own right, requiring expertise from the disciplines of urban planning, transportation planning and civil engineering to develop decision-support systems to guide policy formulation and to evaluate outcomes. It is for this reason that having guidelines for practitioners and developers is vital. An integrated land use and spatial development planning guidelines would spell out a number of considerations that would inform spatial planning and transportation planning at the metropolitan and neighbourhood scales. This could be in the form of estimated impact on transport demand of various land use allocations and/or estimated impact on adjoining transport infrastructure of proposed land developments. This information could then inform the organization of land uses with the aim of shaping emergent trip patterns and travel behaviours in a given locality. Such guidelines would also help practitioners and developers to evaluate the land use and/or transportation impacts of proposed development, right from the design stages and subsequently at the point of application for planning approval.

Thus, at the institutional and policy levels, achieving the goal of integration between spatial development and transportation would require a number of interventions. Firstly, a good understanding of the principles of integration and how these are embedded in the prevailing realities of spatial structures and mobility patterns

---

[6]The National Transport Policy was first introduced in 2008. After close to a decade from this period to when this book was written, no guidelines to facilitate an integrated approach to land use and transportation planning had been issued.

in towns and cities are required in order to articulate workable policy directions at the national level. These policy imperatives could then be pursued at sub-national level of plan-making and implementation. Better integration outcomes would only be possible through an effective spatial planning system. In this regard, the new spatial planning system, with its goal of strengthening land use planning and development management at all levels of decision-making, does offer new opportunities to address spatial development and transportation issues in a holistic manner. Indeed, the new generation of spatial development plans including the NSDF, GKSR-SDF and WRSDF all have proposals that are founded on the importance of linking the distribution of population settlement functions with investment in and development of transportation infrastructure. That notwithstanding, the spatial development—transportation integration strategies advanced in these frameworks and plans are very broad and as such intended to only set the stage for integration at local-level planning and project implementation. At the metropolitan, city, town and neighbourhood scales, where these strategic policy principles and strategies would actually find their expression in space through physical development, a lot more effort and initiative would be required to bring the domains of transportation planning and spatial development planning together. Ultimately, well-resourced public-sector institutions together with a consciously cultivated culture of institutional collaboration and co-ordination would be vital to an integrated approach to spatial development and transportation at all levels.

11.4.2 *Urban Planning and Development to Promote Sustainable, Non-Motorized Mobility*

The way we design and build our towns and cities shape emergent mobility patterns and travel behaviours. Urban planning and design not only have lasting impact on the spatial organization of broad land use functions but also influence structural elements, including development density, diversity of land use activity as well as the layout of and connectivity between road/street systems and activity locations. The evidence suggests that people living in higher housing and employment density locations tend to have lower levels of auto ownership and undertake relatively shorter trips using non-motorized transport options such as walking and cycling (Pinjari et al. 2011; Aditjandra et al. 2013; Naess 2013). Furthermore, a study of the residential-employment-commuting nexus in the context of the Kumasi metropolis (see Acheampong 2017b) found home–work separation distance exceeding half of a kilometre and suburban locations was associated with motorized transport use in general. Also, residence in the suburban and peri-urban locations of the metropolis increased the likelihood of car ownership and private car use as work travel mode compared to residence in the historical-core neighbourhoods of the metropolis that was found to be associated with walking to work.

What the aforementioned evidence suggests is that we could design and build our metropolises, cities and towns to reduce automobile dependence while promoting sustainable mobility options such as bicycling and walking. Before going ahead to this how this could be achieved in practice, it is worth mentioning that there are fundamental attitudinal factors at play at the individual level which prevent the adoption of active, non-motorized travel mode options such as bicycling. For example, in Ghana, while a significant proportion of the population do walk for work and non-work trips over relatively shorter distances, public attitude towards bicycling as a mode of transport is generally negative. There is widespread perception in the population that the bicycle is an inferior mode of transport (see Acheampong 2016; Ministry of Finance and Economic Planning 2010). Consequently, in major towns and cities, bicycling constitutes less than 1% of the total mode share. Bicycling levels are, however, considerably high in the towns and cities in the northern part of the country such as Tamale, where vehicle ownership is very low, probably as a result of the relatively higher incidence of poverty in these areas. Even so, in recent years, motorcycles are gradually replacing bicycles in these areas, signalling a gradual shift towards high levels of motorization in the Tamale metropolis in particular.

Besides the deep-seated attitudinal and sociocultural factors outlined above, physical environment factors have been found to influence bicycle commuting in many cities, including those in Ghana. In the northern regions of Ghana, for example, the reasons for high levels of bicycling include the topography of the area which is fairly flat as well as scattered pattern of settlements which makes the bicycle a perfect substitute as an affordable yet faster mobility option. Moreover, a recent study of the determinants of utility cycling in the Tamale metropolis (see Acheampong and Siiba 2018) found that individuals living closer to the centre of the Tamale township are more likely to cycle to places such as work, shopping, social events and religious activities compared to suburban residents. Also, built-environment factors, which are shaped directly by urban planning design, such as dedicated bicycle lanes, existence of connected routes to get between places and good road surface conditions, all increased the propensity of people to cycle in the metropolis. On the contrary, individuals would not cycle if they perceived challenges and dangers such as difficult road junctions where traffic control systems that priorities non-motorized transport are non-existent and streets with high volumes of traffic, fast-moving traffic and blockages resulting from on-street parking and encroachments.

The forgoing demonstrates how fundamental spatial planning and design at the city and neighbourhood scales are to providing spaces that supports all forms of mobility, but especially non-motorized transport, which is considerably low in urban areas in Ghana. In 2008, the NTP sets an ambitious target of increasing the modal share of non-motorized travel by 10% as way of increasing access to affordable and healthy mobility while reducing travel-induced impact on the environment. The policy identifies a number of strategies to achieve this, although as mentioned previously, the policy is limited in terms of urban design strategies that would fundamentally transform existing and new built environments to bike-friendly and pedestrian-friendly spaces. At the same time, while it has been a decade since the policy came into existence, concrete actions to achieve the set target on non-motorized transport are yet to be seen.

> **Box 11.3: National Transport Policy: Policy Statement and Strategies for Non-motorized Transport**
> **Policy statement**:
> Non-Motorized Transport (NMT) Infrastructure shall be developed to improve affordability and accessibility for urban and rural communities—Aiming for 10% of passenger movement.
> **Strategies**
> - Rehabilitate and free from encroachment existing NMT routes
> - Raise awareness of benefits of NMT especially the use of bicycles and pedestrian safety starting with schools and other educational institutions
> - Carry out surveys to determine user needs and, where required, incorporate NMT facilities in infrastructure planning and development
> - Raise awareness for careful driver attention to pedestrians and cyclists
> - Provide adequate regulation for NMT operations
> - Ensure that schools, universities and shops provide facilities for parking of bicycles and other NMT vehicles
> - Set up credit schemes to allow students to purchase and maintain bicycles at affordable prices
> - Provide strict enforcement and penalties to discourage encroachment on existing NMT facilities.

In the past, the spatial planning system has been weak and largely ineffective in delivering its objectives, including planning for cities that accommodates different mobility types. While some towns and cities still do not have local development plans, for those which such plans were prepared several years ago, provisions for cycling and pedestrian infrastructure were not made. That said, when it comes to planning for non-motorized transport, what the experience from other countries teaches us is that with careful planning, design and traffic control and safety measures, adequate room can be made in the existing road infrastructure to accommodate non-motorized traffic. Indeed, in many cities across Europe, for example, careful planning has made it possible to accommodate motorized and non-motorized traffic on narrow street systems that were built in the medieval period. In fact, these road systems, which were built centuries before the invention of the automobile in the 1920s, would first be adapted to accommodate cars as they became a dominant mode of transport in cities.

Thus, with modest effort at reorganization and space reallocation, existing built environments can be adapted to accommodate societal needs, including making them conducive for cyclists and pedestrians. In the context towns and cities in Ghana, integrated urban land use and transport master planning could provide the vehicle to adapting existing built environments to the needs of non-motorized road users. Road reservations (developed and undeveloped) in most towns and cities are wide enough to allow for sections to be dedicated for cyclists and pedestrians. Plans for future growth areas, including large cities and neighbourhoods, could also reflect the need to promote non-motorized transport options. Such plans would not only make reservations for roads to be built for vehicles, but also dedicate bicycle lanes and/or pedestrian paths that connect non-auto-dependent commuters to existing urban opportunities. Investment in traffic-calming measures and modern traffic control systems that reduces conflicts by prioritizing cyclists would constitute additional measures in retrofitting cities to become bike- and pedestrian-friendly. To bring urban residents closer to urban opportunities, urban development planning should promote high-density, mixed-use development especially in suburban neighbourhoods and peri-urban towns. The potential exists through the urban redevelopment process to gradually transform old inner-city areas to newly built forms that provide outdoor spaces and bicycle/pedestrian paths to promote active mobility forms. In addition to these, softer interventions aimed at shifting attitudes and changing behaviours would be vital to the portfolio of measures needed to decrease automobile dependence and increase active, non-motorized transport in urban areas. Such measures would include bicycling promotion and attitudinal change campaigns aimed at reversing the negative image associated with cycling in the population; enforcement of road safety regulations; and education of road users, especially motorists about ways in which they could contribute to ensuring the safety of vulnerable, non-motorized road users.

11.4.3 *A Portfolio of Complementary Policy Strategies: Public Transit, Shared-Mobility and Intermodal Integration*

An integrated approach to land use and transportation planning ultimately addresses the fundamental question of how to effectively manage travel demand. Growth in travel demand is directly linked to population growth and physical expansion of towns and cities. In corollary of this, as towns and cities grow, a portfolio of policy strategies, complementary to those discussed in the preceding sections, is required to effectively manage the emergent travel demand and evolving travel behaviours in urban areas.

Mass transit services, including city buses, light rail (tram) and passenger trains, can move relatively large number of passengers at any given time, using less energy per passenger mile while reducing pollution and traffic congestion. Therefore, in larger cities, an efficient public transit system is crucial to meeting the travel needs of the growing population. In urban areas in Ghana, for example, over 90% of the population depend on public transport for both work and non-work travel purposes. Yet, the quality of public transport services, provided mainly by 10–19-seater minibuses, known locally as *'Tro-tro'*, is generally very low due to poor vehicle conditions and lax safety standards (Amoh-Gyimah and Aidoo 2013). While the *Tro-tro* system provides relatively flexible means of inter- and intra-urban travel to the larger mass of the population, they are unsafe, unreliable and uncomfortable. Also, given the relatively small capacity of minibuses, the urban transport system is not efficient in terms of energy use per passenger mile and the overall impact on the environment. Thus, besides building urban transport infrastructure that supports non-motorized transport, urban planning and transport policy must address the aforementioned challenges in the public transport sector.

In recent years, new mobility concepts are increasingly becoming popular in cities. In many urban areas across the globe, shared-mobility services have become an important component in the portfolio of urban transport policy strategies required in transition towards sustainable urban futures. Shared-mobility services are being implemented for motorized transport consumers in the form of ride-sharing and car-sharing systems. For non-motorized transport users, bike-sharing systems are also becoming popular. The basic concept of shared-mobility models is that urban citizens are given access to a fleet of cars/bikes to use when they need to instead of the conventional model where individuals own cars or bicycles (see e.g. Cervero et al. 2007; Shaheen and Martin 2010; Efthymiou et al. 2013).

Different models of shared-mobility exist. One of the popular models of shared-mobility is *on-demand taxi* services where customers book a taxi service online using a smartphone of computer to be picked to their destination by a designated driver. In this model, individuals travelling from a common origin to a common destination may share the taxi ride and the associated cost. In another model of motorized shared-mobility, called *ride-sharing*, commuters having common trip destinations, such as working in the same area of the city or the same organization, could travel together in a single vehicle instead of each individual commuting by their own car. This type of travel arrangement could offer a cost-effective and energy efficient means of travel to participants (e.g. work colleagues). The third model of shared-mobility, referred to as *car-sharing,* is one whereby a company provides customers access to a fleet of vehicles to use when needed. To participate in car-sharing, an individual has to register with the company and book to use a car via a smartphone or computer application purposely built for that. Thus, the concept of car-sharing integrates aspects of conventional public transportation and car ownership in the sense that the car(s) are accessible to members of the sharing community, but at the same time, the service offers the benefits of comfort and privacy that is often associated with personal modes of motorized transport.

Elsewhere in this chapter, we have established that vehicle ownership is on the increase in urban areas in Ghana. At the same time, urban areas in Ghana, similar to cities in other developing countries, are in the very early stages of motorization. The potential therefore exists to introduce alternatives personal vehicle ownership through shared-mobility systems. By offering access to a limited fleet of cars, motorized shared-mobility services could meet the travel needs of potential users while helping to reduce travel-related impacts on land use, energy use, environmental pollution, congestion and public health. With no direct purchasing and maintenance costs, shared mobility could also provide cost-effective alternative to car ownership.

Ultimately, integration of various transport modes would contribute to the overall efficiency and effectiveness of the urban transport system. Modal integration means that in a single leg of trip, passengers requiring to switch from one mode to another would be able to do so seamlessly. For example, one may use the bike as a first-mile travel mode to connect to other transport modes such as buses or trains. Modal integration requires co-ordination at two levels. At the policy level, the various transport infrastructures such as roads, bicycle lanes, railways and tram lines must be designed and built as an integrated network. This would require, for example, organizing the network of infrastructure and supporting infrastructures such as bus stops, car parks and bike-sharing stations such that commuters could have easy access to major bus terminals from train stations or from car parks to use bike-sharing services if necessary in a single journey. Secondly, service-level co-ordination is required to ensure effective inter- and intra-modal integration. For example, reliable schedule information showing the departure and arrival times of various modes not only allow travellers to plan their journeys but also makes it easy for them to switch between modes where necessary.

11.4.4 The Role of Integrated Land Use and Transport Decision-Support Systems

An integrated approach to spatial development and transportation involves dealing with multiple sub-systems that are dynamic and inherently complex. For example, the land use component comprises various activity locations such as residence, employment, shopping and recreation. Each of these activity locations is the outcome of an interplay of decisions made by various actors in the urban property and job markets. For example, residential land use, which constitutes the largest share of urban land use in urban areas, emerges from the location decisions of individuals, developers and businesses. Depending on the context and the efficiency of the existing planning systems, these decisions may or may not be shaped by formal plans and the regulations that accompany them. The transportation system also comprises the various physical infrastructures, the supporting software and technologies, traffic flow dynamics as well as individual travel behaviours that are complex to understand and predict with a high degree of certainty.

In view of the foregoing, policy-makers and practitioners in some metropolitan organizations, in making land use and transport policies, tend to complement conventional approaches to planning and decision-making with more sophisticated, computer-based decision-support systems. At the beginning of this chapter, we introduced land use and transport interaction (LUTI) modelling as field of research that has sought to offer decision-support tools to assist practitioners in land use and transportation planning. Over more than 60 years, various LUTI models have been developed for assessing the impacts of land use decision on transportation and vice versa, and to evaluate large-scale transport investments.[7]

A typical LUTI model has two main sub-models that are coupled together. These are the land use sub-model and the travel demand sub-model. The land use model uses data on existing land use activities such as residential locations and employment locations as well as forecasting techniques to predict future home and employment locations. The travel demand model is anchored on the land use system to derive trip production and distribution patterns and the associated travel mode and route choices. In more sophisticated activity and travel demand forecasting models, individual activity-travel choices at different time periods in a typical day are modelled; the emergent activity-time patterns then provide the basis to forecast travel demand and to design strategies to manage the demand.

Operational LUTI models are currently not in use by the various local planning authorities in Ghana. That notwithstanding, in the context of rapid urbanization and the attendant increase in motorization, the forecasting capabilities of such decision-support systems could provide a data-driven approach to long-range land use and transport planning. Some initial attempts have been made to develop such models for metropolitan areas in Ghana, although mainly as an academic research endeavour. For example, Acheampong (2017a) developed the Metropolitan Location and Mobility Patterns Simulator (METLOMP-SIM). METLOMP-SIM has so far been calibrated for the Kumasi Metropolis. The model simulates the residential and employment location choice behaviour of households and individuals and their co-evolution with travel patterns, which is derived from the model as trip origin and destination pairs at the level of TAZs, home–work travel distance and work travel mode choice. Ultimately, the utility of METLOMP-SIM and other decision-support systems that may be developed in the future would depend on the extent to which they can inform practical policy applications in the urban and transportation planning process. Certainty, the availability of reliable data, logistics and expertise from domains including spatial planning, transport engineering and computer science would be crucial to the development and application LUTI decision-support systems in integrated urban and transportation planning in Ghana.

11.5 Conclusion

In this chapter, we have examined the relationship between spatial development and transportation. We have explored how mobility patterns and travel behaviours are fundamentally shaped by the configuration of land use activities at the strategic, city and neighbourhood scales. The central theme we have explored in this chapter is that

---

[7]A comprehensive discussion of the territory is beyond the scope of this chapter. Further reading on LUTI models could be found in Chang (2006), Iacono et al. (2008), Acheampong and Silva (2015). Acheampong (2017a) also presents information on the initial development of the Metropolitan Location and Mobility Patterns Simulator (METLOMP-SIM) and its caliberation for the Kumasi Metropolis.

11.5 Conclusion

an integrative approach to spatial development and transport policy and planning is fundamental to the attainment of long-term sustainable urban development objectives. In view of this, we identified pathways towards achieving the goals and benefits of an integrated approach to spatial development and transportation. Firstly, close collaborations among the various institutions with competences in spatial planning and transportation planning are needed in order to achieve the imperatives of integration for both policy formulation and implementation. Secondly, large metropolis and cities can use integrated land use and transport Master Plans to envisage spatial development patterns and assess the associated mobility outcomes. For purposes of long-term planning and policy evaluation applications, planning agencies could commission the development of integrated land use and transport models for use as decision-support systems. Employing models and decision-support systems could help decision-makers to explore various scenarios of urban spatial structure and required transportation infrastructure investments and travel demand management strategies in an integrative manner. Thirdly, in order to reduce automobile dependence and encourage sustainable mobility choices, urban development planning must prioritize active/non-motorized transport infrastructure such as cycling and walking paths. Fourthly, a shift from exclusionary zoning towards a more inclusionary land use zoning policies that encourages large-scale, mixed-use development would result in urban spatial structures in which the need for long-distance travel using motorized modes is considerably reduced. Moreover, major cities in Ghana are still in the very early stages of motorization, implying that the opportunity exists to promote mass transit for interurban and intra-urban commutes. Doing so would require connecting key urban functions with efficient public transport systems. Finally, as part of wider travel demand management strategies, large cities could explore ways in which they could benefit from emerging ICT-enabled shared-mobility services such as car-sharing, ride-sharing and bike-sharing.

References

Acheampong RA (2016) Cycling for sustainable transportation in urban Ghana: exploring attitudes and perceptions among adults with different cycling experience. J Sustain Dev 9(1): 110–124

Acheampong RA (2017a) Understanding the Co-emergence of urban location choice and mobility patterns: empirical studies and an integrated geospatial and agent-based model. Doctoral Thesis, University of Cambridge. https://doi.org/10.17863/CAM.13849

Acheampong RA (2017b) Towards sustainable urban transportation in Ghana: exploring adults' intention to adopt cycling to work using theory of planned behaviour and structural equation modelling. Transp Dev Econ 3(2): 18

Acheampong RA., Silva EA (2015) Land use–transport interaction modeling: a review of the literature and future research directions. J Transp Land use 8(3): 11–38

Acheampong RA, Siiba A (2018) Examining the determinants of utility bicycling using a socio-ecological framework: an exploratory study of the Tamale Metropolis in Northern Ghana. J Transp Geogr 69: 1–10

Aditjandra PT, Mulley C, Nelson JD (2013) The influence of neighbourhood design on travel behaviour: empirical evidence from North East England. Transp Policy 26: 54–65

Amoako C, Korboe D (2011) Historical development, population growth and present structure of Kumasi. In Adarkwa KK (ed) Future of the tree: towards growth and development of Kumasi University Printing Press, KNUST, Kumasi

Amoh-Gyimah R, Aidoo EN (2013) Mode of transport to work by government employees in the Kumasi metropolis, Ghana. J Transp Geogr 31: 35-43

Batty M (2013) The new science of cities. MIT Press, Cambridge, MA, and London

Beenackers MA et al (2012) Taking up cycling after residential relocation: built environment factors. Am J Prev Med 42(6): 610–615

Brotchie JF (1984) Technological change and urban form. Environ Plann A 16: 583–596

Chang JS (2006) Models of the relationship between transport and land-use: a review. Transp Rev 26: 325–350

Cervero R, Duncan M (2003) Walking, bicycling, and urban landscapes: evidence from the San Francisco Bay Area. Am J Public Health 93(9): 1478–1483

Cervero RJ, Landis J (1997) Twenty years of the Bay area rapid transit system: land use and development impacts. Transp Res A 31: 309–333

Cervero R, Golub A, Nee, B (2007) City CarShare: longer-term travel demand and car ownership impacts. Transp Res Rec J Transp Res Board, 70–80

Cobbinah PB, Amoako C (2012) Urban sprawl and the loss of peri-urban land in Kumasi, Ghana. Int J Soc Human Sci 6(388): 397

Efthymiou D, Antoniou C, Waddell P (2013) Factors affecting the adoption of vehicle sharing systems by young drivers. Transp policy 29: 64–73

Ettema D (2017) Apps, activities and travel: An conceptual exploration based on activity theory. Transportation 1–18

Gim THT (2013) The relationships between land use measures and travel behavior: a meta-analytic approach. Transp Plann Technol 36: 413–434

Hanson S, Giuliano G (2004) The geography of urban transportation, 3rd edn, The Guilford Press, New York; London, pp. xii, 419 p

Iacono M, Levinson D, El-Geneidy A. (2008) Models of transportation and land use change: a Land use transport interaction modeling guide to the territory. J Plan Lit 22: 323–340

Lundqvist L (2003) Land-use and travel behavior. a survey of some analysis and policy perspectives. EJTIR 3: 299–313

Martínez LM, Viegas JM, Silva EA (2007) Zoning decisions in transport planning and their impact on the precision of results. Transp Res Rec J Transp Res Board 1994(1): 58–65

McNally MG (2000) The activity approach. In Hensher DA, Button KJ (eds) Handbook of Transport Modeling. Pergamon, Oxford

Ministry of Finance and Economic Planning (2010) Integrated transport plan for Ghana. Vol. 1. http://www.mrh.gov.gh/files/publications/Integrated_Transport_Plan_for_Ghana_2011___2015.pdf

Ministry of Transport (2016) Vehicle population and growth rate. PowerPoint Presentation (by Daniel Essel). Retrieved from: http://staging.unep.org/Transport/new/PCFV/pdf/2016Ghana_VehiclePopultaionGrowth.pdf (02/11/17)

Mokhtarian PL, Salomon I, Handy SL (2006) The impacts of ICT on leisure activities and travel: a conceptual exploration. Transportation 33(3): 263–289

Naess P (2013) Residential location, transport rationales and daily-life travel behaviour: the case of Hangzhou Metropolitan Area, China. Prog Plann 79: 5–54

Pinjari AR, Bhat CR (2011) Activity-based travel demand analysis. In: A handbook of transport economics, vol 10, pp 213–248

Pinjari AR, Pendyala RM, Bhat CR, Waddell PA (2011) Modeling the choice continuum: an integrated model of residential location, auto ownership, bicycle ownership, and commute tour mode choice decisions. Transportation 38(6): 933–958

Pucher J, Buehler R (2008) Making cycling irresistible: lessons from the Netherlands, Denmark and Germany. Transp. Rev. 28: 495–528

Shaheen SA, Martin E (2010) Demand for carsharing systems in Beijing, China: an exploratory study. Int J Sustain Transp 4(1): 41–55

Simon D, McGregor D, Nsiah-Gyabaah K (2004) The changing urban-rural interface of African cities: definitional issues and an application to Kumasi, Ghana. Environ Urbanization 16(2): 235–248

Songsore J (2003) Regional development in Ghana: the theory and the reality. Woeli Publishing Services, Accra

Waddell P, Wang L, Charlton B, Olsen A (2010) Microsimulating parcel-level land use and activity-based travel: development of a prototype application in San Francisco. J Transp Land Use 3: 65–87

Webster D (2002) On the edge: shaping the future of periurban East Asia. Asia/Pacific Research Center, Stanford

Wegener M (2004) Overview of land-use transport models. In: Handbook of transport geography and spatial systems, vol 5, pp 127–146

Chapter 12
Spatial Planning and the Urban Informal Economy

Abstract Informality is an enduring feature of urbanism in Ghana. One of the visible manifestations of informal urbanism is the informal economy, an important sector of urban economies of developing countries that comprises various economic activities such as street vending and hawking, artisanal works and small-scale enterprises. This chapter focuses on the informal economy and its implications for spatial planning in Ghana. The chapter briefly discusses the origins and contemporary manifestations of the urban informal economy. It develops a typology of informal activities and uses this as the basis to discuss the location decisions of the different types of informal economic activities. The challenges faced by informal workers as they encounter urban land markets are also identified. Finally, the chapter explores possible ways in which the space needs of activities in the informal economy could be provided for in the local spatial planning and land use allocation decisions.

Keywords Informal economy · Informal urbanism · Artisans · Hawkers · Local economic development · Spatial planning · Ghana

12.1 Introduction

Urban spaces are co-produced and shared by different groups of people in society. As emergent and constantly evolving systems, cities reflect the complex interaction among the various forces that shape their construction, and are sites where various agents struggle to share in the benefits of urban agglomeration. As unique spatial resources, cities serve several purposes for different groups of people. For example, the network of social and economic opportunities available in cities makes them thriving sites for various economic activities, on which the livelihoods of their citizens of different socio-economic characteristics depend. Cities are thus arenas of social, political and economic contestation (Lindell 2010; Doe 2015).

Urban scholars tend to adopt typologies as a way of conceptualizing and differentiating the city into component parts and for purposes of analysing the complex interaction among the multiple forces and actors that underpin its emergence. In economic terms, for example, the traditional division of productive activities into public and private sectors, often applied to national economies, has also been used in the context of cities to denote the two main sources of urban economic growth and development. In the context of developing countries, an informal–formal conception has long been applied to explain the unique structure of the private sector of their urban economies. The term 'informal sector' or 'informal economy' has been coined to refer to economic activities by workers and economic units that are not covered at all or are insufficiently covered by formal arrangements of business operation such as registration, regulation and taxation (Becker 2004; Potts 2008).

By definition, activities in the informal economy are often seen by city authorities and urban practitioners as unconventional and failing to adhere to the formal policies, rules and regulations established to bring about desired patterns of spatial development. Consequently, it is not uncommon for operators in the informal economy, including street vendors and artisans to become targets of mass evictions, demolition and police arrests. Informal workers and advocacy groups, on the other hand, have long argued for a more nuanced approach in dealing with this sector of the urban economy. They point to their right of access to urban spaces and the legitimacy of their struggle to secure their livelihoods, especially in contexts where formal sector employment is limited. In Ghana, for example, the private informal economy remains the largest employer, providing jobs to about 90% of the total working population. In urban areas, 84.1% of the labour force is engaged in informal economy employment. This, according to the 2015 national labour force report published by Ghana Statistical Service, amounts to over eight million of the economically active population (i.e. persons aged 15 years and older). The informal economy therefore partly reflects a failure by the state to provide adequate employment for the growing population.

In this chapter, we will look at the informal economy and its implications for spatial planning. We will first situate the discussion on the informal economy within the much broader scholarship on informal urbanism, a phenomenon which is more prevalent in the rapidly urbanizing cities of developing countries. We will also briefly touch on the origins and evolution of the informal economy, highlighting some of the key factors that have fuelled the growth of informal economic activities globally and in the Ghana. Following the aforementioned, we will present a simple typology in which informal economic activities are divided into two broad categories, namely itinerant informal economy activities, which comprises informal workers (e.g. street traders and hawkers) who move from one place to another to undertake their economic activities and informal economy workers who have fixed and sometimes permanent locations (e.g. artisans including carpenters, welders, auto-mechanics hairdressers). We will also discuss the location decisions of informal economy workers and outline the main challenges they face as they encounter urban land markets, formal business operation requirements and spatial planning regulations set by local governments. Next, we will argue that as a result of the recognition of the significant contribution of the informal economy to livelihoods and urban economic growth, attitudes towards this hitherto marginalized sector of the urban economy have shifted over the years. Yet, in the context of Ghana in particular, comprehensive policies to integrate the informal economy into planning policy and ultimately into the urban space are still lacking. Against this backdrop, we will proceed to explore possible ways in which the space needs of different types of informal economic activities could be provided for through local-level spatial planning.

12.2 Informal Urbanism and the Urban Informal Economy: An Overview of Perspectives

The informal economy is part of a much broader phenomenon that has had an unyielding presence in cities in developing and emerging economies and which has recently been captured under the umbrella terminology of '*informal urbanism*'. The term informal urbanism is used broadly to refer to the production of urban transformations outside formal frameworks and mechanisms instituted by the state (see, e.g., Dovey 2012; Jones 2016; Amoako and Boamah 2017). Activities produced by this form of urbanism and the practices that sustain them often deemed disruptive and outside the norm as they do not comply fully with official rules and regulations such as those set and enforced through the planning system. Throughout this book, we have covered a number of behaviours, practices and systems that can be said to be produced outside formally established state planning or administrative systems. The demarcation and sale of land by traditional authorities in areas that do not have formal local plans, the acquisition and incremental development of those lands without recourse to formal development control and land management systems and 'invasion' of public spaces by squatters to either build their homes and/or conduct small-scale economic activities are all manifestations of informal urbanism.

While the existence of the informal economy in most developing countries dates as far back as the 1960s (Potts 2008), the term 'informal sector' first came to use in the 1970s following the original work of Hart (1973) about popular forms of entrepreneurship she observed in Accra, Ghana. The term would subsequently be adopted and popularized by international agencies such as International Labour Organization (ILO) to refer to the unregulated non-formal portion of the market economy that produces goods and services for sale or for other forms of remuneration (ILO 2002). The informal economy has been delineated and examined from different perspectives (see Becker 2004; Portes and Haller 2010; Chen 2012). One of the dominant perspectives takes a *dualist* view by seeing the informal economy as a separate marginal economy that is not linked directly to the formal economy. The dualist conceptualization of the informal economy was founded on traditional notion that the economy of developing countries was characterized by two different sectors, the first characterized by capitalist mode of production and hence considered as modern and progressive and the second being subsistent, characterized by pre-capitalist modes of production, involving self-employment, family labour that was considered less sophisticated in its operations (Potts 2008).

The *structuralist* school perceives the informal economy as subordinated economic units to the formal economy. Leading exponents of this school of thought explain that privileged capitalists subordinate petty producers and traders in order to reduce costs and increase their competitiveness (see, e.g., Castells and Portes 1989; Moser 1978). Structuralists therefore argue that capitalist growth driven by large firms seeking to reduce labour costs and increase profit and competitiveness, and the associated legal requirements created and enforced by the state drive informality. The capitalist mode of production inherently disadvantages small-scale businesses and informal wage workers who become subordinated to the interests of large firms. Others take a *legalist* standpoint and argue that informal work arrangement is a response to cumbersome, often hostile legal systems that lead the self-employed to operate informally with their own informal extra-legal norms (Chen 2012). Leading exponents explain that such businesses operate outside of formal rules and regulation in order to avoid costs, time and effort of formal business registration processes which often requires property rights to convert assets to legally recognized assets (see de Soto 2000).

In addition to the aforementioned traditional notions about the informal economy, a fourth perspective, referred to as *Voluntarist*, has emerged in recent years to explain the continued existence of the informal economy in cities in developing economies. Voluntarists argue that some entrepreneurs, having weighed the costs and benefits of informality relative to formality, choose to operate informally in a deliberate attempt to avoid regulations and taxation weighing. Voluntarists, therefore, do not blame existing business registration procedures per se for the existence of informal businesses. To them, such businesses, by deliberately avoiding formal regulations and taxes, make profits while creating unfair competition for formal enterprises (see Chen 2012).

In any given national or urban context, a confluence of the aforementioned perspectives would explain the existence of informal economy workers. In Ghana, for example, as a result of limited employment opportunities in the formal sectors, individuals resort to self-employment in the informal economy as a major source of livelihood. Indeed, the majority of economically active urban citizens in most developing countries have never had a formal job, implying that self-employment in the informal economy is not only a matter of choice but also a matter of necessity for survival (Chen 2012). Moreover, as we will later show, informal economy workers such as artisans and street vendors tend to be opportunistic in their business location decisions, targeting vantage locations such as busy streets and major road corridors to operate their businesses. Such informal economy workers may be aware or later become aware of the formal rules governing business registration and development control regulations that prohibit economic activities in certain public spaces where they find conducive to operate their businesses. Yet, it is often the case that these informal workers, some of whom would have moved to the city for the first time in search of better living conditions, cannot afford the high cost of land or rents on property at such vantage urban locations. Consequently, informal workers are unable to obtain formal property rights through either renting or ownership of their business premises that is required to officially register a business. Faced with such difficulties presented by urban land markets and local government business registration processes, the majority of workers tend to operate in public spaces where existing development control rules prohibit such activities or on marginal urban lands such as those at risk of perennial flooding or contaminated with toxic pollutants.

The perspectives outlined above help us to delineate the informal economy and to account for some of the possible reasons that explain informality in urban economic activities. In addition to these, it is important to understand the fundamental forces that gave rise to and continue to fuel expansion of the informal economy in developing countries such as Ghana. In the section that follows, we will examine the historical factors that first led to the emergence of the informal economy and explore how contemporary forces linked to rapid population growth, urbanization and national economic policies have led to further expansion of the informal economy by absorbing labour force from a wide range of socio-economic backgrounds.

12.3 Historical Origins and Contemporary Manifestations of the Informal Economy

As the economies of most developing countries including Ghana plunged into macroeconomic crises in the 1980s and 1990s, the implementation of economic restructuring programmes became necessary. Through Structural Adjustment Programmes (SAPs), a portfolio of neoliberal economic reforms was introduced to fundamentally transform the structure of the economy by reducing national budget deficits, liberalizing trade, improving macroeconomic stability and promoting a private-sector-led economic growth (Hilson and Potter 2005; Carmody 1998; Lall 1995). Although the actual impact of SAP is still debated, there is a general consensus in the literature that the pursuit of such neoliberal economic reforms, without cognizance to the prevailing political economies of the beneficiary countries, resulted in a number of negative consequences some of which persists to date.

In Ghana, for example, SAP perpetuated poverty and contributed to persistent economic hardship, unemployment and uncertainty for the larger mass of the population (Briggs and Yeboah 2001; Langevang and Gough 2009). International assistance under SAP carried unrealistic conditions that required a reduction in government expenditure, commitment to privatization and the imposition of cost containment measures in state-owned enterprises in beneficiary countries. Consequently, wage-earning employment in many private and public enterprises would be curtailed. Over a six-year period between 1985 and 1991, it is estimated that formal sector employment fell from 464,000 to 186,000, resulting in 278,000 job losses (Gockel 1998). The majority of those who were retrenched included young workers, labourers, cleaners, drivers, messengers and workers in the lower grades of the public sector. As Gockel further show, women were the hardest hit group over this period, partly because of their particularly low skill levels then; despite women accounting for only 23.5% of the total formal sector workforce, they constituted 31.7% of those who lost their jobs in 1987 alone. With a shrinking public sector, growing population and an underdeveloped private sector, self-employment in small-scale businesses and petty trading became the source of livelihoods for individuals affected by the labour retrenchments as well as new entrants into the labour market. Thus, although self-employment had long existed, the period immediately after SAP saw a significant increase in the size of this sector, thereby establishing the informal economy as the largest sector of non-formal employment in Ghana.

From the latter part of the twentieth century to date, Ghana has at different periods in its economic history experienced macroeconomic challenges that are similar to conditions in the 1980s. In response, governments have sought financial assistance from international organizations such as the IMF under arrangements that are similar to those implemented under SAP. For example, in 2013, just after decade of Ghana receiving a debt relief under the Heavily Indebted Poor Countries (HIPC) Initiative, the government approached the IMF for Extended Credit Facility Arrangement (ECFA)—a three-year bailout arrangement that will see some US$918 million of financial assistance from the IMF in support medium-term economic reforms. Similar to SAP, one of the main targets of the ECFA is 'structural reforms to strengthen public finances and fiscal discipline by improving budget transparency, cleaning up and controlling the payroll, rightsizing the civil service and improving revenue collection' (IMF 2014). While structural reforms under the ECFA have not led to a retrenchment of labour force in the public sector, it has imposed an embargo on new employments as across many sectors as means to reducing the public-sector expenditure on wages. At the same time as ECFA programmes are being implemented, the population of the country is growing (see Chap. 8), while unemployment levels continue to soar. Total unemployment rate in Ghana as of 2015 was estimated at 11.9% (Ghana Statistical Service 2016). This translates into about 1,250,913 unemployed persons aged 15 years and older. In urban areas, unemployment levels are higher. The 2015 labour force report estimates that nearly one in six of the economically active population in urban areas are unemployed, with urban employment rate high at 13.4%.

The current economic climate has not only expanded the size of the informal economy but also diversified the backgrounds of individuals who now engage in informal economic activities. Previously, the informal economy was thought of as comprising mainly of low-skilled workers. In recent years, the proportion of college-educated individuals has increased significantly. Yet, with a rather small manufacturing base, accounting for less than 1% of total employment in urban areas, formal employment opportunities outside of the traditional public and civil service sectors have not matched the large numbers of young adults who graduate annually from the various institutions of higher learning in Ghana. With limited formal sector employment opportunities and rising graduate unemployment, a new form of informal economy in which actors are highly educated and skilled has emerged in urban areas in Ghana (Obeng-Odoom and Ameyaw 2014). Indeed, according to the labour force report, as of 2015, 66% of urban informal economy workers were aged between 25 and 44 years. The same report estimates that about 11.2% of the urban population hold tertiary-level qualifications. While the report does not provide information about the educational backgrounds of informal economy workers, we can infer from the forgoing facts that a large number of highly educated individuals also find employment in relatively low-wage jobs in the informal economy.

In addition to the macroeconomic conditions, local processes that are linked to the urbanization process underway in most developing economies account for the increasing number of their populations resorting to the informal economy for their livelihoods. In Chap. 9, we established how in the context of major metropolitan areas in Ghana, urbanization and associated processes of sub-urbanization and peri-urbanization have reflected in the extensive utilization of hitherto agricultural land for urban land uses such as residential development. The evidence suggests (see, e.g., Korah et al. 2018; McGregor et al. 2011; Cobbinah et al. 2015) that in these peri-urban areas, the ongoing spatial transformations have had major impacts on forest—and agriculture—dependent livelihoods. In some cases, residential development in peri-urban communities has either completely replaced farmland or significantly reduced the amount of farmable land. It is also common in these areas to find developers speculatively acquire large tracks of farmlands. These lands often remain undeveloped several years following acquisition, while access to them for farming may be curtailed for the families and communities that previously owned the land. With their assets and sources of livelihood taken away and often lacking the skills needed to find employment in other sectors, indigenes of rapidly changing peri-urban communities turn to economic activities such as petty trading as the alternative coping strategy.

12.4 Location Decisions and Challenges of Urban Informal Economy Businesses and Workers

Ultimately, informal economic activities manifest spatially. It is therefore important to identify the different types of informal economic activities and to understand the location decisions of business units and individual workers within the informal economy. Previous studies have differentiated between informal economic activities based on the activity types. Boapeah (2001), for example, categorizes activities in the informal economy into service, construction and manufacturing enterprises. Informal service workers include urban food traders, health and sanitation workers, repairers of electrical gadgets, auto-mechanics, welders, hairdressers. Masons, carpenters, steel benders and small-scale plumbers fall under informal workers in the construction sector, while manufacturing activities include small-scale value addition activities such as food processing, textile and garments, wood processing and metal smelting and fabrications.

As we have established at the beginning of this chapter, we are interested in the informal economy and its implications for spatial planning. To this end, a typology of activities that combines both the activity types and locational aspects (i.e. areas in the urban space where these activities tend to locate) is necessary. Understanding the location preferences and behaviours of informal economy businesses and individual workers is also particularly vital to any attempt at integrating their space needs into long-term land allocation decisions in the urban planning and development process. By focusing on their spatial dimensions, we can distinguish between four types of informal economy activities. These are:

i. Home-based informal economic activities;
ii. Itinerant informal economy workers;
iii. Sedentary sales/commercial activities located outside the home;
iv. Sedentary artisanal activities located outside the home.

Within each of the four categories outlined above, a wide range specific activity types could be identified. The location preferences and behaviours of actors as well as the challenges they face as they compete with other activities in the urban land market for space are discussed in the sections below.

12.4.1 Home-Based Informal Economic Activities

In many cities in developing countries including Ghana, the home serves the dual purpose of being a place of residence and place of work. In these urban areas, different types of small-scale activities in the informal economy tend to locate either in the home or within immediate vicinity of the home. Home-based informal economy work locations are more pronounced among low—and intermediate-skilled workers engaged in jobs in the commercial sector of the service industry. While such activities can be found throughout the urban area, in Ghana, home-based workers are more likely to live in traditional compound housing in the historical-core neighbourhoods of towns and cities. Living and working from home in the historical-core neighbourhoods where there is high concentration of commercial activities and a large pool of potential client population offer several benefits to workers including reduced or no home–work distance, relatively cheaper rents for those renting rooms in compound houses and the absence of rents for those living in extended-family-owned housing (see Acheampong 2018).

Invariably, households and individuals make their residential location decisions with the primary objective of providing for their housing needs. However, over time, the place of residence may be adapted to accommodate other activities such as small-scale businesses. For example, it is common to find new extensions and/or alterations to residential units where residents undertake different economic activities such as petty trading, corner shops, saloons, pubs and sachet water production. Such extensions may be constructed from different materials including sandcrete, wood and aluminium sheets. It is also common to find freight containers converted into shops and attached to existing residential units for commercial use. In most cases, such alterations, extensions and attachments proceed without planning permission. In fact, most of the existing residential units to which extensions or alterations are made for commercial uses tend not to have development and building permits from the relevant local government authorities.

Alterations and/or extensions to the home for commercial activities without planning approval flout development control regulations and are therefore considered unauthorized development. This means that, as a matter of the principles of planning and development control regulations, such developments could be demolished at any time by the relevant local authorities. It is not common for demolitions of this kind to happen, yet the fact that such developments remain without threats of demolition does not regularize or legitimize their existence either. It is also possible that businesses operating in 'unauthorized' structures within the premises of the home would have difficulties registering them; operators may, thus, be unable to access opportunities to grow and expand their business for which having a formally registered and recognized business premises is a key requirement. Later in this chapter, we will discuss some of the possible measures that could be taken within the planning system to accommodate physical developments of this kind, as a means of supporting the activities of informal economy businesses and workers in urban areas.

12.4.2 Itinerant Informal Economy Workers

Itinerant informal workers are essentially footloose in the sense that they are not bound to any particular location for purposes of conducting their daily economic activities. They are mobile and tend to visit several locations in a day to sell their goods and/or provide domestic services to clients. For example, in a typical day, cleaners and gardeners may visit several homes to provide services to clients who have requested them. Those who do not have requests for their services may still move around from one neighbourhood to another prospecting for work. Some petty traders also follow a similar mode of operation, finding new customers and/or going to existing customers scattered across the city every now and then to sell their goods. Others undertake their activities within designated hubs such as markets, major transport terminals and industrial areas where they move their goods from one point to another either carrying them on their heads or using wheelbarrows and tricycles modified to suit their needs.

Moreover, for some informal economy workers such as roadside hawkers, street vendors and squeegee merchants, their activities depend largely on vehicular traffic and pedestrian movements generated in the urban area at different times of the day. For commuters and pedestrians, roadside hawkers and vendors offer a convenient means of buying essential household items such as toiletries as well as food and beverages. Consequently, itinerant informal workers tend to be opportunistic in their daily employment location choices, targeting vantage locations such as pavements along congested road corridors, intersections, roundabouts and around traffic signals, where the slow-moving traffic and/or large population concentrations provide ready market for their goods and services. With the knowledge gained from their observation of flows on different road corridors at different times of the day, such as during morning and evening peak hours, street hawkers and vendors are able to adjust their locations periodically to take advantage of the slow-moving traffic and pedestrian flows. Certainly, the availability of mobile phones also makes it easy for hawkers and vendors to exchange information and to converge at the most suitable locations where they can attract the highest patronage for their goods and services.

In their daily struggles to earn a living and to provide for the large share of urban households who depend on their rather low earnings, roadside hawkers and street vendors face several risks and challenges. Although local authorities do collect tolls from them on a daily basis, hawkers and vendors are often the main targets of routine 'decongestion' exercises. In most cases, decongestion exercises only affect them temporarily and/or displace their activities to other locations in the urban area. Notwithstanding, routine evictions in the name of decongesting inner-city areas constitute a major source of uncertainty for itinerant informal economy activities. Encounters between street hawkers and enforcement officers deployed by local authorities to carry out decongestion exercises sometimes result in destruction and/or confiscation of their goods; operators also sustain minor to life-threatening injuries in instances where their encounters with enforcement officers turn out violent and/or result in a stampede as people attempt to escape arrest. Fatalities and injuries from collision between motorized traffic and street hawkers are also very common.

12.4.3 Sedentary Sales/Commercial and Artisanal Activities Located Outside the Home

In this section, we examine the location decisions and challenges of the third and fourth categories of urban informal economy operators identified as sedentary sales and commerce workers and sedentary artisanal activities located outside the home, respectively. Activities in the former group include retailers, food vendors, petty traders and vehicle spare part dealers who require. Operators in the latter group tend to come mainly through the apprenticeship training sector and included artisanal workers such as auto-mechanics, welders and metal fabricators, vulcanizers, manufactures of construction inputs such as sandcrete blocks, bricks and culverts, hairdressers and dress-makers.

Sedentary activities in the informal economy tend to locate within or closer to existing agglomeration of similar enterprises. Such agglomerations may come about as a result of land being designated purposely for them such as in neighbourhood markets, urban satellite markets and artisanal enclaves. While formal planning may designate land for these activities, for example, in Structure Plans or local plans, in most cases, the site itself is either not planned or receives minimal planning involving the designation of lots for stalls and shops and local access roads. Actual construction of sheds and stores may be left to the operators. As more and more activities are attracted to these enclaves beyond the initially planned capacity, road reservations and other public spaces become developed into more stores and sheds, often without any permission from the relevant local authorities. Over time, vehicular access to parts of the enclave may become restricted or completely curtailed, making it difficult for timely response in emergencies such as a fire outbreak.

Moreover, some sedentary informal economy enterprises tend to be attracted to existing uses such as transport terminals, lorry parks, large vehicle repair enclaves and light industrial areas where they can find patronage for their goods and services. Activities in the artisanal sector, in particular, such as auto-mechanics, spare part dealers, metal workers and vulcanizers, tend to be complementary and interdependent. The need for the artisans themselves to benefit from these complementarities in their daily operations while presenting their enclaves as one-stop shops where clients can receive a wide range of services underpins the large concentrations of informal artisanal enclaves that can be found in many cities in developing countries. In addition, operators such as food vendors depend on the customers and workers in these artisanal enclaves. Moreover, large concentrations of informal artisanal workers could sometimes be a strategy on the part of operators to achieve strength in numbers, especially in instances where they are constantly threatened by evictions by local governments or landowners.

Yet again, for some small- and medium-scale artisanal enterprises, the nature of their businesses means that they find locations along major, often busy road corridors suitable. They consider these locations for a number of reasons. Firstly, among those involved in the processing of raw materials into, for example, building construction materials such as sandcrete blocks, culverts, door and window frames, locations along major road corridors are easily accessible and convenient for delivery of the bulky raw materials they require as inputs, as well as for loading finished products to various supply destinations. Also, busy road corridors can offer some of the most suitable locations for strategically locating showrooms, where finished products such as furniture can be exhibited to attract patronage from road users. Artisans such as welders and metal fabricators, carpenters and sandcrete block manufactures choose locations in emerging residential neighbourhoods in peri-urban areas to take advantage of the booming construction industry.

The aforementioned location preferences and strategies of sedentary informal economy enterprises imply that operators must compete with other developers and businesses for land at prime urban locations. The general accessibility value provided by major road corridors means that such locations have some of the highest land prices in the urban area. Operators must therefore compete with other investors involved in large-scale commercial developments such as petrol-filling stations, hotels apartment blocks and warehouses. Landowners tend to prefer to lease to investors who can afford to pay the market price of the subject land while bringing about upliftment in the value of adjacent land by virtue of their development. While a large number of informal workers could pool resources together to enable them to compete for the land, their bid for the land may be rejected on the basis that their activities could be a nuisance and potential source of blight for such prime urban locations. In the end, land invariably goes to the highest bidder, implying that following the dictates of the urban land market, informal economy businesses stand very little to no chance of accessing land at the locations most suitable for their businesses. In fact, some landowners may initially lease their land to these small- and medium-scale informal businesses but evict them to either sell to other developers or develop the site into other uses that yields higher returns to the landowner. Artisans displaced by the workings of the land market end up as squatters on public spaces for which they have no permission and security of tenure. It is common in urban areas in Ghana, for example, to find artisans in areas within road reservations; locations directly under high-voltage power lines despite the potential insidious health impacts; and as squatters on unprotected land belonging to the state and other public institutions.

As a result of lacking tenure over the land they occupy, artisans in the informal economy are exposed to poor and in some instances, hazardous conditions in the work environment. Poor environmental conditions that are often associated with slum settlements, such as the lack of on-site water and sanitation facilities, also typify artisanal enclaves in urban areas across Ghana. Operators also find it difficult to connect to existing utilities such as electricity required to power their activities. Utility service providers such as the Ghana Electricity Company Limited require operators to produce permits certifying approval of their premises by the relevant local authority in order to be able to connect to existing electricity distribution lines. Moreover, as a result of not having their businesses registered and/or lacking tenure over the land on which they operate, informal economy enterprises are unable to insure their businesses against potential risks such as those arising from burglary or damages from fire. While having a registered business premise at an authorized location may not be a requirement in demonstrating creditworthiness especially to informal money lenders and small credit and loans establishments, informal business operators would not be able to access loans in instances where land may be required as collateral.

Furthermore, constant threats and fear of eviction constitute a major source of uncertainty which affects the extent to which small-scale informal businesses could grow and expand to make significant contributions to the economic development of their respective local government areas. Indeed, it would not be prudent for operators to invest in expanding their businesses in premises over which they have no secure tenure. For those who risk and invest in their businesses, evictions could result in major loses from which businesses often take several years to recover or are unable to recover at all. Finally, being unable to register businesses as a result of operating in unauthorized premises exclude small and medium-scale businesses in the informal economy from being able to participate in competitive bidding for public sector projects. Without business registration certificates and operation permits from the appropriate institutions, such businesses cannot meet the formal requirements of public procurement that public institutions are expected to follow in awarding contracts for the purchase of goods and services.

12.5 Planning to Accommodate the Informal Economy

Informal urbanism and associated economic activities such as street trading and small-scale artisanal enterprises have longed been perceived as problematic and disruptive in the pursuit of the ideals of order, efficiency and 'modernity' in spatial planning. Several attempts to rid cities in developing countries of informal economic activities have stemmed from the generally negative attitude that planners and local governments have had towards this sector as being chaotic, backward and not conforming to formally established development control regulations (Potts 2008). However, as we have established elsewhere in this book, in the context of Ghana, for example, the planning system until recently had been generally weak and ineffective in delivering its objectives. At the same time, with high levels of urbanization and limited employment opportunities in the formal sectors, the number of urban citizens engaged informal sector jobs has increased significantly in recent years. This means that even when practitioners and local governments have acknowledged the contributions of the informal economy and expressed the need to integrate them into planning policies, the formal system and mechanisms that were needed to make urban development planning responsive to the informal economy were essentially non-existent. Thus, to a larger extent, unresponsive urban planning has been responsible for some of the chaos and the resulting urban blight that has accompanied the expansion of the informal economy in towns and cities in Ghana.

12.5.1 Shifting Attitudes and Contemporary Policy Responses

In recent years, attitudes towards the informal economy are shifting. In part, policymakers at the national and local levels now recognize the significant contribution the informal economy makes to employment creation, local economic development and Gross National Product (GDP). Indeed, in periods of macroeconomic instability and decline, the informal economy has remained resilient and continued to absorb a large share of the urban labour force. In addition, like private formal sector businesses, enterprises in the informal economy continue to leverage technology and innovation in their operations, providing new opportunities for them to adapt their businesses to the changing times while increasing productivity. The emergence and rapid expansion of the mobile money transfer industry, which is used by the majority of a hitherto unbanked business in the informal economy, are a case in point.

At the same time, informal economy businesses and workers continue to innovate and adapt to a competitive business environment, and academic research and advocacy from international organizations and local organizations have ensured that their interests and concerns are represented in policy debates. The National Urban Policy (NUP), for example, is indicative of the policy shifts towards accommodating the informal economy into urban planning and development. One of the central objectives of the NUP is promoting urban economic development. Four of the eight initiatives identified to help realize this policy objective focus on the urban informal economy. These are outlined as follows:

- Change official attitude towards informal enterprises from neglect to recognition and policy support;

- Ensure that urban planning provides for the activities of the informal economy;
- Build up and upgrade the operational capacities of enterprises in the informal economy;
- Improve funding support for informal economic activities.

Furthermore, in order to make urban planning responsive to the informal economy, the NUP has identified the need to: incorporate in planning legislation, standards and zoning regulations, provisions that protect and facilitate informal economic activities; involve small-scale businesses in the informal economy in the urban planning process; designate industrial and business estates in strategically selected places and provide essential facilities and services for informal economy businesses; and provide businesses currently operating in authorized locations alternative serviced sites for relocation.

Moreover, the new zoning guidelines and planning standards, which were introduced as part of the reforms that established the new spatial planning system in Ghana, have made an attempt to address the space needs of informal economic activities in land use zoning policies. The zoning guidelines and planning standards document have, for the first time, introduced a sub-class development zone called informal business zone (BL) to be designated in spatial development frameworks and Structure Plans to accommodate informal economy businesses. A description of the intent of this sub-class of land use and an outline of the permissible and non-permissible uses within it are provided in Box. 12.1.

Box 12.1: Informal Business Zone (BL)

Intent

Informal business zone has for the first time in the history of spatial planning in Ghana been introduced new zoning guidelines and planning standards which was published in 2011 by MESTI and the TCPD.

This zone is intended for local community business activities, incorporating small shops, sub-offices and agencies and a community market. Local business centres should be sited in locations which are convenient for pedestrian and vehicular access. The zone is intended to be developed progressively, with adequate provision for services, waste disposal and public amenities. The market is intended to be the focus of day-to-day community activity around which will be constructed more permanent retail and community facilities.

Permitted uses

The under listed uses are permitted in the informal business zone:

- Local market (1 ha or less.);
- Local shops;
- Service agencies;
- Light industries with floor area less than 20 m^2;
- Restaurant and bars;
- Cottage industries;

- Motor sales points;
- Taxi and bus terminals;
- Car park.

Prohibited uses

- Industrial development;
- Major commercial and retail development;
- Cemetery and crematorium;
- Extractive industries;
- Animal husbandry;
- Animal slaughterhouse;
- Warehousing.

12.5.2 Translating National Policies into Local Strategies and Actions

The recognition at the national level of the need to integrate informal economic activities into local land use planning decisions in Ghana is commendable. This, however, is only a first step in the portfolio of measures, policy strategies and actions that are needed at the local level to achieve this policy imperative. The policy intent of designating informal business zones in spatial development plans, for example, is laudable. However, land in such zones would not necessarily become affordable for small—and medium-scale informal economy businesses should its allocation be left to the dictates of the formal land market. In order to ensure that informal economy businesses actually benefit from land allocated in spatial development plans to accommodate their activities, local authorities should acquire and protect such lands for their designated purposes. The experience over the past decades has shown that where the planning system has failed to provide for the space needs of informal economic activities, operators have eventually found ways of meeting this need, often in spaces where such activities are not permitted. In order to avert this, local governments should as far as possible endeavour to protect land designated for informal businesses from being reallocated to other uses, even when proposals for rezoning are initiated through the formal development management procedures.

Moreover, as we established from the analyses of the location preferences of informal economy businesses, it is important that informal business zones are designated in easily accessible areas such as along existing road corridors or planned arterials where businesses would have adequate patronage for their goods and services. Such locations would particularly be ideal for non-home-based sedentary businesses such as artisans and retailers. Also, to ensure home–work balance while contributing to sustainable commuting, informal business zones should be located close to existing residential areas and/or near areas planned for future residential development. In instances where land is only available in areas quite remote from existing built-up areas, such as in Greenfield areas in the urban periphery, informal business zones should be connected to other urban functions by public transit to facilitate easy commuting for informal workers and the urban population who would patronize their goods and services. Ultimately, creating opportunities for informal economy workers and business units to actively participate in the plan-making process would ensure that their location needs would be adequately catered for in the urban development process.

Similar to other sectors of planning, data would be vital to assessing and providing for the location requirements of informal businesses. Given that the vast majority of informal businesses are not registered, most local governments do not have any reliable data on the number of such businesses in their jurisdictions. While enumerating all informal economy businesses would not be possible, sedentary businesses can easily be identified and counted. With knowledge of the locations informal businesses find suitable in the urban area, TCPDs, working with the Business Advisory Units (BAUs) of their local governments could make use of digital information resources such as Google Earth and Open-Street Maps as part of identifying and enumerating sedentary informal economy businesses in their jurisdictions. The enumeration process could, for example, begin by identifying from these online geospatial data sources, structures along major road corridors and neighbourhood streets that are likely to be used by informal economy operators in the urban area.

Informal enterprises in the artisanal sector also provide apprenticeship training to individuals outside of the existing formal apprenticeship training system. Apprentices, on completing their training, tend to open their own businesses independent of the centres where they received their training. The apprenticeship sector is therefore one of the main channels by which the number of informal businesses in an urban area increases over time. This also implies that by identifying and enumerating informal businesses providing apprenticeship training, local governments would also be able to estimate new activities that are likely to emerge in the informal sector, and to plan accordingly to accommodate their space needs. Over time, a system could be created, whereby local government would periodically receive up-to-date data on the number of individuals undergoing apprenticeship training in various informal economy enterprises to inform land allocation decisions.

Making flexible, development control systems that were instituted under exclusionary land use zoning regimes would also complement measures needed to effectively address the space needs of informal economy businesses in the urban development process. We have already established how home-based informal economy businesses respond to their space needs by making alterations and/or extensions to existing residential units, often without recourse to existing zoning regulations. How should the planning system deal with developments of this nature that have already taken place as well as new ones that will occur in future? Certainly, some level of discretion on the part of local governments is needed in regularizing alterations and extensions that have already taken place. In fact, one may argue that while many extensions and alterations do not conform to zoning regulations, most of the new uses that emerge in the process tend to be compatible with the existing residential uses. Authorizing these uses retrospectively could therefore constitute one of the strategies towards achieving desired levels of mixed uses in hitherto neighbourhoods zoned exclusively for residential development.

Beside alterations and extensions to existing residential units, it is becoming increasingly common in recent years for developers to submit designs for mixed residential–commercial development for planning approval, albeit the subject sites tend to be exclusively zoned originally for residential purposes. In such instances, the planning system is not clear as to how such applications could be approved when they deviate from the prevailing zoning policies. That notwithstanding, planning officers could identify various scenarios under which exercising reasonable amount of discretion by, for example, relaxing the existing zoning regulations would allow such development proposals to be approved. One such scenario is recommending for approval, proposals for mixed-use development in areas exclusively zoned for residential development if there is a clear case that the proposed development would not fundamentally conflict with adjacent uses. To facilitate a transparent development approval system under such circumstances, local authorities could also set and publish thresholds against which such development applications could be assessed. For purpose of illustration, let us consider a scenario where a developer submits planning application for a site originally zoned exclusively for residential use. There could be a flexible caveat that if the proposed development has between 80 and 90% conforming to the original allocated use (i.e. residential) while light commercial uses constitute about 10–20%, then such a development may be approved subject to meeting other requirements, without going through the formal procedures of change of use or rezoning, which we outlined in Chap. 4. Also, by storing plans in digital databases, information regarding the subject site could be updated by a matter of adding an attribute to the parcel in question, to reflect the approved mixed-use development without having to revise the entire planning scheme. It may even be possible, under certain circumstances, for certain types of alterations and extensions carried out for purposes of undertaking home-based economic activities to be permitted at a later stage at the time of granting development approval to an original application. In such instances, an on-site inspection by a building inspector may be required for the sole purpose of ensuring that the alteration or extension meets basic building regulations.

Increasingly, cities are experimenting with innovative ways of making urban spaces adaptable to multiple often temporary uses. Reimaging public spaces as adaptable spaces could provide yet another effective way of creating spaces for informal economic activities such as street vending to thrive in towns and cities in Ghana. To do so would ultimately require a shift in current urban planning and design ethos from seeing urban spaces such as streets as serving a single purpose as stipulated in zoning codes towards reimagining them as adaptable spaces that could be put to multiple uses. For example, through careful planning and coordination, it may be possible to open up public spaces such as open spaces, busy streets in inner-city locations as well principal streets in neighbourhoods for purposes of street trading on certain days for a given period of time. In addition, existing and new pedestrian zones could be designed by integrating spaces for trading activities with paths for non-motorized traffic in inner-city locations.

12.6 Conclusion

The informal economy constitutes one of the enduring features of many cities in emerging economies including Ghana, serving as a source of livelihood and escape from poverty for the larger mass of the urban labour force. In this chapter, we have focused on the informal economy in relation to spatial planning at the local level. Focusing on the spatial dimensions of the informal economy, we developed a typology of activities in this sector comprising home-based activities, itinerant informal economy workers, sedentary sales/commercial activities and sedentary artisanal activities located outside the home. We also examined the location preferences and strategies of activities within the above categories and highlighted the challenges they face in seeking to access spaces in the urban area to conduct their activities. What the analyses of the location preferences and challenges reveal is that, for informal economy businesses, a number of factors including landowners and city officials regarding such activities as nuisance and competition from large-scale developers for prime urban locations make it difficult for them to find spaces that are suitable and safe for their activities.

In addition, we highlighted how through a combination of advocacy and a realization of the significant role the informal economy plays to economic development at the local and national levels, and attitude towards informal business operators have shifted in recent years. In the context of Ghana, for example, this attitudinal shift is reflected in the recognition given in the National Urban Policy to the need to integrate informal economy businesses into local economic development strategies and land use allocation decisions. Despite this recognition at the national level of policy-making, specific policy strategies and actions needed at the local level to address the space needs of enterprises in the informal economy are still rare. In response to this, we have proposed a number of ways by which spatial planning at the local level could make the necessary space provisions to accommodate the different types of informal economy activities in urban areas. We argued that while designating informal business zones in spatial development plans is a step in the right direction, local governments would have to ensure that such lands are acquired, protected and made accessible to small—and medium-scale businesses in the informal economy. Equally important is the need for local government to begin to build a database of informal economy businesses in their respective jurisdictions to inform both long-term land allocation decisions and policy responses that may be required over relatively short-term horizons. In particular, we highlighted the value that periodic information on the artisanal activities providing apprenticeship training would have in helping local governments to estimate new activities that are likely to emerge and to factor them into land allocation decisions. Last but not least, relaxing development control procedures based on exclusionary zoning ethos while experimenting with new ways of

12.6 Conclusion

putting public spaces such as streets to multiple, adaptable uses would also provide opportunities for informal economy businesses to operate in towns and cities.

Ultimately, a spatial planning system that is responsive to the needs and challenges of the informal economy not only plays a significant role in enabling urban citizens to conduct economic activities on which their livelihoods depend, but also helps in creating the conditions necessary for entrepreneurship and for local economic development.

References

Acheampong RA (2018) Towards incorporating location choice into integrated land use and transport planning and policy: a multi-scale analysis of residential and job location choice behaviour. Land Use Policy. https://doi.org/10.1016/j.landusepol.2018.07.007

Amoako C, Boamah EF (2017) Build as you earn and learn: informal urbanism and incremental housing financing in Kumasi, Ghana. J Housing Built Environ 32(3): 429–448

Boapeah S (2001) The informal economy in Kumasi. In: Adarkwa KK (ed) The fate of the tree: planning and managing in the development of Kumasi, Ghana, Woeli Publishing Services, Accra, pp 59–77

Becker KF (2004) The informal economy. Swedish International Development Cooperation Agency. Retrieved from http://rru.worldbank.org/Documents/PapersLinks/Sida.pdf

Briggs J, Yeboah IE (2001) Structural adjustment and the contemporary Sub-Saharan African city. Area 33(1): 18–26

Carmody P (1998) Constructing alternatives to structural adjustment in Africa. Rev Afr Polit Econ 25(75): 25–46

Castells M, Portes A (1989) World underneath: the origin, dynamics and effects of the informal economy. John Hopkins University Press, Baltimore

Chen MA (2012) The informal economy: definitions, theories and policies, vol 1, no 26, pp 90141-90144. WIEGO working paper

Cobbinah PB, Gaisie E, Owusu-Amponsah L (2015) Peri-urban morphology and indigenous livelihoods in Ghana. Habitat Int 50: 120–129

de Soto H (2000) The mystery of capital: why capitalism triumphs in the west and fails everywhere else. Basic Books, New York

Doe B (2015) Urban land use planning and the quest for integrating the small-scale informal business sector. Doctoral dissertation, TU Dortmund University

Dovey K (2012) Informal urbanism and complex adaptive assemblage. Int Dev Plann Rev 34(4): 349–368

Ghana Statistical Service (2016) 2015 labour force report

Gockel AF (1998) Urban informal sector in Ghana: features and implications for unionisation. In: Paper presented within the framework of the GTUC/ICFTU-AFRO new project approach to structural adjustment in Africa, Oct 1998

Hart K (1973) Informal income opportunities and urban employment in Ghana. J Mod Afr Stud 11(1): 61–89

Hilson G, Potter C (2005) Structural adjustment and subsistence industry: artisanal gold mining in Ghana. Dev Chang 36(1): 103–131

Internal Labour Organization (2002) Women and men in the informal economy: a statistical picture. Employment Sector Paper. Geneva, Switzerland

International Monetary Fund (IMF) (2014) Heavily Indebted Poor Countries (HIPC) Initiative And Multilateral Debt Relief Initiative (MDRI)— Statistical Update. Washington, D.C. https://www.imf.org/external/np/pp/eng/2014/121214.pdf

Jones P (2016) Informal urbanism as a product of socio-cultural expression: insights from the Island Pacific. In: Dynamics and resilience of informal areas. Springer, Cham. pp. 165–181

Korah PI, Nunbogu AM, Akanbang BAA (2018) Spatio-temporal dynamics and livelihoods transformation in Wa, Ghana. Land Use Policy 77: 174–185

Lall S (1995) Structural adjustment and African industry. World Dev 23(12): 2019–2031

Langevang T, Gough KV (2009) 'Surviving through movement: the mobility of urban youth in Ghana', Soc Cult Geogr 10(7): 741–756

Lindell I (2010) Informality and collective organising: identities, alliances and transnational activism in Africa. Third World Q 31(2): 207–222

McGregor DF, Adam-Bradford A, Thompson DA, Simon D et al (2011) Resource management and agriculture in the periurban interface of Kumasi, Ghana: problems and prospects. Singap J Trop Geogr

Moser CN (1978) Informal sector or petty commodity production: dualism or independence in urban development. World Dev 6

Obeng-Odoom F, Ameyaw S (2014) A new informal economy in Africa: the case of Ghana. Afr J Sci, Technol, Innov Dev 6(3): 223–230

Portes A, Haller W (2010) The Informal economy (chap. 18). In: The handbook of economic sociology, p 403

Potts D (2008) The urban informal sector in sub-Saharan Africa: from bad to good (and back again?). Dev South Afr 25(2): 151–167

Chapter 13
Epilogue: Perspectives on Pathways Towards a Responsive Spatial Planning System

Abstract In the preceding chapters, matters that relate specifically to the spatial planning system as well as various developmental challenges that reflect the scope of issues embraced by spatial planning have been discussed. This chapter provides a summary of the main issues discussed in the previous twelve chapters of this book. This chapter synthesizes the various perspectives advanced in this book on these issues into a portfolio of pathways towards making spatial planning effective in delivering social, economic and environmental development objectives at all levels.

Keywords Planning system · Local planning · Urbanization
Growth management · Housing · Transportation · Informal economy
Public participation · Ghana · Spatial planning

13.1 A Synopsis of the Story so Far

We set out in this book to explore the subject of spatial planning from the unique sociocultural, economic and political context of Ghana. To set the stage for examining spatial planning in this unique context, we first examined the evolving meaning and scope of the activity, identified the rationale for spatial planning and discussed the normative theories that underpin the discipline of spatial planning and its professional practices globally. Among other things, we have established that the nature, scope and purpose of spatial planning have evolved to encompass its traditional function of land use allocation and development control as well as how these regulatory functions are connected to the realisation of wider economic, social and environmental development goals at different spatial scales. In its contemporary conception and practice, therefore, spatial planning is meant to embody a promotional, interventional and regulatory purpose, support the co-production of socio-economic transformations and reconcile the essential tensions among various development goals across spatial scales and among policy domains. It is the vehicle through which various goals, including achieving orderly organization of population and activities, promoting healthy and liveable towns and cities, providing adequate and affordable shelter, promoting business growth and economic development, and protecting biodiversity and the environment are articulated and pursued.

On the subject of spatial planning in Ghana—the substantive focus of the book—we explored the key issues under two broad themes. Under the first theme, we focused on issues that relate specifically to the spatial planning system—the milieu of evolving institutional and legal arrangements, processes, mechanisms and practices, as well as the policy instruments used to articulate and achieve the goals of spatial planning.[1] The second theme explored in the book covered a range of cross-cutting issues, which reflect some of the key developmental challenges that spatial planning is meant to address. In other words, the issues identified and discussed here, together, reflect the scope and purpose of the activity of spatial planning. In corollary of this, we covered a range of topics including: the multiple scales of spatial planning and policy (i.e. national, regional and local levels); public participation in spatial planning; spatial planning and housing development; urbanization and growth management; the nexus between spatial development and transportation and spatial planning and the urban informal economy. For each of the issues identified under these two broad thematic areas, we systematically analysed past trends and their contemporary manifestations, with the aim to gain a firm understanding of the forces at play and the prevailing challenges.

On the basis of a strong understanding of past trends as well as the contemporary expressions of the issues interrogated, we proceeded to explore possible pathways through which the key challenges identified could be addressed. In relation to the spatial planning system, we established that if it would become effective and responsive to the developmental needs of the country, a number of fundamental changes must be made in the existing system of legislations and institutional arrangements that determines the competencies for planning. These changes, in turn, will have implications for the ways in which the range of instruments that are deployed at the national, regional and locals of spatial planning and policy are formulated and implemented to achieve social, economic and environmental development priorities. Furthermore, we explored a number of perspectives on possible pathways towards leveraging the spatial planning system to respond to the key socio-economic and environmental development issues, which we identified and explored as single issues in spatial planning in Part III of the book. In the sections that follow, we will synthesize the various perspectives into a comprehensive portfolio of approaches and pathways that could be adopted and implemented to strengthen the spatial planning system and to make spatial planning a vital vehicle of public policy to bring about socio-economic transformation of towns, cities and regions in Ghana.

---

[1] See Chaps. 3–7 for the range of issues covered under this theme of the book.

13.2 Strengthening the Spatial Planning System

Throughout the chapters in Part II of this book, we subjected the planning systems of Ghana to critical analyses. Taking a historical perspective, we discovered, among other things, the origins and evolution of the planning system from when it was formally instituted in the mid-1940s, through the early years following the declaration of independence in 1957, the decades of political instability and economic uncertainty that followed from the mid-1960s, all the way to the late 1980s when the process of democratization and decentralization would start and continue to the present day. Since the inception of democracy and decentralization, the national governance structure has transformed gradually from a centralized unitary system to a three-tier decentralized structure, comprising the national government, administrative regions and local governments.

Alongside the aforementioned transformation of the governance structure, the institutional and legal arrangements for planning at all levels have evolved. We established that the Town and Country Planning Ordinance (CAP 84, 1945), the first spatial planning legislation introduced by the colonial government in the then Gold Coast, instituted a planning system that was explicitly spatial and comprehensive in ethos, scope, purpose and practice. However, at the beginning of the twentieth century, the pursuit of decentralization accordance with the tenets of democratic governances would fundamentally change the landscape of spatial planning at all levels of decision-making. Two major legislations which accompanied the democratic reforms, namely the Local Government Act (Act 462, 1933) and the National Development Planning (System) Act (Act 480, 1994), helped to establish the apparatus for a new culture of planning, which we identified as the development planning system. The ethos of development planning instituted by these legislations pursued the strategic goals of achieving macroeconomic stability and alleviating poverty, using medium-term development plans formulated at the national and local levels as the main vehicles to articulate and realize development goals.

As we discovered throughout the analyses, in an attempt to respond to developmental challenges which were largely perceived and understood in their socio-economic manifestations, the newly found ethos of development planning, which endures to date, would ultimately neglect the principles and strategic objectives spatial planning. Consequently, the spatial planning system—the system of institutions, policy instruments, strategies, tools and interventions that would address the wider challenges of resource allocation within and between places, as well as the spatial organization of population and activities in towns, cities and regions and their environmental implications—gradually became weak and ineffective to contribute to the nations socio-economic transformation agenda in a meaningful way. Essentially, the development planning system in its conception and functioning failed to connect in any meaningful way the nexus between poverty reduction and macroeconomic stability goals on the one hand, and the geographical scales of policy conception and implementation needed to achieve those strategic goals on the other hand. Spatial planning which hitherto had been the main vehicle for governments to anchor national visions and development priorities to towns, cities and regions lost its standing and relevance under the new paradigm of development planning.

With an established development planning system in place for nearly two-and-a-half decades, the promulgation of the Land Use and Spatial Planning Act (Act 925) subsequently in 2016, as part of reforms to reintroduce spatial planning at all levels of decision-making, has emerged two separate planning systems—the development planning system and the new spatial planning system. As the analyses showed, these traditions of planning, each drawing on the institutional structures embodied in the three-tier decentralized governance structure, deploy separate institutional arrangements and policy instruments to accomplish the single task of planning at the national and sub-national levels. The resulting situation of '*a one nation, two planning systems*' has compartmentalized the authority and competences into the two competing planning traditions. While in theory these traditions of planning overlap in many respects, in practice, they are operated by the actors involved in accordance with a rather narrow conception that suggests that what ought to be embraced within the scope of spatial planning as a *type* of planning is unique and distinguishable from the ethos and practices of development planning.

Strengthening the planning system and making it responsive to the development needs of the country would require fundamental changes in the existing institutional and legal systems that underpin the activity. Indeed, as we have already established, given the obvious overlaps in scope, purpose and functioning of the two competing planning systems, the current distinction is not only confounding intellectually but also creates unnecessary compartmentalization, duplication of functions and resource wastage as institutions attempt to delineate and deliver their respective mandates. Sober reflections on the prevailing system would ultimately lead us to seek new reforms which would establish a unified planning system. As well as the institutional restructuring that would become necessary, institutionalizing a unified system of planning that is comprehensive and integrated into scope and purpose would require harmonizing the existing laws that grant the institutional powers and competences to undertake the task of planning at all levels.

In addition to addressing the internal contradictions in the contemporary planning systems, a number of issues are worth addressing at the various scales of planning. At the national level, we found that the established culture under the development planning system has been the formulation of medium-term development planning articulated in NPFs. This is, however, inherently inconsistent with the long-range, strategic and comprehensive imperatives espoused under the new spatial planning system to underpin the formulation of SDFs at the national, regional/sub-regional and local (i.e. MMDAs) levels, respectively. Pragmatic approaches would be required if strategic spatial planning instruments can reconcile their long-term goals on the one hand with the short-term priorities articulated in election manifestoes of politicians and political parties, which get translated into NPFs on the one hand. In the absence of comprehensive development plans at the national level, institutions with the authority and competences for spatial planning would have to engage with the political process in addition to carrying out their technical mandate. Through constant engagements and dialoguing with governments, the ensuing consensus could

13.2 Strengthening the Spatial Planning System

provide new opportunities to bridge the gap between strategic imperatives put forward in long-term spatial planning instruments and development priorities as defined and set from the perspectives of politicians who tend to have a relatively short-term outlook.

With respect to regional-level planning, we established that the tradition of regional spatial planning has only recently been instituted, implying that the concept of regional spatial planning and the associated practices is still being experimented with. Besides, the limited institutional experience with regional policy formulation and implementation, the existing institutional arrangements for the activity require streamlining to enable the system to deliver its objectives effectively. At the same time, the geographies of regional spatial planning including existing administrative regions, new administrative regions that are planned to be delineated from the existing ones and the newly designated city-regions and urban networks (see Chap. 4) yield very complex spatial governance structures for achieving the objectives of regional spatial planning. Amidst the emerging fuzzy and fragmented geographies for regional spatial planning, comprising layers of administrative regions, sub-regions, city-regions and urban networks, the concept of special Development Zones is also becoming increasingly popular. In particular, politicians in recent years have been designating such zones as the vehicle to bring into focus the perceived developmental challenges of specific geographical areas in the country.

A number of measures are worth pursuing in order to avoid these emerging complex layers of geographies and the plethora of accompanying institutional actors obfuscating the regional spatial development planning process. Firstly, resolving the internal contradictions in the present-day planning systems would also streamline the scope and institutional settings for spatial planning and governance of the different types of regions that may be conceived and designated. Also, having a multi-stakeholder platform, where the relevance of designating Special Development Zones, in addition to the administrative regions, multi-district functional regions, city-regions and urban networks that currently exist, would help build consensus and ultimately avoiding the chaotic outcomes which the system appears to be headed. Indeed, designating any type of region is not an end in itself. Instead, given the complex institutional settings that accompany such declarations, using these geographies as vehicles to address developmental challenges at a scale above local government jurisdictions, would depend, to a larger extent, on cooperation among the relevant in stakeholders. Thus, conscious effort to promote and institutionalized a culture of co-operation among multiple stakeholders comprising local governments and their citizens would be crucial to making regional spatial planning effective in delivering expected economic, social and environmental development objectives.

13.3 Responding to Developmental Challenges Through Spatial Planning

One of the central focuses of this book has been to examine developmental challenges facing town, cities and regions and the state and to identify ways in which spatial planning can contribute to addressing these challenges. To this end, we identified and analysed a range of developmental issues in Part III of the book. Using these developmental challenges to illustrate the scope of spatial planning and the functions expected of this activity in practice, we also identified the technical competences, practical strategies and tools that spatial planning offers, and their relevance in bringing about sustainable solutions to socio-economic and environmental challenges in towns, cities and regions. In the sections that follow, syntheses of the core arguments advanced in relation to responding to key developmental issues through spatial planning are outlined.

13.3.1 Local-Level Spatial Planning and Development Control

One of the key mandates of spatial planning is to bring about orderly physical development by using various types of frameworks and plans (indicative and/or binding) and policy instruments to guide spatial development and manage land in the best interest of the public. To be effective, spatial planning must coordinate decisions of various actors in the urban development process in accordance with policies and standards articulated in a plan. The quality of physical development outcomes in neighbourhoods, towns and cities, therefore, depends mainly on the ability of local spatial planning and development management to guide parcel-level development towards desired outcomes stated in the plan. We have, however, revealed that historically, local spatial planning in Ghana—spatial planning for areas including a neighbourhood, town, city, metropolis, district or part of a district comprising multiple settlements—has been weak and ineffective in delivering the aforementioned objectives.

Throughout this book, we have explored the multiple factors that account for the failure of the spatial planning system to accomplish its forward planning and development management objectives. One of the primary reasons we identified is the crippling of the spatial planning system at all levels, following the inception of the tradition of development planning. In recent years, however, reforms implemented through the Land Use Planning and Management Project (LUPMP) has contributed to a strengthening of the national spatial planning system. The renewed interest in spatial planning at national level is also reinvigorating local-level spatial planning. More importantly, the 1945 Town and Country Planning Ordinance (CAP 84), which albeit being obsolete, remained in force for several decades, has now been replaced with the Land Use and Spatial Planning Act (Act 925). Following promulgation in

2016, the new planning law has effectively institutionalized the new three-tier spatial planning system.

While the aforementioned reforms have laid down the foundations needed for an effective spatial planning, the planning system must continue to innovate and implement policies and strategies to address major challenges at the local level. Despite there being a new spatial planning system in place, local plans and Structure Plans for most of the major towns and cities are outdated. As a result of decades of rapid urbanization, many suburban and peri-urban areas have experienced massive influx of population, resulting in significant increases in built-up land. Most of these peri-urban and suburban settlements are developing without any local plans, indicative or binding, to guide and coordinate the process. We also established that even in places where up-to-date local plans exist, the general culture of non-compliance to planning and building regulations constitutes one of the major challenges that local spatial planning and development management must continue to grapple with. Outdated plans are essentially of limited use to the urban development process, but the process of revising them do actually present new opportunities to consolidate hitherto unplanned and fragmented development while reimaging the futures of existing towns, cities and their surrounding areas in comprehensive and coherent frameworks. In doing so, spatial planning would have to address the crucial question of how to integrate existing developments, the majority of which have occurred without any formal planning, into new urban development visions that would be required to transform existing towns and cities.

As previously highlighted in the discussions in Chap. 6, in the absence of planning and/or failure by developers to comply with formal development management regulations and procedures, emergent patterns of physical development tend to be chaotic and of poor quality. While this is true, it is also true that in newly emerging peri-urban and suburban neighbourhoods, the most pressing problem is not actually with the quality of individual developments but rather with the absence of supporting facilities and services. One of the pragmatic approaches in such situations, therefore, would be for local governments (i.e. MMDAs) to regularize the existing development retroactively and to devise innovative strategies to finance the infrastructure deficit in these areas. In order to avert the unsustainable patterns of physical development which have occurred previously, local governments, using their planning powers, must act proactively in identifying and prioritizing spatial development plans for areas that are currently dominantly rural or peri-urban, but have the potential to become hotspots of physical development in the future. Ultimately, a streamlined development management process that removes unnecessary and costly administrative bureaucracies for prospective developers could help install public confidence in the development and building permitting system and contribute to the overall quality of physical development outcomes.

Furthermore, we have discovered that spatial planning at the local level in Ghana must navigate a complex customary land tenure system and the associated land market practices which do not always take place within the formally established procedures of land acquisition and physical development. We have discovered that under the prevailing customary land tenure system, more than 80% of all land in Ghana that would become the subject of spatial planning at the national, regional and local levels is vested in traditional authorities. However, spatial planning has not found ways to effectively harness the authority and powers vested in the chieftaincy institution in the local land use planning and development management process. Instead, historically, planning legislations, which have deep roots in both the colonial and contemporary governance systems, have mainly sought to entrench the mandate of the public-sector agencies of local planning and only regarded traditional authorities as stakeholders in the spatial planning process. It has been argued that because spatial planning at the local level is fundamentally about land and its allocation for various uses, the powers and interests of traditional authorities cannot be taken for granted in land allocation decisions. Indeed, since its inception, the planning system has struggled to fully utilize the determination of land use rights vested in it by planning legislations and other statutes that affect land governance. That said, the possibility exists for the spatial planning system to leverage the institution of chieftaincy to help accomplish the goals of local spatial planning. It would be worth exploring the possibility of traditional authorities becoming empowered by law to become an integral part of the institutional apparatus of spatial planning at the local level. In addition to local governments (i.e. MMDAs) leading the spatial planning process, traditional authorities could be granted basic planning powers that make them responsible for undertaking land use planning and development control functions in areas under their jurisdiction. Within this arrangement, traditional authorities would rely on the expertise of professional planners working in the public or private sectors to prepare plans. The final plan would then be approved for implementation by the relevant local government authority.

13.3.2 Urbanization Impact Monitoring and Settlement Growth Management

The analyses of urbanization trends in Chap. 9 revealed that in the last seven decades, major towns and cities across the country have experienced a significant increase in population and the size of built-up land. At the same time, integrated and coherent strategies to manage the urbanization process towards sustainable outcomes are lacking. We have recognized that innovative spatial strategies would be required in order to optimize the benefits of urbanization while addressing the associated negative challenges. Among other things, spatial planning must deal with the spatio-environmental consequences of years of rapid physical expansion of major towns and cities. As the discussion on spatio-temporal urban expansion trends in the three major urban agglomerations (i.e. Accra, GKSR and Sekondi-Takoradi) have shown, modern technology has now made it not only possible but also relatively easy and cheaper to monitor the footprints of urban growth. Using remotely sensed data such as Landsat Satellite data and existing geospatial software programmes—as was demonstrated in Chap. 9—spatial planners can now retrospectively examine land use transitions at the smallest possible spatial resolution in order to understand settlement growth patterns from the past to the present and to measure the associated spatial and socio-environmental impacts. With a solid understanding of past trends and impacts, planners can now design evidence-based policies to respond to the existing challenges and manage future growth towards sustainable outcomes. In order to effectively respond to the challenges of unfettered urban expansion, sprawl and fragmented settlement development patterns, the spatial planning system must design and implement settlement growth management strategies. As we discovered in Chap. 9, a range of growth management instruments exists to manage the spatio-environmental impacts of urbanization on towns and cities: the spatial planning system could deploy a combination of regulatory, fiscal and incentive-based growth management instruments to shape the timing and location of planned development and to raise revenue through the urban development process to finance infrastructure and services. Definitely, the adoption and implementation of innovative growth management strategies would face several challenges. That said, spatial planners must be confident and willing to experiment with new ideas and strategies as doing so could be one of the effective pathways towards adapting growth management strategies to the urban contexts of Ghana.

13.3.3 *Housing Supply—The Role of Spatial Planning*

Access to decent and affordable shelter is a major problem faced by many households and individuals of different socio-economic status in Ghana. As a result of the growing mismatch between housing need and housing supply, housing is increasingly becoming unaffordable to the larger mass of the urban population. The number of people sleeping rough on streets, slums and/or makeshift, sub-standard structures such as kiosk and containers continue to increase in urban areas. Given that land constitutes one of the critical inputs for housing development and spatial planning is fundamentally about land and its allocation for different uses, we have explored ways in which the spatial planning system can contribute to addressing the urban housing crises. To this end, two housing supply models, which draw on the principles of land value appreciation and value capture to unlock housing supply, have been advanced. In the first model, we explored the possibility of leveraging the spatial planning system to make land available for mass housing initiatives. As we discovered in Chap. 10, the key strategy under the land supply model would involve MMDAs using their spatial planning powers secure land at suitable locations for the development of housing and supporting amenities. Whereas the first model aims to make land available for housing development, the second model advanced in this book explores ways in which housing supply and/or finance for new housing could secure at various stages of the development process. Under this model, local governments would design and implement various development exaction instruments to raise revenue and/or set affordable housing quotas to be met in new, medium-scale and large-scale developments proposed by private developers. One possibility would be to apply these instruments at the point of granting planning permission to the proposed development. While these models have never been explored in the history of spatial planning in Ghana, existing legislations, including the Local Government Act (Act 462) and Land Use and Spatial Planning Act (Act 925), contain provisions founded on the principles of betterment and value capture from which local governments would be able to design and implement these housing supply models.

13.3.4 Integrated Spatial Development and Transport Planning for Sustainable Mobility

Spatial planning can play a significant role in designing and building cities, towns and neighbourhoods that promote sustainable mobility. From the discussions in Chap. 11, we know that urban land use patterns and the overall emergent structure of towns and cities provide the structural conditions, which in turn shape patterns of flow and spatial interaction as well as individuals' travel behaviours. Spatial planning therefore has an important role to play in reducing the negative impacts on society of automobile dependence, travel-related energy use and greenhouse gas emissions. Moreover, we have established that land use configurations, once they emerge, can remain unchanged for several decades, making it extremely difficult, in the short to medium terms, to bring about any significant reorganization of existing structures that underline travel patterns. This implies that in order to achieve the goals of sustainable urban development, land use planning and design decisions on the one hand and transportation planning and investment decisions on the other hand must be formulated and implemented in an integrated manner. Adopting an integrated approach to spatial development and transportation planning and policy would require coordination at the various institutional levels, where programmes, policies and strategies are articulated. Models and decision support systems could offer new opportunities for planning authorities to gain useful insights on which policy decision could be made with a much better degree of certainty with respect to expected outcomes. Using bespoke planning support systems, decision-makers can explore possible sustainability outcomes of various scenarios of spatial development, transportation infrastructure investment and travel demand strategies in an integrated manner.

Moreover, spatial planning must prioritize active non-motorized transport (i.e. walking and cycling) in existing built environments as well as in future growth areas. Instead of low-density exclusionary zoning policies, spatial planning must promote high-density mixed-land uses to reduce the need for motorized travel. Rather than planning to accommodate cars, local governments in their spatial development plans should provide for well-connected dedicated bicycle lanes and/or pedestrian paths to enable more people to access to urban opportunities using non-motorized transport. Planning for efficient public transit systems would also help connect population to urban opportunities in the existing monocentric spatial structure of towns and cities, which is characterised by the segregation between dominantly residential functions in suburban and peri-urban areas and commercial functions in inner-city areas. As we have already discovered, an efficient public transit system provides the benefits of moving a relatively larger number of passengers at any given time while reducing energy use, congestion and pollution. Acting together, institutions with competences in spatial planning and transportation could explore ways in which towns and cities could also benefit from emerging ICT-enabled shared-mobility systems such as car-sharing and bike-sharing.

13.3.5 *Supporting Livelihoods and Local Economic Development—Spatial Planning and the Urban Informal Economy*

One of the visible manifestations of informal urbanism in towns and cities in Ghana is the informal economy. In the penultimate chapter of this book, we focused on the urban informal economy, discussing the spatial dimensions of informal economic activities and the challenges businesses faces in finding suitable locations for their activities. A major highlight of the discussion was that if urban areas would obtain maximum benefits from the informal economy while addressing the negative impacts on the quality of urban spaces, spatial planning must continuously evolve and implement lasting measures to accommodate the space needs of activities in this sector. We discovered that the spatial planning system has introduced the concept of informal business zones which would be designated in towns and cities to accommodate economic activities in the informal sector. It was argued that while the idea of designating informal business zones constitutes a major step towards supporting the informal economy in urban areas, local governments, as part of their spatial planning functions, must explore strategies that would enable them to acquire land designated in local plans to be developed into informal business zones and ensure that such land is protected from invasion by other uses. Maintaining a database on informal economy businesses, especially artisanal activities that provide apprenticeship training to would-be informal economy business operators, would also enable local governments to plan accordingly for new activities that are likely to emerge in the future. As part of the portfolio of measures needed to promote home-based informal economy activities, the planning system must revise its development control regulations that are based on exclusionary zoning principles and encourage mixed-use development, especially in inner-city locations. Finally, by experimenting with new ways of putting public spaces such as streets to multiple adaptable uses, spatial planning at the local level can provide new opportunities for informal economy conduct business in designated public spaces on a temporary basis.

13.3.6 Public Engagement for the Co-production of Urban Transformations

Ensuring that the outcomes of socio-economic development are equitably distributed across space and between different groups of people in society remains one of the core objectives of spatial planning. To achieve its objectives, therefore, spatial planning must embrace the democratic principles of inclusivity, dialogue, collaboration and consensus building. Spatial planning must engage with existing power structures that affect the distribution of resources. Planners must actively seek ways of accessing the interests and needs of various stakeholders and use them to transform cities, towns and neighbourhoods in inclusive and sustainable ways. In Chap. 8, where we examined the topic of public participation in spatial planning, we discovered that over the years, statutory requirements and institutional practices regarding stakeholder engagement in Ghana's spatial planning system have evolved. In principle, the contemporary spatial planning system in Ghana advocates a decentralized place-making process that fosters co-creation of communities through the active involvement of relevant stakeholders at any given spatial scale. At the same time, we have discovered that planners continue to rely on traditional methods of engagement, whereby elected representatives of communities and local governments are the main mediums through which public participation is achieved in practice. A major shortcoming of this method is that it lacks any effective mechanism of ensuring that inputs into plans and policies by elected representatives actually reflect the development needs and priorities of the communities they represent. There is therefore the need for the public sector charged with the responsibility of planning to explore innovative methods of stakeholder engagement at all levels. We have discussed the prospects of spatial planning leveraging ICT and participatory planning support systems to offer citizens the opportunity to actively participate in the planning process. Over the years, overall educational attainment and literacy levels in the population have increased while Internet penetration and smartphone adoption rates continue to expand rapidly particularly in urban areas. These developments therefore provide the basic conditions for the spatial planning system to begin to institute technology-mediated public participation systems. ICT-enabled participatory tools have the potential to facilitate easy mobilizing of ideas for policy formulation and differing perspectives in the evaluation of alternative courses of action. Important tasks such as prioritizing community development goals and voting on issues could be achieved efficiently with limited resources. Finally, leveraging ICT in the spatial planning process could help establish the hitherto missing feedback link that is required between public officials and the communities they serve.

13.4 Conclusion

Since its formal inception in the 1945, spatial planning in Ghana has had moments of relevance when it played a leading role in formulating and delivering programmes, policies and plans to bring about socio-economic transformation. There have also been difficult times of political instability and uncertainty from the mid-1960s to the late-1980s when the weakened machinery of state negatively impacted all systems of governance, including spatial planning. The return to democratic governance at the beginning of the twentieth century provided new opportunities for decentralized governance and place-based policy articulation and implementation. Yet, the new ethos of development planning that accompanied the reforms and persists to date deemphasised spatial planning at the national level and effectively rendered the spatial planning system weak and ineffective to achieve its objectives at the local level. Fortunately, spatial planning has in recent years received a renewed impetus that can enable it to assert its role in spearheading the delivery of solutions to social, economic and environmental challenges at multiple scales. As we have discovered elsewhere in this book, a number of challenges still remains, especially with the fragmentation of institutional competences under the two existing competing planning systems that ought to be addressed. Notwithstanding the challenges, the new spatial planning system instituted by the 2016 Land Use and Spatial Planning Act (Act 925) provides a foundation for the strategic platform needed to bring together actors to formulate and implement solutions to socio-economic and environmental development problems. With its mandate firmly grounded in various legislative instruments and an entrenched decentralized governance structure, the spatial planning system in Ghana can provide a coherent multi-level governance system to deliver integrative solutions to major problems, including housing, transportation, rapid urban expansion, uncoordinated physical development, inadequate social services and poor quality of built environments, environmental degradation, unemployment and poverty. To remain relevant, spatial planning must constantly evolve the capabilities to dealing with emerging challenges. One of such challenges is reducing the impacts on towns and cities on climate change. With its ability to act between spatial scales and across policy domains, spatial planning can provide the strategic framework to design and deliver both mitigation and adaptation strategies. Increasingly, urban areas are becoming arenas of experimentation with and diffusion of technological innovations. While the intersection of technology and urban spaces could aid a transition towards sustainable futures, there would be wider equity issues that spatial planning must consider in order to make innovation beneficial to different groups in society. Ultimately, the success of spatial planning in the years to come would not be judged on the number of programmes, policies and plans that would be produced. Rather, a justification for the relevance and continued existence of the activity can only derive from the actual improvements it can bring to society. For spatial planning in Ghana to be responsive to the evolving needs of society, it must co-create transformations with its stakeholders. Spatial planning has to be ambitious yet realistic in its undertakings.

Planners should be willing to experiment with new ideas and have the confidence to make permanent the solutions that have been proven to work in the process.

References

Land use and spatial planning Act (2016) Act 925 The nine hundred and twenty fifth act of the parliament of the Republic of Ghana
Local Government Act (1933) Act 462 The four hundred and sixty two act of the parliament of the Republic of Ghana
National Development Planning (System) Act (Act 480, 1994) The four hundred and eightieth Act of the parliament of the Republic of Ghana. http://urbanlex.unhabitat.org/sites/default/files/urbanlex//gh_national_development_planning_act_1994.pdf. Accessed 24/07/2018
Town and Country Planning Act (1945) CAP 84. http://www.epa.gov.gh/ghanalex/acts/Acts/TOWN%20AND%20COUNTRY%20PLANNING%20ACT,1945.pdf